Conceptual Academy

Physics

Written and Illustrated by

Paul G. Hewitt

City College of San Francisco

Editor in Chief: John Suchocki
Operations Specialist: John Wright
Senior Marketing Manager: Graham Orr
Composition: Tracy Suchocki, Ian Suchocki
Photo Research: Conceptual Productions, LLC
Copyediting: Lillian Hewitt
Indexing: Ian Suchocki
Cover Image Credit: Silent Yachts

Credits and acknowledgments appear with images.

Library of Congress Cataloging-in-Publication Data available upon request from Publisher.

Printed in China
4 5 6 7 8 9 10

Conceptual Academy

Conceptual Academy, PBC www.ConceptualAcademy.com

ISBN: 979-8-9862706-2-3

About Conceptual Academy

Conceptual Academy is a public benefit company of science teachers, media producers, and authors known for their conceptual science textbooks, such as *Conceptual Physics*, *Conceptual Chemistry*, *Conceptual Biology*, and others. Conceptual Academy provides a comprehensive library of video lessons that, altogether, create the "video textbook"—just as comprehensive and cohesive as the traditional textbook but featuring video. Conceptual Academy also produces traditional textbooks, such as the one you are now holding, as well as many other learning resources, such as lab activities, worksheets, podcasts, test banks, an automated encouragement-based assessment system, and resources for an instructor's professsional development all now hosted by the PocketLab Notebook learning and lesson plan system. Variety is the key to serving the varying needs of our diverse students and helping to keep each classroom truly interactive and learner-centered. Toward this, our programs follow a double helical model as described here:

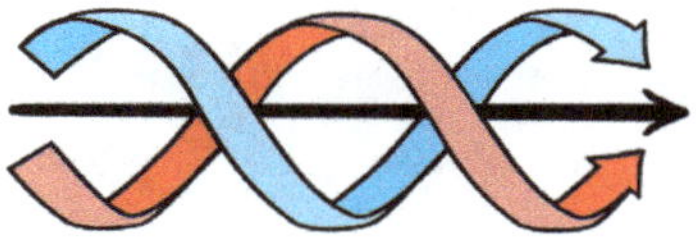

About PocketLab

PocketLab is a hands-on, classroom-ready, learning and lesson plan system that brings STEM labs and lessons to life. This learning system features hand-held data sensors and digital lab software where students collect, visualize, and analyze real-time experimental data. PocketLab hosts hundreds of comprehensive lesson plans available for instructors to customize for their own classroom needs. Lesson plan developers include PhET Interactive Simulations, Kesler Science, TeacherGeek, Green Ninja, Open SciEd, Hooked on Science with Jason Lindsay, StarTalk with Neil deGrasse Tyson, and Conceptual Academy with Paul Hewitt. PocketLab also sponsors the community-building "Science is Cool" virtual unconferences for science teachers across the world. Known for a relaxed and engaging atmosphere, each webinar unconference allows teachers to connect with science celebrities and, just as celebrated, science educators doing amazing things in their classroom.

TEXAS EDITIONS

About the Conceptual Academy Science Series

SBOE approved

The Conceptual Academy authors have revived their physical science and integrated science "Explorations" titles once published for middle and secondary level students. The content has been 1) updated and reorganized into distinct module lessons, 2) integrated with the Conceptual Academy video library and other resources including PocketLab sensor activities, 3) made available through the dynamic Pocket Lab Notebook platform that allows each instructor to organize and implement this library of content according to their classroom needs, and 4) aligned to the TEKS and ELPS Texas state standards. To both teach and reach our modern students effectively, we need not just curricula of high quality but also of great variety and adaptability. Combine this with affordability, then we have the means to pay it forward to a sustainable future built by compassionate, curious, and scientifically literate citizens who love and respect the natural universe for what it truly is. That's good energy!

To the Instructor and Student

In most American schools, physics has usually been the last course in a science sequence. This policy goes back to the 1800s when farming was dominant and it was enormously important that personal hygiene be learned in school. Biology was the course that taught hygiene. A second year was devoted to chemistry, at a time before the electron was known, when students learned to mix and measure substances. For those reaching the senior year a more rigorous course in applied mathematics was taught—physics.

Today we view this sequence differently. Physics is the simplest of the sciences. How so? Look to the number of vocabulary terms found within any science textbook. By far, a textbook in physics shows the fewest terms, followed by chemistry, and then biology, which is the most complicated of these sciences. When physics is presented not as applied mathematics but as the study of the fundamental rules of nature, it becomes quite relevant to everyday life—while serving as a building block to the more complex sciences of chemistry and biology. The *Conceptual Academy Science* sequence is physics first, chemistry second, and then followed by biology. Physics deals with such basic topics as motion, energy, electricity and magnetism, light, sound, and nuclear processes all providing a necessary foundation from which a student can reach for his or her personal stars.

I present physics as a study of nature's rules. Learning to comprehend the basic rules (laws) of physics is a tall order. Honing that comprehension further can be the goal of a follow-up physics course. A grand mistake is trying to "do it all" in a first course, which results in the dreaded INFORMATION OVERLOAD. As a teacher, I have vowed not to inflict on my students undue exhaustion that results from much-too fast acquisition of knowledge. Suitable pacing of knowledge acquisition is a defining feature of Conceptual Academy.

Within physics, mathematical equations can be used as recipes for problem solving. But, more importantly, they can also be used as guides to clear thinking. The conceptual approach emphasizes the latter. Just as a music student learns to hear the music behind a musical score, a physics student learns to discern the relationships depicted within a mathematical equation. An example is Newton's second law of motion, $a = F/m$, which relates the acceleration, a, acquired by a body of mass, m, to the net force, F, exerted on it. Remarkably, this one simple equation was fundamental to humans reaching the Moon in 1969. Interestingly, Galileo understood acceleration, force, and mass, but he failed to recognize how these three concepts are related. Why? Because he didn't have the equation $a = F/m$! This relationship, even today, is one of the most complex concepts a student faces in a study of physics. Again, suitable pacing of knowledge acquisition is vital. Once $a = F/m$ is understood, the student has achieved a respectable milestone. Applying this law to numerical outcomes is small by comparison and can await a follow-up course.

Good Energy!

Paul G. Hewitt

Conceptual Academy Physics

Table of Contents

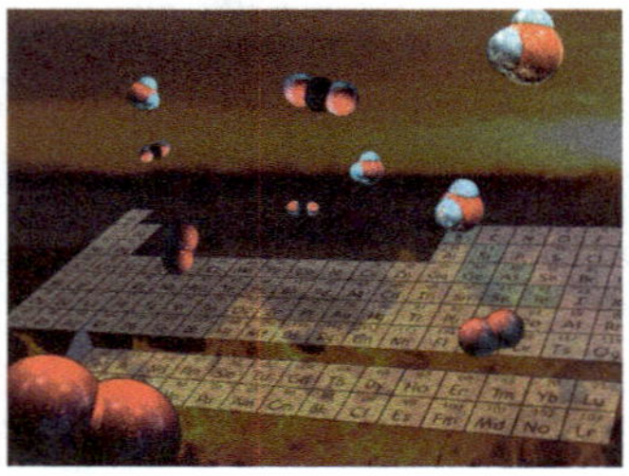

Conceptual Academy

TEXAS EDITION

PocketLab
SCIENCE. EVERYWHERE.

Introduction

About Science

THE MAIN IDEA

Science is the study of nature's rules

Science has given us much. Our modern world is built on it. Nearly all forms of technology—from medicine to space travel—are applications of science. But what exactly is this amazing thing called science? How should science be applied? Where did science come from? And what would the world be like without it?

Science is an organized body of knowledge about nature. It is the product of observations, common sense, rational thinking, and (sometimes) brilliant insights. Science has developed from group efforts as well as individuals' discoveries. It has accumulated over thousands of years and gathered from places all around the world. It is a huge gift to us today from the thinkers and experimenters of the past. Yet, science is not just a body of knowledge. It is also a method, a way of exploring nature and discovering the order within it. Importantly, science is also a tool for solving problems.

Science began back before recorded history, when people first discovered repeating patterns in nature such as star patterns in the night sky, weather patterns, and animal migration patterns. From these patterns, humans gradually learned to make predictions that gave them some control over their surroundings. Now *you* can learn more about science. Let's begin!

0.1 Copernicus and Galileo

When a sore throat keeps you home from school, you probably ask yourself, "How did I catch this cold?" You then speculate about how you could have been exposed to germs, and how you might avoid exposure next time. When a light goes out in your room, you ask, "How did that happen?" You might check to see if the lamp is plugged in, check the bulb, or even look at your neighbors' houses to see if there has been a power outage. When you think like this, you are searching for *cause and effect* relationships—trying to find out what events cause what results. This type of thinking is *rational thinking.* Rational thinking is basic to science.

Today we use rational thinking so much and so often that it's hard to imagine other ways of interpreting our experiences. But it wasn't always this way. At times, people have relied more on superstition and magic to interpret the world around them—or have simply failed to ask, "Why?"

Rational thinking became very popular in Greece in the 3rd and 4th centuries BC. From there it spread throughout Rome and other parts of the Mediterranean world. When the Roman Empire fell in the 5th century AD, advancements in science came to a halt in Europe. Rational thought was replaced by the Dark Ages. But during this time science continued to advance in other parts of the world. The Chinese and Polynesians were charting the stars and the planets. Arab nations developed mathematics and learned to make glass, paper, metals, and certain chemicals. Finally, the Greek philosophy of rational thinking was brought back into Europe by Islamic people who entered Spain during the 10th to the 12th centuries. Then universities emerged. When the printing press was invented in the 15th century, science made a great leap forward. This invention did much to advance scientific thought (just as computers and the Internet are doing today).

Up into the 16th century most people thought the Earth was the center of the universe. They thought the Sun circled the stationary Earth. This thinking was challenged when the Polish astronomer Nicolaus Copernicus quietly published a book proposing that the Sun is stationary and that the Earth revolves around it. These ideas conflicted with religious teachings and were banned for the following 200 years.

Nicolaus Copernicus
(1473 - 1543)

Galileo Galilei
(1564 - 1642)

READING CHECK

How did Galileo study nature's behavior?

But about 100 years after Copernicus, the Italian physicist Galileo Galilei revived the Copernican view. Galileo used experiments, rather than speculation, to study nature's behavior. Galileo was arrested for popularizing the Copernican theory and for his other contributions to scientific thought. Yet a century later his ideas and those of Copernicus were accepted by most thinking people.

Scientific discoveries are often opposed, especially if they conflict with what people want to believe. Every age has its intellectual rebels who are persecuted, condemned, or suppressed at that time. But later they seem harmless and are often essential to the elevation of human conditions. To quote the Belgian poet Count Maurice Maeterlinck: "At every crossway on the road that leads to the future, each progressive spirit is opposed by a thousand men appointed to guard the past."

0.2 Scientific Methods

In the 16th century the Italian physicist Galileo and the English philosopher Francis Bacon developed a formal method for doing science—the **scientific method**. Although their method is not the only one, all scientific methods are based on rational thinking and experimentation. Some or all of the following steps are likely to be found in the way most scientists perform their work:

1. **Observe** Closely observe the physical world around you. Recognize a question or a puzzle—such as an unexplained observation.

2. **Question** Make an educated guess—a hypothesis—that might resolve the puzzle.

3. **Predict** Predict consequences of the hypothesis.

4. **Test predictions** Do experiments to see if the consequences you predicted are present.

5. **Draw a conclusion** Formulate the simplest general rule that organizes the three main ingredients: hypothesis, predicted effects, and experimental findings.

A scientific **hypothesis** is an educated guess. When a hypothesis has been tested over and over again and has not been contradicted, it may become known as a **law** or *principle.* A scientific **fact**, on the other hand, is something that competent observers can observe and agree to be true. For example, it is a fact that amputated limbs of certain crabs can grow back. Anyone can watch it happen. It is not a fact—yet—that a severed limb of a human can grow back.

Scientists use the word theory in a way that differs from everyday speech. In everyday speech a theory is the same as a hypothesis—a statement that hasn't been tested. But in science, a **theory** is a vast combination (synthesis) of facts and well-tested hypotheses. Physicists speak of the quark theory of the atomic nucleus. Chemists have the theory of chemical bonding. Geologists use the theory of plate tectonics, and astronomers speak of the theory of the Big Bang.

Theories are a foundation of science. They are not fixed, but evolve. They pass through stages of refinement. For example, since the theory of the atom was proposed 200 years ago it has been refined many times in light of new evidence. Some people argue that scientific theories can't be taken seriously because they change. Those who understand science, however, see that theories grow stronger as they evolve to include new information.

READING CHECK

What do all scientific methods have in common?

FOR YOUR INFORMATION

Here's an analogy for how science works. You're in a closed room with no windows and no light. It's pitch dark. You're **wondering** if there might be a cat somewhere in the room. So, being **clever**, you open a can of smelly tuna. No cat arrives. Does this mean there is no cat in the room with you? (No.) Or might it be better to say the **probability** of there being no cat is high? (Yes.) Say, a cat shows up. How then might you **determine** its color? Or do we accept that some things just remain **unknowable**? Or do we invent new **technology** to help us see past our **limitations**? At the heart of all science is a **curiosity** to learn the nature of things for what they are and not for what we might wish them to be. Better understanding those things helps us to care for those things. This include cats and all creatures of this planet, including ourselves.

0.3 Scientific Hypotheses

A scientific hypothesis is one that is testable. A test or series of tests can determine whether a hypothesis is valid. A hypothesis that is not testable is not scientific. It is **pseudoscience**—fake science with no tests for validity In scientific work, most hypotheses turn out to be wrong. Scientists have to be patient though, and keep testing their ideas for accuracy.

A well-known scientific hypothesis that proved to be incorrect was that of the greatly respected Greek philosopher Aristotle (384–322 BC). Aristotle claimed that heavy objects naturally fall faster than light objects. This hypothesis was considered true for nearly 2000 years—mainly because of respect for this great man. Also, air resistance was not recognized as an influence on how quickly objects fall. We've all seen that stones fall faster than leaves fluttering in the air. Without investigating further it is easy to accept false ideas.

Galileo very carefully examined Aristotle's hypothesis. Then he did something that caught on and changed science forever. He *experimented.* Galileo showed the falseness of Aristotle's claim with a single experiment—dropping heavy and light objects from the Leaning Tower of Pisa. Legend tells us that they fell at equal speeds. In the scientific spirit, one experiment that can be reproduced outweighs any authority, regardless of reputation or the number of advocates. Albert Einstein put it well when he stated, "No number of experiments can prove me right; a single experiment can prove me wrong." In science, the test of knowledge is experiment.

READING CHECK
What is the test of knowledge in science?

Although Aristotle was wrong about falling objects, his hypothesis was scientific. That's because it was testable. In Aristotle's time, experimentation hadn't caught on. Even the claim that the Moon is made of green cheese is a scientific hypothesis, because the claim is *testable*. The hypothesis was tested and proved wrong with the lunar landings that began in 1969.

The scientific method is effective in advancing knowledge. But many scientific advances involve trial and error, experimenting without guessing, or just plain accidental discovery. The trained eye, however, is better at noticing questions in the first place and making sense of evidence. More than a particular method, the success of science has to do with an attitude common to scientists. This attitude involves inquiry, experimentation, and being humble before the facts. A good scientist can admit to being wrong when wrong.

CONCEPT **CHECK**

Which statements are *scientific* hypotheses?

a. Better stock market decisions are made when the planets Venus, Earth, and Mars are aligned.
b. Atoms are the smallest particles of matter that exist.
c. Albert Einstein was the greatest physicist of the 20th century.

CHECK YOUR ANSWER All statements are hypotheses, but only statements *a* and *b* are scientific hypotheses—because they are testable. Statement *a* can be tested (and proven wrong) by researching the performance of the stock market during times when these planets were aligned. Statement *b* not only can be tested, but has been tested. Although it is untrue (many particles smaller than atoms have been discovered), the statement is nevertheless a scientific one. Lastly, statement *c* is an assertion or opinion. Greatness is a quality that cannot be measured in an objective way.

0.4 Scientific Attitude

The curious mind works hard to make sense of what it experiences. How then to satisfy that curiosity? We can make up fanciful explanations. Or we can do as Galileo did and rely on the results of experimentation. When we follow Galileo's lead, we are satisfying our curiosity through science. In following this path, we quickly come to see value in a set of attitudes.

For example, we must accept our experimental findings even when we wish they were different. We must strive to distinguish between the results we see and those we wish to see. This is not easy. Most people can easily fool themselves. Furthermore, none of us has the time or resources to test every idea. Thus, we tend to accept the word of someone we trust. But who do we trust? To reduce the likelihood of error, the science-thinking individual trusts those whose findings are testable—if not in practice, then at least in principle. Ideas that cannot be tested are regarded as "unscientific."

The fact that scientific statements will be thoroughly tested helps keep the science-thinking individual honest. Sooner or later, mistakes (or deception) are discovered. A scientist exposed for cheating doesn't get a second chance in the community of scientists. Honesty, so important to the progress of science, thus becomes a matter of self-interest. There is little bluffing in a game where all bets are called.

READING CHECK

What kind of ideas are unscientific?

We have discovered that the physical universe is guided by a relatively small set of rules, such as gravity, electromagnetism, and nuclear forces. The application of these few rules to the gadzillion atoms and molecules of the universe, however, results in significant complexity. A living organism is a prime example. Science has done a remarkable job at unraveling many of the mysteries arising from this complexity. But given the vast complexity of the universe, what we know today is pale to what remains to be learned.

This leads to another key aspect of the scientific attitude. We can be proud of what we have learned, but it's important to remain *humble* for all that we still do not understand. This includes not just the unknowns, but the even greater number of unknown unknowns. An "unknown unknown" is that which you don't even know you don't even know. A good example is the idea of dark energy. From astronomical observations starting in the 1990s we learned of a universal force we now call *dark energy*. It permeates the universe but we still don't know what it is—it's a known unknown. Prior to the 1990s, dark energy was an unknown unknown, which means we were less than clueless. It's as if the more we learn, the more we learn we don't know. This pushes the mindful scientist to remain humble.

Lastly, there is this idea of *uncertainty*. It's human nature to want things to be simply black or white. Is it this? Or is it that? Given the complexity of the natural universe, however, most discoveries fall within a gray spectrum of fairly certain to most uncertain. Climate scientists, for example, have identified carbon dioxide to be a powerful greenhouse gas, which means it helps to keep the planet warm. This is fairly certain. As humans produce more carbon dioxide, this means the planet will become warmer, which brings a change in climates. This is fairly certain too. But how exactly might climates change? This is less certain. To the scientist, uncertainty is inspiration for further study. The scientific attitude embraces uncertainties because they are ever-present and part of the fabric of this complex universe.

FOR YOUR INFORMATION

To many who influence public opinion, including politicians, uncertainty is cause for ignoring the issue. Why? Because uncertainty is uncomfortable, and who would vote for a politician who makes them uncomfortable? The politician might say: "If people want black and white, then give them black and white." This attitude is NOT scientific.

0.5 Science and Technology

Science and technology are different from each other. Science is concerned with gathering knowledge and organizing it. **Technology** is the engineering of that knowledge for practical purposes. Plus, technology provides the instruments scientists need to conduct their investigations.

Technology is a double-edged sword. It can be both helpful and harmful. We have the technology, for example, to extract fossil fuels from the ground and then burn the fossil fuels to produce energy. Energy production from fossil fuels has benefited society in countless ways. On the flip side, the burning of fossil fuels damages the environment. It is tempting to blame technology itself for problems such as pollution and resource depletion. These problems, however, are not the fault of technology any more than a stabbing is the fault of the knife. It is humans who use the technology, and humans who are responsible for how it is used.

Remarkably, we already possess the technology to solve many environmental problems. Today we are posed to see a switch from fossil fuels to more sustainable energy sources. We recycle waste products in new and better ways. In some parts of the world, progress is being made toward limiting the human population explosion, which unchecked could worsen almost every problem faced by humans today. Difficulty solving today's problems results more from social inertia than failing technology. Technology is our tool. What we do with this tool is up to us. The promise of technology is a cleaner and healthier world. Wise applications of it *can* improve conditions on planet Earth.

READING CHECK
What is the promise of technology?

We bath in radiation whenever we venture outside. Some places more so than others.

Risk Assessment

Technology comes with risks as well as benefits. When the benefits are seen to outweigh risks, a technology can be accepted and applied. X-rays, for example, continue to be used to diagnose disease despite their potential risk for causing cancer. The benefits outweigh the risks. Of course, when the risks of a technology outweigh its benefits, it should be used sparingly or not at all.

A difficult ethical problem arises when a technology benefits one group of people but poses risk to a different group of people. Dumping raw sewage into the local river may pose little risk for a town located upstream, but for towns downstream the untreated sewage is a health hazard. Technologies involving different risks for different people raise questions that are often hotly debated. Which medications should be sold to the general public over-the-counter and how should they be labeled? Should food be irradiated to eliminate the food poisoning that has consistently killed thousands of Americans each year? Or are the unknown potential hazards of food irradiation sufficient cause for banning food irradiation?

People seem to have a hard time accepting the fact that zero risk is impossible. Airplanes cannot be made perfectly safe. Processed foods cannot be completely free of toxicity, for all foods are toxic to some degree. You cannot go to the beach without risking skin cancer no matter how much sunscreen you apply. You cannot avoid radioactivity, for it is naturally in the air you breathe and the foods you eat, and has been that way before humans first walked the Earth. Even the cleanest rain contains radioactive carbon-14; our bodies do as well. You might hide in the hills, eat the most natural foods, practice obsessive hygiene and still die from cancer caused by radioactivity. The probability of eventual death is 100%. So far, no one is exempt.

Science helps to determine the most probable results. As the tools of science improve, risks can be evaluated more and more accurately. Acceptance of risk, on the other hand, is more of a social issue than a scientific one. If society were to demand zero risk from its technology, this goal would not only be impractical, but selfish. Any society striving toward a policy of zero risk would consume its present and future economic resources. A society that accepts no risks receives no benefits.

0.6 Skepticism and Denialism

Skepticism is being doubtful over unsubstantiated claims. Skepticism is integral to the scientific process. A good scientist is a good skeptic. Sorting out the few kernels of wheat from the large pile of chaff requires much effort and cautious inference.

Denialism is being unfairly resistant to substantiated claims, often motivated by ideology or strong beliefs, where commitment to the belief takes precedence over the evidence. Belief comes first, reasons for belief follow, with reasons skewed to ensure that the belief survives intact.

There have always been groups of people skeptical about issues such as the Holocaust, the 1969 Moon landing, tobacco, climate change, and vaccines. The list is seemingly without end. Keep in mind the word *some* when identifying skeptics and deniers. Some people hold that the Earth is flat, while other people consider Earth as round—in fact, spherical. Importantly, what is the *percentage* of each? More often than not, deniers make up a small percentage of people.

Emotional motivation often fosters denial when a policy is seen as distressing or distasteful. As an example, if your dentist tells you that you need a root canal, you may not want to believe the evidence and reasoning. You are emotionally motivated to deny the report. So, you go for a second opinion, and perhaps a third—which is quite reasonable. You are gathering data to evaluate. But at a certain point, if all of the credible dentists you visit are in consensus about your diagnosis, you will be wise to set your emotions and short-term preferences aside, and try to understand the dental science of your tooth.

Think of a less wise approach: You interview 1000 dentists until you find one who declares no root canal is needed. You then heed the advice that no root canal is needed. Further, you praise and elevate that dentist's credentials because finally you found someone who knows how to listen to your underlying needs. You convince yourself there are always two sides to a story. You think of yourself as a crafty renegade for bucking a system that's clearly corrupt and self-interested. What you would be doing here is known as "cherry picking" the evidence. Cherry picking evidence means that you knowingly or unknowingly only consider evidence supporting your case, no matter how outlying that evidence is. Cherry picking evidence is a common way to deny the validity of scientific claims.

READING CHECK

What is meant by "cherry picking"?

The physical world, knowable to a large extent through science, offers *real* rewards, joys, sustenance, and plenty of hazards too. The human population is booming like never before. This means we are taxing the environment that sustains us like never before. Can we ultimately use what we learn from science to address our modern challenges and to accomplish great things? If we've been to the Moon and back, then the answer is a resounding yes.

CONCEPT **CHECK**

Why are long term problems, like over-population, easier to deny than short term problems?

CHECK YOUR ANSWER It's common for people today to think short term. I'm hungry. I want to buy that car. A problem that develops over a longer time scale, such as too many people in too small a space—overpopulation—is more difficult to grasp, especially if you live in a sparsely populated area. Addressing long-term problems requires a focus on the future and commitment to a sustainable way of life. This is a greater challenge.

Concept Review

Introducttion

Summary of Terms

Denialism Being unfairly resistant to considering substantiated claims.

Fact A phenomenon about which competent observers can agree.

Law A general hypothesis or statement about the relationship of natural quantities that has been tested over and over again and has not been contradicted. Also known as a *principle*.

Hypothesis An educated guess; a reasonable explanation that is not fully accepted as factual until tested over and over again by experiment.

Pseudoscience Fake science that has no tests for its validity.

Science The collective findings of humans about nature, and a process of gathering and organizing knowledge about nature based upon experimental evidence.

Scientific Method An orderly method for gaining, organizing, and applying new knowledge.

Skepticism Being doubtful over unsubstantiated claims.

Technology Method and means of solving practical problems by applying the findings of science.

Theory A synthesis of a large body of information that encompasses well-tested hypotheses about certain aspects of the natural world.

Review Questions

1. Throughout the ages, has acceptance or resistance usually been the general reaction to new ideas about established "truths"?
2. Outline the steps of the scientific method.
3. Distinguish among a scientific fact, a hypothesis, a law, and a theory.
4. How many experiments are necessary to invalidate a scientific hypothesis?
5. Must a statement prove correct to be a scientific hypothesis?
6. In science, what kind of ideas are generally accepted?
7. Why is honesty a matter of self-interest to a scientist?
8. How do scientists regard "not knowing" in general?
9. Clearly distinguish between science and technology.
10. Who uses and who is responsible for how technology is applied?
11. In daily life people are often praised for maintaining some particular point of view, for the "courage of their convictions." A change of mind is seen as a sign of weakness. How is this different in science?
12. In daily life we see many cases of people who are caught misrepresenting things and who soon thereafter are excused and accepted by their contemporaries. How is this different in science?
13. In answer to the question, "When a plant grows, where does the material come from?" Aristotle hypothesized by logic that all material came from the soil. Do you consider his hypothesis to be correct, incorrect, or partially correct? What experiments do you propose to support your choice?
14. What is probably being misunderstood by a person who says, "But that's only a scientific theory"?
15. Make an argument for bringing to a halt the advances of technology. Make an argument that advances in technology should continue. Contrast your two arguments.

Answers to Odd-Numbered Questions

1. Resistance.

3 Whereas a fact is generally a close agreement by competent observers of a series of observations of the same phenomenon, a hypothesis is an educated guess, which when verified may become a law. A scientific theory is a synthesis of a large body of information that encompasses well-tested and verified hypotheses.

5. No. It need only be testable.

7. The scientist is in an arena where falsehood will sooner or later be found, and discredit that scientist.

9. Science has to do with gathering and organizing knowledge; technology puts that knowledge into practice and provides the tools needed for finding more.

11. A change of mind is a strength of science, enabling refinements in theories.

13. Aristotle's hypothesis was partially correct, for material that makes up the plant comes partly from the soil, but mainly from the air and water. An experiment would be to weigh a pot of soil with a small seedling, then weigh the potted plant later after it has grown. The fact that the grown plant will weigh more is evidence that the plant is composed of more material than the soil offers. By keeping a record of the weight of water used to water the plant, and covering the soil with plastic wrap to minimize evaporation losses, the weight of the grown plant can be compared to the weight of water it absorbs. How can the weight of air taken in by the plant be estimated?

15. The answers to these are open ended.

Chapter 1

Newton's First Law

THE MAIN IDEA

No net force means no change in motion

If you want to move something, you apply a force—you either push it or pull it. Why then is a large boulder so difficult to move? There are a number of reasons. Consider a boulder floating free in outer space. Apply a force and it will indeed start moving. But because the boulder has so much mass, the degree to which it starts moving is relatively small. We say that the massive boulder has a lot of inertia, which means it has a resistance to change how it's moving. The greater the mass, the more the inertia. Now place that same boulder on Earth. Gravity presses the boulder downward against the ground and this can create significant friction to oppose any attempt to move the boulder. To get the boulder moving you have to overcome both inertia and friction. You are likely not strong enough to accomplish this, so the boulder doesn't move. It remains in what we call *equilibrium*, which means a steady state. In this chapter we discuss the basic ideas behind moving as well as not moving.

1.1 Motion

The idea that motion requires a force (a push or a pull) goes back to the 4th century BC when the Greeks were developing scientific ideas. This was the time of Aristotle, the most famous Greek scientist. Aristotle classified motion into two kinds: natural motion and unnatural motion.

Natural motion, Aristotle believed, occurred without force. For example, motions of the Sun, Moon, and other objects in the sky were considered natural. These celestial objects moved continuously without the need for any force to push or pull them. He taught that natural motion on Earth's surface was directed either up or down. He said, for example, that it was natural for a boulder to fall down toward Earth and a puff of smoke to rise in air. He believed that objects have resting places that they naturally seek. He went on to say that it was natural for heavy objects to fall faster than light objects.

READING CHECK
How did Aristotle classify motion?

Unnatural motion, on the other hand, required forces such as pushes or pulls by people or animals. For example, the only way to get a cart

moving across the ground was to push or pull on it. Unnatural motion required forces.

Aristotle's ideas were taken as fact for nearly 2000 years. But in the early 1500s the Italian scientist Galileo demolished Aristotle's belief that heavy things fall faster than light things. According to legend, Galileo dropped a heavy object and a light object from the Leaning Tower of Pisa. He showed that except for the effects of air friction, objects of different weights fell to the ground at the same time.

Galileo made another huge discovery. He showed that Aristotle was wrong about forces being necessary to keep objects moving. Galileo discovered that a force is required to start an object moving, but once moving, no force is required to keep it moving—except for the force needed to overcome friction. (We'll study friction, a force that opposes motion, in the next chapter). When friction is absent, a moving object does not need a force to keep it moving.

Galileo found a simple and powerful way to test his revolutionary idea. He rolled balls along flat plane surfaces tilted at different angles. He noted that a ball rolling down an inclined plane picks up speed. This is shown in **Figures 1.2** and **1.3**. Gravity increases the ball's speed. He also noticed the ball slows down when rolling up an inclined plane. Then gravity decreases the ball's speed. What about a ball rolling on a level surface? While a ball rolls on a level surface, it doesn't roll with nor against gravity. Galileo saw that a ball rolling on a smooth horizontal plane doesn't speed up and doesn't slow down. It has a constant speed. Galileo reasoned that a ball moving horizontally would move forever if friction were entirely absent. A ball would move of itself.

All objects show the same property of motion as the balls rolling on Galileo's planes. The tendency of things is to remain as they are. If moving, they tend to remain moving. If at rest, they tend to remain at rest. This property of objects is called **inertia.**

∧ Figure 1.1
Galileo's famous demonstration.

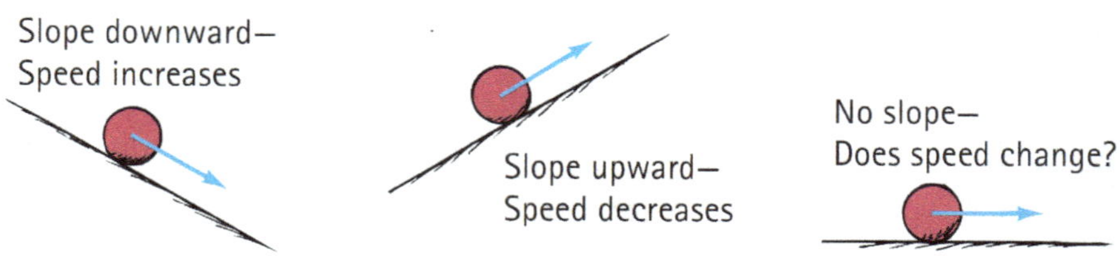

Figure 1.2 >
Motion of balls on various planes.

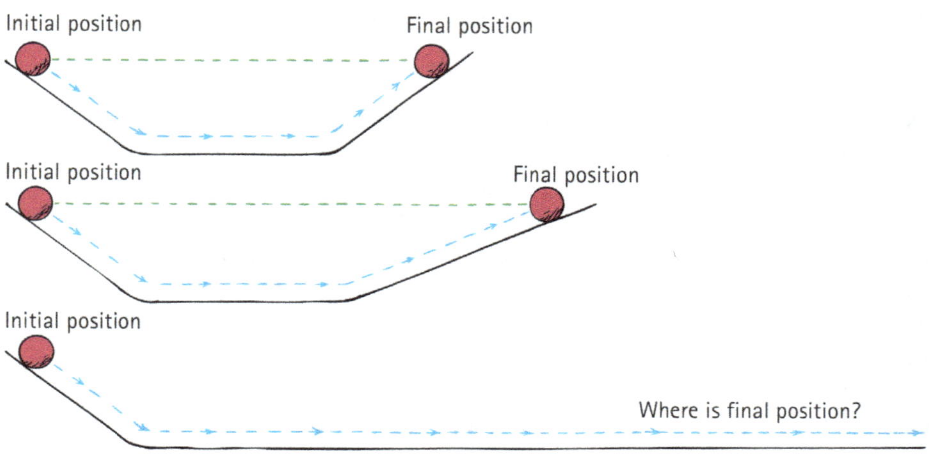

Figure 1.3 >
A ball rolling down an incline on the left tends to roll up to its initial height on the right. The ball must roll a greater distance as the angle of incline on the right is reduced.

CONCEPT CHECK

A ball rolling on a pool table slowly comes to a stop. How would Aristotle explain this behavior? How would Galileo explain it? How would you explain it?

CHECK YOUR ANSWER *Did you think about the questions and arrive at your own answers before reading this? Please do so, and you'll find yourself learning more. Much more!*

Aristotle would probably say that the ball stops because it seeks its natural state of rest. Galileo would probably say that friction overcomes the ball's natural tendency to continue rolling. Only you can answer the last question!

Motion is Relative

Everything is always moving. Even when you think you're standing still, you're actually speeding through space. You're moving relative to the Sun and stars—though you are at rest relative to Earth. Right now, your speed relative to the Sun is about 100,000 kilometers per hour. And you're moving even faster relative to the center of our galaxy.

When discussing motion, we mean *relative motion*—motion relative to something else. When we say a space craft moves at 30,000 kilometers per hour, we mean relative to Earth below. When we say a racing car reaches a speed of 300 kilometers per hour, we mean relative to the track. Unless stated otherwise, all speeds discussed in this program are relative to the surface of Earth. Motion is relative.

< Figure 1.4
When you sit on a chair, your speed is zero relative to the Earth but 30 km/s relative to the Sun.

FOR YOUR INFORMATION

Galileo did all his work before the advent of mechanical clocks. He timed some of his experiments with his pulse, and others with the dripping of water drops. Einstein called Galileo the father of modern physics, and modern science.

1.2 Speed and Velocity

Λ Figure 1.5
The greater the distance traveled each second, the faster the horse gallops.

Speed

Before the time of Galileo, people described moving things as simply "slow" or "fast". Such descriptions were vague. Galileo was the first to measure speed by considering the distance covered and the amount of time it takes. He defined speed as the distance covered per unit of time. The word per means "divided by."

$$\text{Speed} = \frac{\text{distance}}{\text{time}}$$

For example, if a horse covers 20 kilometers in a time of 1 hour, its speed is 20 km/h. If you run 6 meters in 1 second, your speed is 6 m/s. These values are average speeds. The speed at any instant, instantaneous speed, may be different.

With distance in the numerator and time in the denominator, any combination of units can be used for speed—kilometers per hour (km/h), centimeters per day (the speed of a sick snail), or whatever is useful and convenient. The slash symbol (/) is read as "per." In science the preferred unit of speed is meters per second (m/s). **Table 1.1** shows some comparative speeds in different units.

Table 1.1 Approximate Speeds in Different Units

12 mi/h	=	20 km/h	=	6 m/s (bowling ball)
25 mi/h	=	40 km/h	=	11 m/s (very good sprinter)
37 mi/h	=	60 km/h	=	17 m/s (sprinting rabbit)
50 mi/h	=	80 km/h	=	22 m/s (tsunami)
62 mi/h	=	100 km/h	=	28 m/s (sprinting cheetah)
75 mi/h	=	120 km/h	=	33 m/s (batted softball)
100 mi/h	=	160 km/h	=	44 m/s (batted baseball)

Λ Figure 1.6
Is a speed limit road sign the law for your maximum average speed or instantaneous speed?

Λ Figure 1.7
This speedometer gives readings in both miles per hour and kilometers per hour.

Velocity

When we know both the speed and direction of an object, we know its velocity. For example, if a car travels at 60 km/h, we know its speed. But if we say it moves at 60 km/h to the north, we have specified its velocity. Speed is a description of how fast; velocity is how fast and in what direction.

Constant speed means steady speed. Something with constant speed doesn't speed up or slow down. Constant velocity, on the other hand, means both constant speed and constant direction. Constant direction is a straight line—the object's path doesn't curve. So constant velocity means motion in a straight line at constant speed.

So, think about what happens when a car turns a corner at a constant speed. The speedometer will show the same number throughout the turn. Is the velocity also constant? The answer is NO! The velocity is actually changing, but not by virtue of a change in speed. Rather, the velocity is changing by virtue of a change in direction. Speed and velocity are related, but they are not the same thing.

READING CHECK

Distinguish between speed and velocity.

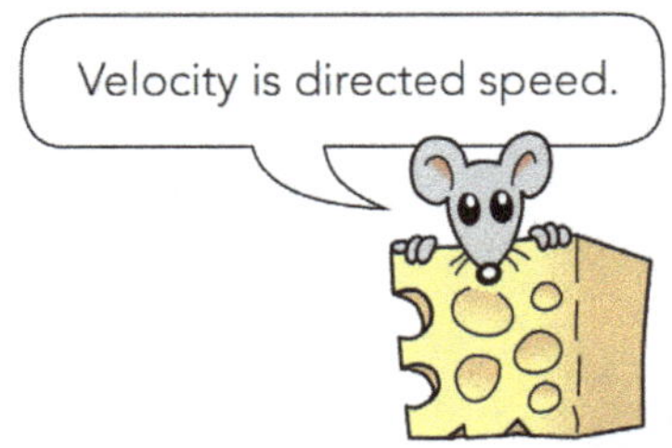

Scalar and Vector Quantities

There are two types of quantities: scalar and vector. A scalar quantity only includes the magnitude of the quantity. Magnitude tells us 'how much".

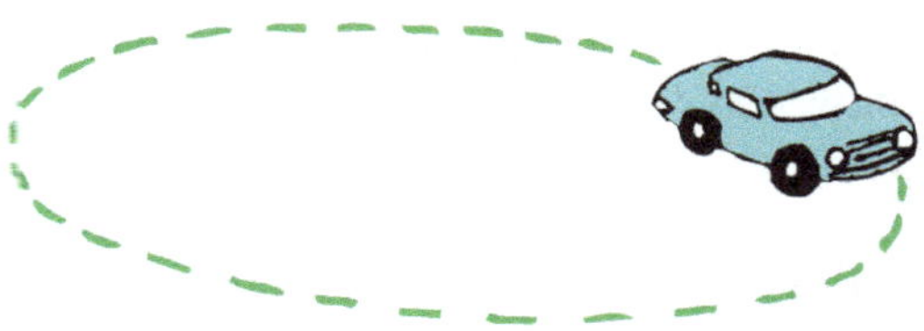

< Figure 1.8
The car on the circular track may have a constant speed, but its velocity is changing every instant. Why?

Mass is an example of a scalar quantity. Consider a 10 kg rock and ask yourself "how much" do you have? The answer is 10 kg. The magnitude is "10". You can think of magnitude as the number associated with the unit. Speed is also an example of a scalar quantity. Consider traveling at 50 km/s. How much speed do you have? The answer is 50 km/s. The magnitude of your speed is "50".

A vector quantity is any quantity that includes magnitude (the number) but necessarily also includes direction. Velocity is an example of a vector quantity. Why? Because velocity, by definition, includes how fast an object is moving (the magnitude) and in what direction.

CONCEPT CHECK

1. What is the average speed of a cheetah that sprints 100 m in 4 s? How about if it sprints 50 m in 2 s?
2. The speedometer on a bicycle moving east reads 50 km/h. It passes another bicycle moving west at 50 km/h. Do both bikes have the same speed? Do they have the same velocity?
3. "The bird flies at a constant speed in a constant direction." Say the same sentence in fewer words.

CHECK YOUR ANSWER

1. In both cases the answer is 25 m/s.
2. Both bicycles have the same speed, but they have opposite velocities because they move in opposite directions.
3. "The bird flies at constant velocity."

As we study physics further, we'll be coming across various types of quantities. If the quantity is fully expressed with only magnitude, then it's scalar. If the quantity also requires direction, then it's vector. For example, energy only has magnitude and is thus a scalar quantity. By contrast, a force is a push or a pull, and any push or pull can only be made in a particular direction. Try pushing on a table without doing so in a particular direction. This can't be done! Thus, force is an example of a vector quantity—it necessarily includes direction.

1.3 The Law of Inertia

Aristotle used logic to establish his ideas of motion. Galileo used experiment. Galileo showed that experiments are better than logic in testing knowledge. Galileo was concerned with how things move rather than why they move. The path was clear for Isaac Newton (1642–1727) to make further connections on concepts of motion.

READING CHECK

What scientist had ideas that Newton incorporated into his first law of motion?

Galileo had set the stage for Newton with his concept of inertia, and his thoughts about motion without friction, which he as a force akin to an ordinary push or pull. That meant a friction-free moving object would slow down only if a force acted on it. At age 24, Isaac Newton elevated Galileo's ideas to the status of a fundamental law that underlies all motion. (He then developed two other laws of motion that we'll study in following chapters.)

Newton's first law, usually called the *law of inertia*, is a restatement of Galileo's idea.

> **Every object continues in its state of rest or of uniform speed in a straight line unless a net force acts on it.**

This says that things tend to keep on doing what they're already doing. Objects at rest tend to remain at rest. A *force* is needed to set them in motion, where, by definition, a **force** is a push or a pull. For example, dishes on a table are in a state of rest. The dishes tend to remain at rest even if you snap a tablecloth from beneath them. Again, for emphasis: it is the tendency of the dishes NOT to move. Also, the heavier those dishes the greater that tendency not to move (don't try this with light plastic ware). And, importantly, when you snap the table cloth, you need to do so in just the right fashion so that frictional forces between the heavy dishes and the table cloth are properly overcome.

Figure 1.9 > Inertia in action.

Quick Activity

Can You Pull a Piece of Paper Out From Beneath a Coin?

1. Place a penny on top of a smooth sheet of paper on a table.
2. Pull the paper horizontally with a quick snap.
3. Repeat using a quarter.
4. Try with a slow pull. Why the difference?

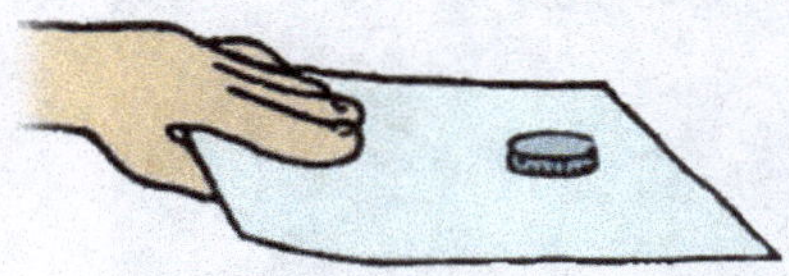

The law also says that when an object is moving, its tendency is to remain moving, along a straight-line path. For example, if you slide a hockey puck on the surface of slippery ice, it moves a long way until ice and air friction finally stops it. If you slide it along the surface of a city street, it is quickly brought to rest by the force of friction with the ground. It tends to keep sliding but friction acts against it. If you toss an object where there is no friction, such as in the vacuum of deep outer space, it will continually move in a straight-line path. An object's tendency to continue in its state of motion is called its *inertia*.

While the ancients thought continual forces were needed to maintain motion, the law of inertia provides a completely different way of thinking about motion. We now know that objects continue to move by themselves. If an object is at rest you'll have to apply force to get it moving, but once in motion no force is needed (except that to overcome any friction). The object moves in a straight line indefinitely. In the next chapter we'll see that forces are needed to change the speeds and directions of objects, but not to maintain motion if there is no friction.

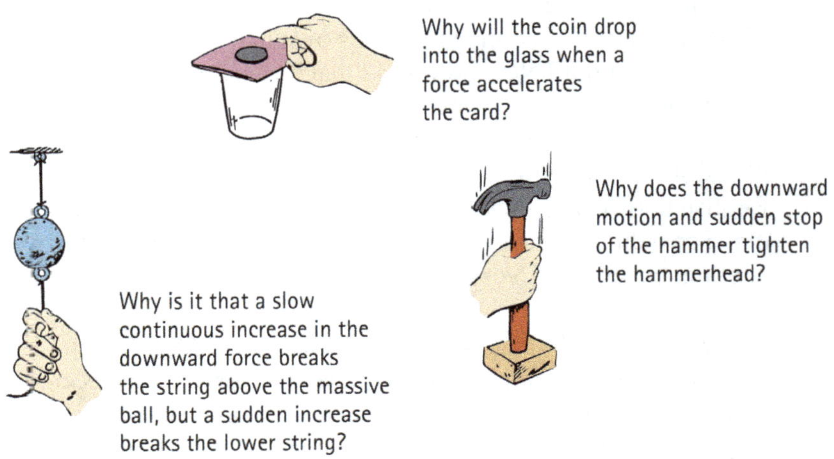

< Figure 1.10
Examples of inertia.

CONCEPT **CHECK**

When a space craft moves in a nearly circular orbit around Earth, is a force needed to maintain its high speed? If suddenly the force of gravity were cut off, what type of path would the craft follow?

CHECK YOUR ANSWER No force in the direction of the craft's motion exists. The space craft coasts by its own inertia. The only force acting on it is the force of gravity, which acts at right angles to its motion (toward Earth's center). We'll see later that this right-angled force holds the space craft in a circular path. If it were cut off, the craft would move in a straight line at constant speed (constant velocity), which means it would leave Earth.

1.4 The Life of Isaac Newton

Λ Figure 1.11
Isaac Newton at 47 years of age.

Isaac Newton was born prematurely and barely survived on Christmas Day, 1642, the same year that Galileo died. Newton's birthplace was his mother's farmhouse in Woolsthorpe, England. His father died several months before his birth, and he grew up under the care of his mother and grandmother. As a child he showed no particular signs of brightness, and as a young teen he was taken out of school to work on his mother's farm. As a farmer he was a failure, preferring to read books he borrowed from a neighboring druggist. An uncle sensed the scholarly potential in young Isaac and prompted him to study at the University of Cambridge, which he did for 5 years, graduating without particular distinction.

A plague swept through London, and Newton retreated to his mother's farm—this time to continue his studies. At the farm, at age 23, he laid the foundations for the work that was to make him immortal. Seeing an apple fall to the ground led him to consider the force of gravity extending to the Moon and beyond. He formulated the law of universal gravitation (which he later proved). He invented calculus, a very important mathematical tool in science. Newton extended Galileo's work and formulated the three fundamental laws of motion. He also formulated a theory of the nature of light and showed with prisms that white light is composed of all colors of the rainbow. It was Newton's experiments with prisms that first made him famous.

READING CHECK
What type of mathematics did Newton invent?

When the plague subsided, Newton returned to Cambridge and soon established a reputation for himself as a first-rate mathematician. His mathematics teacher resigned in his favor and Newton was appointed the Lucasian professor of mathematics. He held this post for 28 years. In 1672 he was elected to the Royal Society, where he exhibited the world's first reflector telescope. It can still be seen, preserved at the library of the Royal Society in London with the inscription: "The first reflecting telescope, invented by Sir Isaac Newton, and made with his own hands."

It wasn't until Newton was 42 that he began to write what is generally acknowledged as the greatest scientific book ever written, the *Principia Mathematica Philosophiae Naturalis.* He wrote the work in Latin and completed it in 18 months. It appeared in print in 1687 and wasn't printed in English until 1729, 2 years after his death. When asked how he was able to make so many discoveries, Newton replied that he solved his problems by continually thinking very long and hard about them—and not by sudden insight.

At the age of 46 he was elected a member of Parliament. He attended the sessions in Parliament for 2 years and never gave a speech. One day he rose and the House fell silent to hear the great man. Newton's "speech" was very brief; he simply requested that a window be closed because of a draft.

READING CHECK
What did Newton do after resigning his professorship?

A further turn from his work in science was his appointment as warden and then as master of the mint. Newton resigned his professorship and directed his efforts toward greatly improving the workings of the mint, to the dismay of counterfeiters who flourished at that time. He maintained his membership in the Royal Society and was elected president, then was re-elected each year for the rest of his life. At the age of 62, he wrote *Opticks,* which summarized his work on light. Nine years later he wrote a second edition to his *Principia.*

Although Newton's hair turned gray at 30, it remained full, long, and wavy all his life. Unlike others in his time, he did not wear a wig. He was a modest man, very sensitive to criticism, and never married. He remained healthy in body and mind into old age. At 80, he still had all his teeth, his eyesight and hearing were sharp, and his mind was alert. In his lifetime he was regarded by his countrymen as the greatest scientist who ever lived. In 1705 he was knighted by Queen Anne. Newton died at the age of 85 and was buried in Westminster Abbey along with England's kings and heroes.

Newton showed that the universe ran according to natural laws—a knowledge that provided hope and inspiration to people of all walks of life and that ushered in the Age of Reason. The ideas and insights of Isaac Newton truly changed the world and elevated the human condition.

< Figure 1.12
Replica of reflecting telescope invented and built by Isaac Newton. Photo Credit: Andrew Dunn.

1.5 Net Force

Changes in motion are produced by a force or combination of forces (in the next chapter we'll refer to changes in motion as acceleration). When more than one force acts on an object, we consider the **net force**, which is the combination of all the acting forces. It is the net force that changes an object's state of motion.

READING CHECK
What, exactly, changes an object's state of motion?

For example, suppose you pull on a box of chocolate bars with a force of 5 pounds. If your friend also pulls with 5 pounds in the same direction, the net force on the box is 10 pounds. If your friend pulls on the box with the same force as you but in the opposite direction, the net force on it is zero. Now if you increase your pull to 10 pounds and your friend pulls oppositely with 5 pounds, the net force is 5 pounds in the direction of your pull. We see this in **Figure 1.13**, (on the next page) where instead of pounds, the scientific unit of force is used—the *newton*, abbreviated N.

In Figure 1.13, forces are shown by arrows. Arrows are used because forces are vector quantities. As mentioned earlier, a vector quantity has both magnitude (how much) and direction (which way). When an arrow represents a vector quantity, the arrow's length represents magnitude and its direction shows the direction of the quantity. Such an arrow itself is called a *vector*.

Figure 1.13 >
Net force is the sum of all the applied forces.

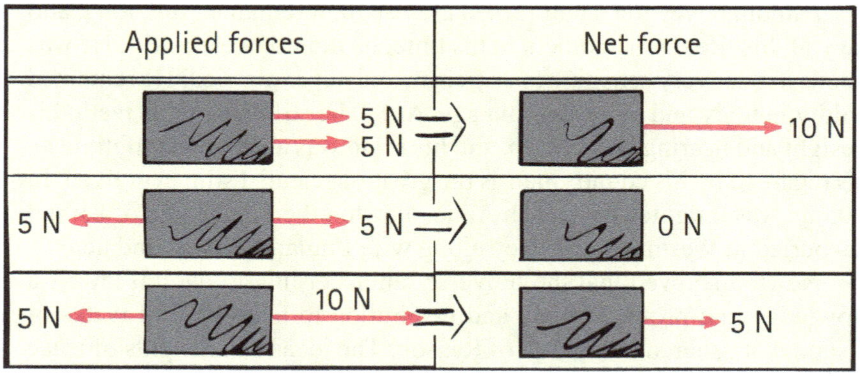

1.6 Static Equilibrium

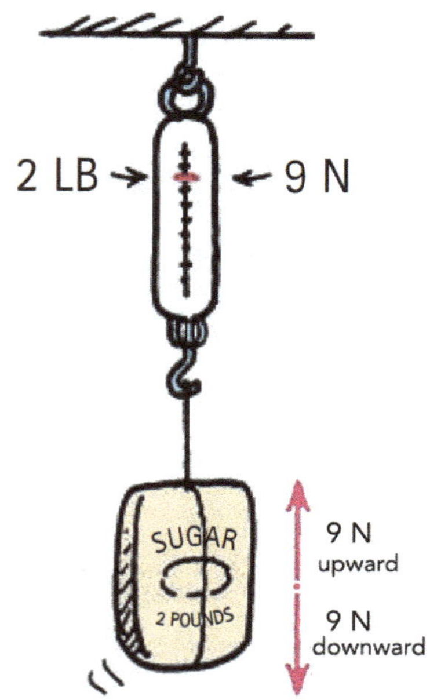

∧ Figure 1.14
The upward tension in the string has the same magnitude as the weight of the bag, so the net force on the bag is zero.

Suppose you tie a string around a 2-pound bag of sugar and hang it from a spring-based weighing scale (**Figure 1.14**). The spring in the scale stretches until the scale reads 2 pounds. The stretched spring experiences a "stretching force" called tension. The same scale in a science lab is likely in units of newtons. This scale will show the weight of the bag of sugar as about 9 newtons rather than 2 pounds. Both pounds and newtons are units of weight. Units of weight, in turn, are units of force (the force due to gravity). The bag of sugar is attracted to Earth with a gravitational force of 2 pounds—or equivalently, about 9 newtons. Hang twice as much sugar from the scale and the reading will be about 18 newtons.

Note there are two forces acting on the bag of sugar—tension force acting upward and weight acting downward. The two forces on the bag are equal and opposite, and cancel to zero. Hence the bag remains at rest. We say it is in *mechanical equilibrium*, which means it remains in a steady state. More specifically, we say the bag is in *static mechanical equilibrium,* or more simply, **static equilibrium**, which means there is no net force while the object is also not moving. The word "static" indicates a lack of motion.

Mechanical equilibrium is a state wherein no physical changes occur; it is a state of steadiness. When the net force on an object is zero, the object is said to be in mechanical equilibrium—this is known as the *equilibrium rule*. We express the equilibrium rule mathematically as:

$$\sum F = 0$$

The symbol $\sum$ stands for "the vector sum of" and F stands for "forces." For a suspended object at rest, like the bag of sugar mentioned earlier, the rule states that the forces acting upward on the object must be balanced by other forces acting downward to make the vector sum equal zero. (Vector quantities take direction into account, so, if upward forces are positive, downward ones are negative, and when summed they equal zero.)

In **Figure 1.15** we see the forces of interest to sign painters on a scaffold. The sum of the upward tensions is equal to the sum of their weights plus the weight of the scaffold. Note how the magnitude sum of the two upward vectors equal the magnitude of the three downward vectors. Net force on the scaffold is zero. The scaffold is in static mechanical equilibrium, which we can tell because it remains steady and also not rising nor falling.

READING CHECK

What does it mean to say that something is in mechanical equilibrium?

< Figure 1.15
The sum of the upward vectors equals the sum of the downward vectors. $\Sigma F = 0$ and the scaffold is in equilibrium.

CONCEPT **CHECK**

Consider the gymnast hanging from the rings.

1. If she hangs with her weight evenly divided between the two rings, how would scale readings in both supporting ropes compare with her weight?
2. Suppose she hangs with slightly more of her weight supported by the left ring. How would a scale on the right read?

CHECK YOUR ANSWER (Again, are you reading this before you have thought about and formulated your reasoned answers? If so, do you also exercise your body by looking at others do push-ups? Exercise your thinking: think before you look at the answers!)

1. The reading on each scale will be half her weight. The sum of the readings on both scales then equals her weight.
2. When more of her weight is supported by the left ring, the reading on the right is less than half her weight. No matter how she hangs, the sum of the scale readings equals her weight. For example, if one scale reads two-thirds her weight, the other scale will read one-third her weight.

1.7 Support Force

Consider a book lying at rest on a desk. It is in equilibrium. What forces act on the book? One is the force due to gravity—the *weight* of the book. Since the book is in equilibrium, there must be another force acting on it to produce a net force of zero—an upward force opposite to the force of gravity.

Where is the upward force coming from? It is coming from the desk that supports the book. We call this the support force, which is the upward force that balances the weight of an object on a surface. Technically, a support force can be called the *normal force*, which means it is a force that acts at a right angle away from the surface. For now, we only consider horizontal surfaces, which means the support force is straight upward—opposite of gravity, which is straight downward.

Notice that Newton's law of inertia and the equilibrium rule say the same thing: When $\Sigma F = 0$, objects don't change their states of motion.

For any object at rest on a horizontal surface, the support force must equal the object's weight. So, in this case, the support force must equal the weight of the book. If we say the upward support force is positive, then the downward weight is negative. The two forces add mathematically to zero. So, the net force on the book is zero. Another way to say the same thing is $\Sigma F = 0$.

READING CHECK

For an object at rest on a horizontal surface, what is the support force equal to?

To better understand how the desk pushes upward on the book, think about a spring being compressed (**Figure 1.16**). If you push the spring down, you can feel the spring pushing up on your hand. Similarly, the book lying on the desk compresses atoms in the desk, which behave like tiny springs. The weight of the book squeezes downward on the atoms, and the atoms squeeze upward on the book. In this way the compressed atoms produce the support force. If you removed those atoms there would no longer be a support force. There would only be the downward force of gravity and the object would fall. The support force of atoms within the ground you stand upon prevent you from falling to the center of the Earth.

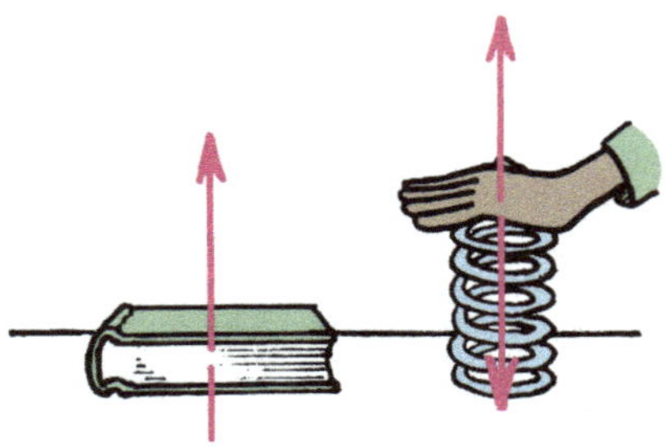

Figure 1.16

(Left) The desk pushes up on the book with as much force as the downward weight. (Right) The spring pushes up on your hand with as much force as you push down on the spring.

When you step on a bathroom scale, two forces act on the scale. One force is the downward pull of gravity, your weight. The other is the upward support force of the floor. These forces compress a mechanism (in effect, a spring) that is calibrated to show your weight (**Figure 1.17**). So, the scale shows the support force. When standing on a bathroom scale at rest, the amount of support force and the amount of your weight are equal. We say they have the same magnitude.

CONCEPT **CHECK**

1. What is the net force on a bathroom scale when a 110-pound person stands on it?
2. Suppose you stand on two bathroom scales with your weight evenly distributed between the two scales. What will each scale read? How about if you lean with more of your weight on one scale than the other?

CHECK YOUR ANSWER

1. Zero, for the scale remains at rest. The scale reads support force, which has the same magnitude as weight—not the net force.
2. The reading on each scale is half your weight. If you lean more on one scale than the other, more than half your weight will be read on that scale but less on the other. In this way they add up to your weight. Like the example of the gymnast hanging by the rings, if one scale reads two-thirds your weight, the other scale will read one-third your weight.

Figure 1.17

The upward support is as much as your weight.

1.8 Dynamic Equilibrium

As described earlier, an object at rest and not moving is in a state of static equilibrium. The forces on it add up to zero. But the forces acting on an object moving at constant speed in a straight-line path also add up to zero. We call this **dynamic equilibrium**, which is a form of mechanical equilibrium where there is motion. Note, the word "dynamic" indicates movement.

In accordance to Newton's first law, the forces on a moving object in dynamic equilibrium add up to zero. Although the moving object is changing locations, the rate at which it moves through space in a particular direction remains the same. For example, a bowling ball rolling at constant velocity is in equilibrium—until it hits the pins. Whether at rest or steadily rolling in a straight-line path, the sum of the forces on the bowling ball is zero: $\Sigma F = 0$.

It follows from Newton's first law that an object under the influence of only one force cannot be in equilibrium. Net force couldn't be zero in such a case. Only when there is no force at all or two or more forces combining to zero, can an object be in equilibrium. We can test whether or not something is in equilibrium by noting whether or not it undergoes changes in motion.

Consider a crate of canned food being pushed horizontally across a storeroom floor. If it moves at constant velocity, it is in equilibrium. This tells us that more than one force acts on the crate—friction. So the net force on the sliding crate is the forceful push by the person minus the opposing force of friction between the crate and the floor. The fact that the net force on the crate equals zero means that the force of friction must be equal in magnitude and opposite in direction to our pushing force.

In summary, objects at rest are said to be in static equilibrium; objects moving at constant velocity are said to be in dynamic equilibrium. Both of these situations are examples of mechanical equilibrium. Interestingly, there are other types of equilibrium. For example, when we study heat, we'll talk about thermal equilibrium, where temperature doesn't change.

Λ Figure 1.18
When the push on the crate is as great as the force of friction between the crate and the floor, the net force on the crate is zero and it slides at an unchanging speed.

READING CHECK
Distinguish between static and dynamic equilibrium.

CONCEPT **CHECK**
An airplane flies at constant velocity. In other words, it is in dynamic equilibrium. Two horizontal forces act on the plane. One is the thrust of the propeller that pushes it forward. The other is the force of air resistance (drag) that acts in the opposite direction. Which force is greater?

CHECK YOUR ANSWER Being in dynamic equilibrium tells us that both forces have the same magnitude. Call the forward force exerted by the propeller positive, then the air resistance is negative. Since the plane is in dynamic equilibrium, can you see that these two forces combine to equal zero?

1.9 Moving Together

∧ Figure 1.19
Can the bird drop down and catch the worm if Earth moves at 30 km/s?

READING CHECK

How does the law of inertia apply to objects in motion?

∧ Figure 1.20
When you flip a coin in a high-speed airplane, it behaves as if the airplane were at rest. The coin keeps up with you—inertia in action!

Before the 16th century, it was believed that Earth was the center of the universe. People then believed that the Sun circles Earth. As mentioned in the introduction chapter, Galileo advanced the idea that Earth moves around the Sun, instead of the other way around. There was much arguing and debate about this idea. People thought like Aristotle, and the existence of a force big enough to keep Earth moving was beyond their imagination.

One of the arguments against a moving Earth was the following. Consider a bird sitting at rest at the top of a tall tree. On the ground below is a fat, juicy worm. The bird sees the worm and drops vertically below and catches it. It was argued that this would be impossible if Earth were moving. A moving Earth would have to travel at an enormous speed to circle the Sun in one year. While the bird is in the air descending from its branch to the ground below, the worm would be swept far away along with the moving Earth. It seemed that catching a worm on a moving Earth would be an impossible task. The fact that birds do catch worms from high tree branches seemed to be clear evidence that Earth must be at rest.

Can you see the mistake in this argument? You can if you use the concept of inertia. You see, not only is Earth moving at a great speed, but so are the tree, the branch of the tree, the bird that sits on it, the worm below, and even the air in between. All move at the same speed. The law of inertia states that objects in motion remain in motion whenever no net force acts. So, objects on Earth move with Earth as Earth moves around the Sun. When the bird drops from the branch, its initial sideways motion remains unchanged. It catches the worm quite unaffected by the motion of its total environment. All the objects within this system are moving together, each of them following the same law of inertia.

We live on a moving Earth. Stand next to a wall. Jump up so that your feet are no longer in contact with the floor. Does the moving wall slam into you? Why not? It doesn't because you are also traveling as fast as Earth, with the same speed, before, during, and after your jump. Both you and the wall are moving together on that Earth.

Four hundred years ago, people had difficulty with ideas like these. One reason is that they didn't yet travel in high-speed vehicles. Rather, they took slow, bumpy rides in horse-drawn carts. People couldn't notice the effects of inertia as much. Today we flip a coin in a high-speed car, bus, or plane and catch the vertically moving coin as we would if the vehicle were at rest. We see evidence for the law of inertia when the horizontal motion of the coin before, during, and after the catch is the same. The coin keeps up with us.

Our ideas of motion today are very different from those of our distant ancestors. Aristotle did not recognize the idea of inertia, because he did not see that all moving things follow the same rules. He imagined different rules for motion in the heavens and motion on Earth. He saw horizontal motion as "unnatural," requiring a steady force. Galileo and Newton, on the other hand, saw that all moving things follow the same rules. To them, moving things require no force to keep moving if friction is absent. We can only wonder how differently science might have progressed if Aristotle had recognized the unity of all kinds of motion.

Concept Review

Chapter 1

Summary of Terms

Dynamic equilibrium A form of mechanical equilibrium where the object is in motion.

Equilibrium Rule $\Sigma F = 0$.

Force A push or a pull.

Inertia The property of things to remain at rest if at rest, and in motion if in motion.

Magnitude The amount of a quantity

Mechanical Equilibrium A state in which there is no net force acting on an object.

Net Force The combination of all forces that act on an object.

Newton The scientific unit of force.

Newton's First Law of Motion—The Law of Inertia: Every object continues in a state of rest, or in a state of motion in a straight line at constant speed, unless a net force acts on it.

Scalar Quantity A quantity fully described by its magnitude.

Speed The distance traveled per time.

Static Equilibrium A form of mechanical equilibrium where the object is at rest.

Support Force The force that supports an object, usually against gravity.

Vector Quantity A quantity that specifies direction as well as magnitude.

Velocity The speed of an object and specification of its direction of motion—a vector quantity.

Review Questions

1. What were two types of motion that Aristotle taught?
2. According to Aristotle, what kind of motion requires no force?
3. What two main ideas of Aristotle did Galileo discredit?
4. What is the name of the property of objects to maintain their states of motion?
5. Distinguish between speed and velocity.
6. Why do we say velocity is a vector quantity and speed is not?
7. How can you be both at rest and also moving at 100,000 km/h at the same time?
8. Between Aristotle and Galileo, who relied on experiments?
9. Who was the first to discover the concept of inertia, Galileo or Newton? Who incorporated it into a law of motion?
10. What is the tendency of a moving object when no forces act on it?
11. When only a pair of equal and opposite forces act on an object, what net force acts on it?
12. Why do we say that both velocity and force are vector quantities?
13. Name the force that occurs in a rope when both ends are pulled in opposite directions?
14. How much tension is in a rope that holds a 20-N bag of apples at rest?
15. What is the meaning of $\sum F = 0$?

16. Why is the support force on an object often called the normal force?

17. When you weigh yourself, is the scale reading the support force, or is it your weight?

18. Give an example of something moving when a net force of zero acts on it.

19. If we push a crate at constant velocity, how do we know how much friction acts on the crate compared to our pushing force?

20. If you're in a smooth-riding bus that is going at 50 km/h and you flip a coin vertically, what is the horizontal velocity of the coin in midair?

Quantitative Questions

21. Calculate Emily's average walking speed when she steps 1 meter in 0.5 second.

22. Calculate your average speed if you run 100 meters in 10 seconds.

23. Calculate the average speed of a mouse who runs across a 4-m long room in 0.4 seconds.

24. Calculate the speed of a bowling ball that moves 8 meters in 4 seconds.

Answers to Odd-Numbered Questions

1. Artistotle taught there were two types of motion: natural and unnatural.

3. Galileo discredited the idea that heavy objects fall faster than light ones, and that a force is necessary to maintain motion.

5. Speed is magnitude (how much) and velocity is speed with direction (how much and which way).

7. You can be at rest relative to the Earth, but moving at 100,000 km/h relative to the Sun.

9. Galileo. Then Newton advanced it to a fundamental law that underlies all motion.

11. Zero.

13. Tension.

15. It means that the vector sum of all the forces that act on an object in equilibrium equals zero. Forces cancel.

17. You actually read the support force acting on you, which is the same as your weight when the scale is stationary.

19. Since it slides in equilibrium (constant velocity), we know the friction must be equal and opposite to our push. That way they cancel and the crate slides without changing speed.

21. Average speed = total distance covered/time interval = 1 m/0.5 s = 2 m/s.

23. Average speed = total distance/time interval and this equals 4 m/0.4 s = 10 m/s.

Chapter 2

Newton's Second Law

THE MAIN IDEA

An object accelerates when a net force acts upon it

Kick a soccer ball and it moves. Does the ball's motion depend on its mass? Does its motion depend on how hard it's kicked? Or does how far the ball moves depend both on its mass and on how hard you kick it? What about air resistance? Does that also affect how far the soccer ball goes?

Consider some more examples. Suppose that you drop two soccer balls from a cliff, one ball with air inside and the other filled with heavy sand. Which hits the ground first? Does air resistance make a difference? Suppose you and a much-heavier friend sky dive from a high-flying plane. You both open your same-size parachutes at the same time. Who reaches the ground first? Does air resistance make a difference in this case? The rules guiding us to these answers are the same that got humans to the Moon—*Newton's Second Law.*

2.1 Acceleration

In addition to speed and velocity (previous chapter), Galileo experimented with inclined planes to develop the concept of *acceleration*. He found that balls rolling down inclines rolled faster and faster. Their velocity changed as they rolled. Further, the balls gained the same amount of velocity in equal time intervals. In other words, their velocity increased by a given amount each second.

Galileo defined the rate of change of velocity as acceleration:

$$\textbf{Acceleration} = \frac{\textbf{change of velocity}}{\textbf{time interval}}$$

You observe acceleration every time you ride in a bus. When the driver steps on the gas pedal the bus gains speed. We say it accelerates. We see why the gas pedal is called the "accelerator"!

When the brakes are applied, the car slows. This is also acceleration because the velocity of the car is changing. When something slows down, we often call this deceleration.

When a car makes a turn, even if its speed does not change, it is accelerating. Can you see why? Acceleration occurs because the car's

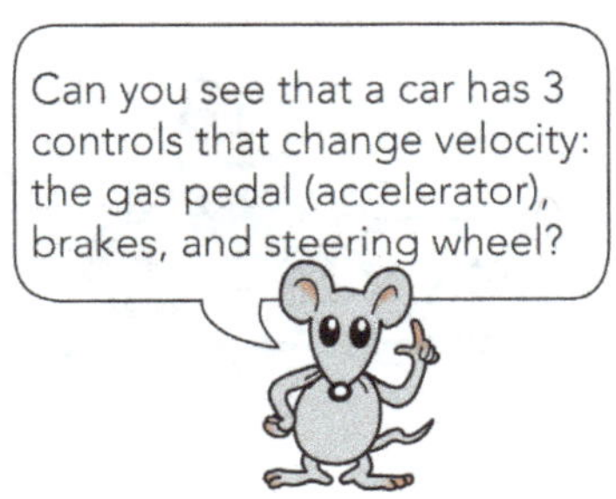

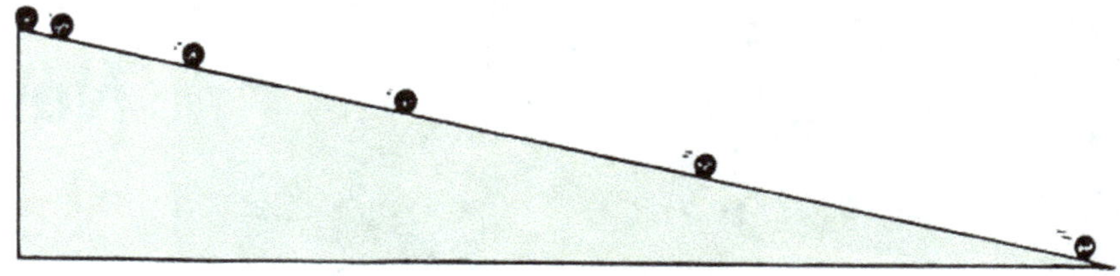

Figure 2.1 >
A ball gains the same amount of velocity each second as it rolls down an incline.

direction is changing. Acceleration is a change in velocity. So acceleration is a change in speed, a change in direction, or a change in both speed and direction. **Figure 2.2** illustrates this.

READING CHECK
Name three kinds of changes in motion that occur with acceleration.

Consider Nellie Newton riding downhill on her skateboard while making these measurements. In 1 second her velocity increases from 2 m/s to 4 m/s. In the next second her velocity increases from 4 m/s to 6 m/s. During this time frame she increases her velocity by 2 m/s for each second of travel. In equation form:

$$\textbf{Acceleration} = \frac{\textbf{change of velocity}}{\textbf{time interval}} = \frac{\textbf{2 m/s}}{\textbf{1 s}} = \textbf{2 m/s}^2$$

In this example Nellie's acceleration is 2 meters per second per second. Note that a unit for time enters twice: Once for the unit of velocity, and again for the interval of time in which the velocity is changing. Also note that acceleration is not just the change in velocity; it is the *change per second* of velocity. If either speed or direction (or both) changes, then the velocity changes. The *rate* at which velocity changes is the concept of acceleration.

It's important to emphasize that zero acceleration does not mean zero velocity. Zero acceleration means that the object will maintain the velocity it happens to have—neither speeding up nor slowing down nor changing direction.

Hold an apple above your head and drop it. It accelerates during its fall. When air resistance doesn't affect its motion, the apple is in free fall. Interestingly, the amount of acceleration is the same for all freely falling objects in the same vicinity. Experiment shows that a freely falling object gains a speed of 9.8 m/s during each second, which is usefully rounded off to 10 m/s each second.

$$\textbf{Acceleration} = \frac{\textbf{change of speed}}{\textbf{time interval}} = \frac{\textbf{10 m/s}}{\textbf{1 s}} = \textbf{10 m/s}^2$$

We read the acceleration of free fall as 10 meters per second squared. This is the same as saying that acceleration is 10 meters per second per second. Note again that the unit of time, the second, appears twice—once for the unit of velocity and again for the time during which velocity changes.

In **Figure 2.4**, we imagine a freely-falling boulder with a speedometer attached. As the boulder falls, the speedometer shows that the boulder falls 10 m/s faster each second. This 10 m/s gain each second is the boulder's acceleration.

Λ Figure 2.2
The vehicle undergoes acceleration when there is a change in its state of motion.

Λ Figure 2.3
Rapid deceleration is sensed by the driver who lurches forward (inertia in action!).

If you drop a ball it gains speed because of the force of gravity. How about a ball thrown straight upward? Once it leaves your hand, the ball continues moving upward for a while and then comes back down. While going up, it moves against gravity and loses speed. Guess how much speed it loses each second while going upward? And guess how much speed it gains each second while coming down? That's right: The change in speed per second is 10 m/s—whether moving up or down! At the highest point, when the ball changes direction from upward to downward, its instantaneous speed is zero. Then it starts downward just as if it had been dropped from rest at that height. It will return to its starting point with the same speed it had when thrown.

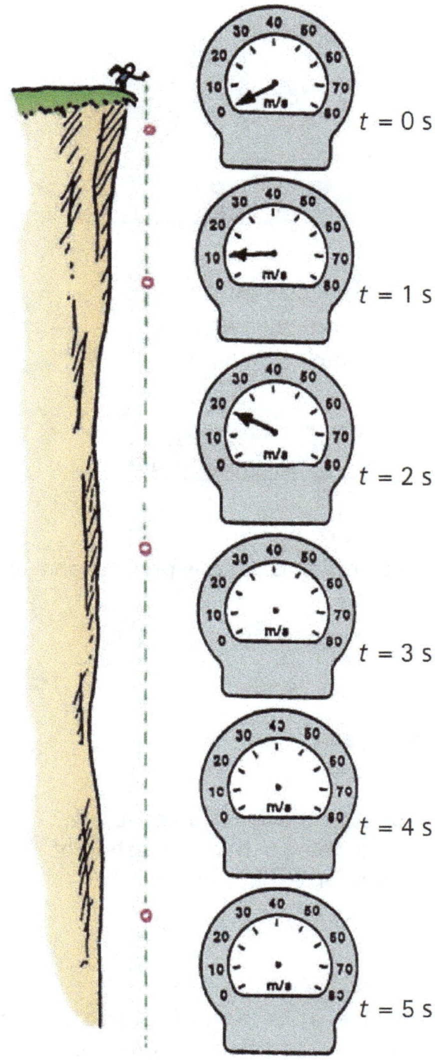

Figure 2.4 >
Imagine a falling boulder equipped with a speedometer. In each succeeding second of fall, you'd find the boulder's speed increasing by the same amount; 10 m/s. What would be the missing speedometer needle at *t* = 3 s, 4 s, and 5 s?

2.2 Net Force Causes Acceleration

Acceleration involves the *changes* in velocity that objects can undergo. Kick a ball and its state of motion changes and in a particular direction. You'll see that your kick has caused the ball to accelerate (Figure 2.5).

In order for any object to accelerate, a net force must be applied to that object. In other words, if you see an object accelerating, then you know that a net force is being applied to that object. You see a car going faster and faster down the street. You can be sure that there is some net force acting on that car.

Recall that by *net force*, we mean the sum of all forces acting upon that object. For example, if you push with 25 N on an object, and somebody else pushes in the opposite direction with 15 N, the net force applied to the object is 10 N. The object will accelerate as if a single 10-N force acts on it.

Suppose you pull a toy wagon with a net force of 20 N. The wagon accelerates. Now double the force and pull with 40 N. How much more will the wagon accelerate? There is a general rule here. If the net force is doubled, the acceleration also doubles. Three times the net force produces three times

Λ Figure 2.5
Kick the ball and it accelerates

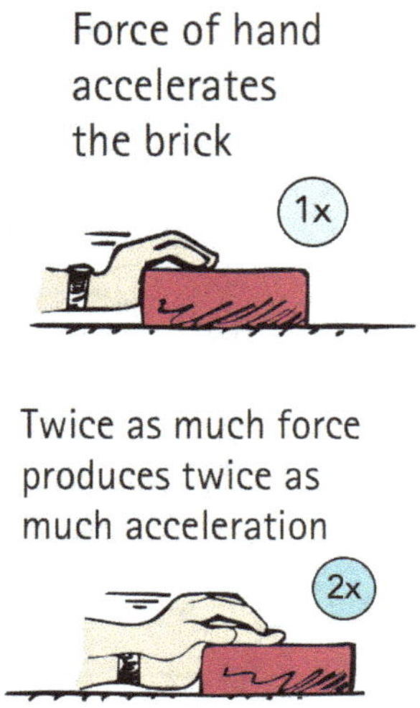

∧ Figure 2.6
Acceleration is directly proportional to net force.

READING CHECK
How does the direction of acceleration compare with the direction of the net force that produces it?

the acceleration. We say that the acceleration is directly proportional to the net force. We write:

$$\textbf{Acceleration} \sim \textbf{net force}$$

The symbol ~ stands for "is directly proportional to." That means any change in one quantity matches the same amount of change in the other quantity.

Acceleration is the rate at which velocity changes with time. This includes going faster, and also includes slowing down. You're in a car cruising at 20 mph. But with a stop sign just ahead, the driver gently applies the brakes and the car slows down. The net force of the brakes causes the car to accelerate. This acceleration, however, is in the opposite direction, so we can call this a negative acceleration, or, usually, a deceleration. Either way, slowing down is a form of acceleration because it involves a *change* in motion.

So, we see that the direction of acceleration need not be the same as the direction of the moving object. The direction of acceleration, however, is *always* in the direction of the net force. There are no exceptions to this. Recall that acceleration and net force are directly proportional to each other. What this means is that if you increase the net force, then acceleration will increase in proportion to that net force and be in the same direction as that net force.

We find that when a force is applied in the direction of the object's motion, the speed increases. This happens to a car when the driver presses the accelerator pedal. When a force is applied in the opposite direction, the speed decreases. This happens to a car when the driver presses the brake pedal (Figure 2.7). What then happens when the force is applied at some angle to the object's motion? Answer: The moving object will change direction as dictated by the direction of the net force. This happens to a car when the driver rotates the steering wheel. No matter what, acceleration is always in the direction of the net force.

Figure 2.7 >
The brakes of a car apply a force in the opposite direction of the car's motion. This causes the car to slow down. Think of it this way, if you wanted to slow down the car yourself, in which direction would you push on the car: from behind or from the front?

Quick Activity

Net Force and Acceleration

The net force acting on an object and the resulting acceleration are always in the same direction. You can demonstrate this with a spool. If the spool is pulled horizontally to the right, in which direction will it roll? Don't overthink this. Just try it and see.

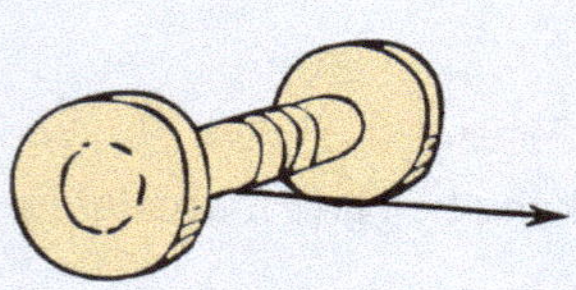

CONCEPT **CHECK**

1. If you push on a shopping cart it will accelerate. If you apply four times the net force, how much greater will the acceleration be?
2. If the net force acting on a sports car is reduced to half, how will the acceleration change?

CHECK YOUR ANSWERS

1. Acceleration will be four times as much.
2. Acceleration will be half as much..

2.3 Mass, Inertia, Volume, and Weight

So, we see that net force produces acceleration. How much acceleration, however, also depends on something else. It depends on the *mass* of the object being pushed or pulled. Consider what happens when you kick an empty can? It accelerates—changes its state of motion. Now kick the same can filled with cement. What happens? It doesn't accelerate as much as when it was empty. If the can is full of something really massive, like solid lead, then the can will very much resist moving, which will very much hurt your toe. Ouch!

The more massive full can has more resistance to a change in motion. We find that the mass of an object relates to its inertia. The greater an object's mass, the greater its inertia. This is why powerful engines are required in tractor trailers that pull massive loads—and why they have powerful brakes for stopping. Heavy loads have lots of inertia, which makes them hard to get moving and hard to stop.

Figure 2.8
The greater the mass, the greater the force needed for a given acceleration.

Mass, however, is more than an indication of an object's inertia. Mass is also a measure of how much material an object contains. Mass depends on the number and kinds of atoms making up the object. We define **mass** as follows:

> **Mass is the amount of matter in an object. Also, a measure of the inertia, or "laziness," that an object shows when you try to change its state of motion.**

A dense material such as lead is made up of many tightly packed atoms. So, lead has a lot of mass for a given volume.

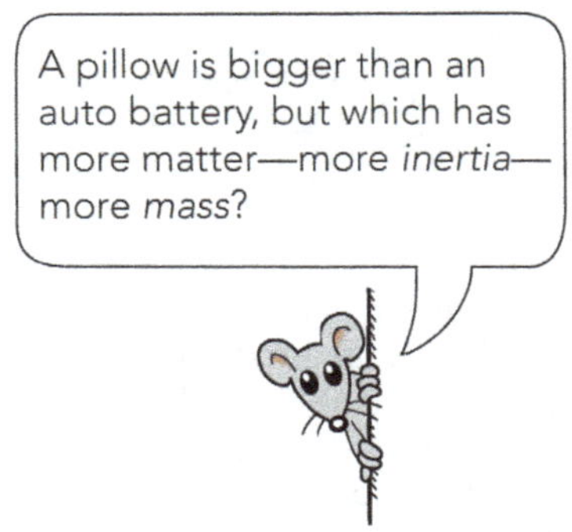

Mass Is Not Volume

Do not confuse mass and volume. **Volume** is a measure of space. It is measured in units such as cubic centimeters, cubic meters, or liters. Mass is measured in **kilograms**. If an object has a large mass, it may or may not have a large volume. For example, equal size bags of popcorn and jellybeans may have equal volumes, but very unequal masses. How many kilograms of matter an object contains and how much space the object occupies are two very different quantities. Mass is a measure of matter, and volume is a measure of space.

Mass Is Not Weight

Mass and weight are often confused with each other. We sometimes think, "Which is which?" The main reason for confusion is that mass and weight are *directly proportional* to each other. An object with twice as much mass also has twice as much weight. Mass refers to how much *stuff* an object is made of. Weight is the *gravitational pull* on that stuff. So know that weight depends on gravity, whereas mass does not. You'd weigh less on the Moon even though your mass would be the same in both locations. The Moon's pull is less because its gravity is less. Weight depends on where you are, while mass stays the same wherever.

Figure 2.9
An anvil in outer space, between the Earth and Moon for example, may be weightless, but it is not massless.

You can sense how much mass is in an object by feeling its inertia. When you shake an object back and forth you can sense its inertia. If it has a lot of mass, it's hard to change the object's direction. If it has a small mass, shaking the object is easier. To-and-fro shaking requires the same force even in regions where gravity is different—on the Moon, for example. The rock's inertia, or mass, is a property of the object itself and not its location.

We define **weight** as follows:

Weight is the force due to gravity on an object.

READING CHECK
Distinguish between mass and weight.

Objects with a great mass have a great weight; objects with less mass have less weight. In the same location, twice the mass weighs twice as much. That's what we mean when we say that mass and weight are proportional to each other. Remember that mass has to do with the amount of matter in the object and its inertia, while weight has to do with how strongly that matter is attracted by gravity.

Figure 2.10
The astronaut in space finds it is just as difficult to shake the "weightless" anvil as if it were on Earth. If the anvil is more massive than the astronaut, which shakes more—the anvil or the astronaut?

One Kilogram Weighs About 10 Newtons

The standard unit of mass is the kilogram, abbreviated kg. The standard unit of force is the newton as discussed in Chapter 1. The standard symbol for the newton is N. The abbreviation is written with a capital letter because the unit is named after a person. A 1-kg bag of any material has a weight of about 10 N in standard units. Away from the Earth's surface where the force of gravity is less, the bag would weigh less.

CONCEPT CHECK

1. Does a 2-kilogram iron block have twice as much *inertia* as a 1-kilogram iron block? Twice as much *mass*? Twice as much *volume*? Twice as much *weight* when weighed in the same location?

2. Does a 2-kilogram iron block have twice as much *inertia* as a 1-kilogram bunch of bananas? Twice as much *mass*? Twice as much *volume*? Twice as much *weight* when weighed in the same location?

3. How does the *mass* of a bar of gold vary with location?

CHECK YOUR ANSWERS

1. The answer is yes to all questions. A 2-kilogram block of iron has twice as many iron atoms, and therefore twice the amount of matter, mass, and weight. The blocks are made of the same material, so the 2-kilogram block also has twice the volume.

2. Two kilograms of anything has twice the inertia and twice the mass of one kilogram of anything else. Since mass and weight are proportional in the same location, two kilograms of anything will weigh twice as much as one kilogram of anything. Except for volume, the answer to all the questions is yes. Volume and mass are proportional only when the materials are the same—when they have the same density. Iron is much more dense than bananas, so two kilograms of iron must occupy less volume than one kilogram of bananas.

3. Not at all! It is made of the same number of atoms no matter what the location. Although its weight may vary with location, it has the same mass everywhere. This is why mass is preferred to weight in scientific studies.

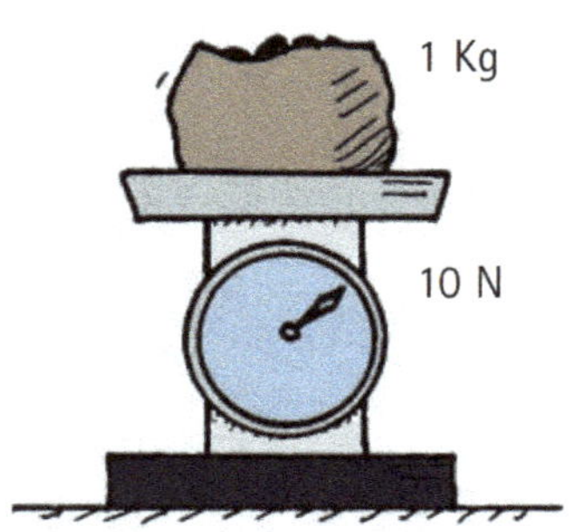

Figure 2.11
One kilogram of nails weighs about 10 newtons, which is equal to about 2.2 pounds.

So, 1 kilogram of something weighs about 10 newtons. If you know the mass in kilograms and want weight in newtons, multiply the number of kilograms by 10. Or, if you know the weight in newtons, divide by 10 and you'll have the mass in kilograms. Weight and mass are proportional to each other. In cases where precision is needed, for example in some lab experiments, 1 kilogram of matter actually weighs 9.8 N. For convenience we round 9.8 N off to 10 N.

The relationship between kilograms and pounds is that 1 kg weighs 2.2 lb at Earth's surface. (That means 1 lb is the same as 4.45 N.)

CONCEPT CHECK

Why is it okay to say a 1-kg bag of sand weighs 10 N, but a 1-kg bag of gold weighs 9.8 N? Don't they weigh the same?

CHECK YOUR ANSWERS

Both 1-kg bags have the same weight, and both weigh 10 N. However, since gold is more valuable, saying precisely 9.8 N rather than 10 N is usually a good idea. Except for in the lab, we won't make a big deal in this program about whether or not rounding off is okay. It's more important to learn the concepts first. Fidgeting over numerical details can come later.

2.4 Mass Resists Acceleration

More massive objects are more difficult to accelerate. Experiments show that for the same force, twice as much mass results in half as much acceleration; three times the mass results in one-third the acceleration, and so forth. In other words, the acceleration produced by a given force is inversely proportional to the mass. We write:

$$\text{Acceleration} \sim \frac{1}{\text{mass}}$$

By inversely we mean that the two values change in opposite ways. When one value gets larger, the other gets smaller. Or when one gets smaller, the other gets larger. More mass means less acceleration, because more mass means more resistance to changes in motion. (Mathematically we see that as the denominator increases, the whole quantity decreases. For example, the quantity 1/100 is less than the quantity 1/10.)

READING CHECK

For a given force, how does the acceleration of an object relate to its mass?

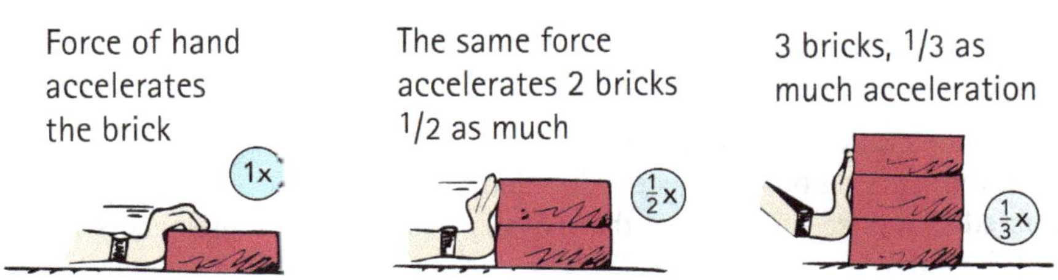

Figure 2.12
Acceleration, shown in blue circle, is inversely proportional to mass.

CONCEPT **CHECK**

Suppose you apply the same amount of force to two carts, one cart with a mass of 4 kg, and the other with a mass of 8 kg. Which cart will accelerate more? How much greater will the acceleration be?

CHECK YOUR ANSWER

The 4-kg cart will have more acceleration. The 4-kg cart will have twice the acceleration because it has half as much mass—which means half as much resistance to changes in motion.

2.5 Newton's Second Law

Isaac Newton was the first to realize the connection between force and mass in producing acceleration. He discovered one of the most important rules of nature ever proposed—his second law of motion. Newton's second law states:

> **The acceleration produced by a net force on an object is directly proportional to the net force, is in the same direction as the net force, and is inversely proportional to the mass of the object.**

Or in shorter notation,

$$\text{Acceleration} \sim \frac{\text{net force}}{\text{mass}}$$

By using consistent units such as newtons (N) for force, kilograms (kg) for mass, and meters per second squared (m/s^2) for acceleration, we get the exact equation.

$$\text{Acceleration} = \frac{\text{net force}}{\text{mass}}$$

In briefest form, where a is acceleration, F is net force, and m is mass:

$$a = \frac{F}{m}$$

The acceleration of an object equals the net force divided by the mass. If the net force acting on an object is doubled, the object's acceleration will be doubled. Suppose instead that the mass is doubled. Then the acceleration will be halved. If both the net force and the mass are doubled, then the acceleration will be unchanged.

Figure 2.13
Acceleration depends on the mass being pushed.

READING CHECK

How does the acceleration of an object relate to the net force on it and its mass?

CONCEPT **CHECK**

If you push on a shopping cart, it will accelerate.

1. If you push with the same force, but the cart is loaded with groceries so it has five times as much mass, what happens to the acceleration?
2. If you push five times harder on the loaded cart, what happens to the acceleration?

CHECK YOUR ANSWER

1. Acceleration will be less—only one-fifth as much.
2. It will have the same acceleration as it had without the load.

2.6 Friction

Friction occurs when one object tends to rub or does rub against something else. Friction occurs for solids, liquids, and gases. An important rule of friction is that it acts in a direction to oppose motion. If you drag a bag of cement along a floor to the left, the force of friction on the bag will be to the right. A boat propelled to the east by its motor experiences water friction to the west. When an object falls downward through the air, the friction due to air resistance (**air drag**) acts upward. Friction acts in a direction to oppose motion.

READING CHECK
When does friction occur?

The amount of friction between two surfaces depends on the kinds of material and how much they are pressed together. Friction between a crate and a rough wooden floor is greater than between the same crate and a polished marble floor. And if the surface is inclined, friction is less because the crate doesn't press as much on the inclined surface.

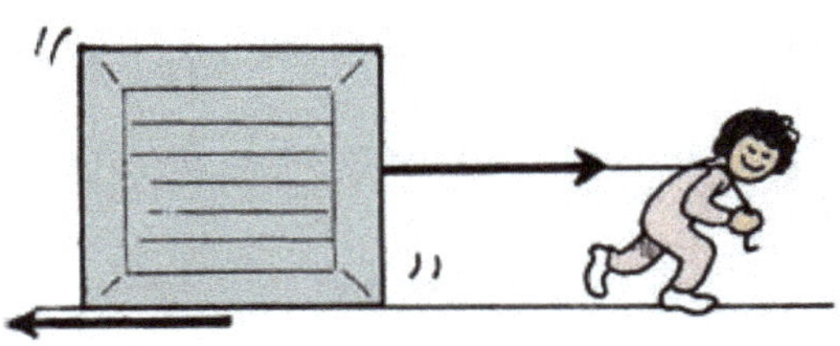

Figure 2.14
Applied force just overcomes friction so the crate slides at constant velocity.

When you push horizontally on your textbook and it slides across the table, both your force and the opposite force of friction affect the motion. When you push hard enough to match friction, the net force on the book is zero and it slides at constant velocity. Notice that we are talking about what we learned earlier—no change in motion occurs when $\sum F = 0$.

Practicing Physics

A crate filled with delicious peaches rests on a horizontal floor. The only forces acting on the crate are gravity and the support force of the floor. We show these forces with vectors. F_w is the weight vector, and F_N is the support force. We use the subscript N to indicate the force is normal to the floor. Normal means at a right angle with respect to the floor. Support forces, interestingly, are always "normal" to the surfaces of support.

1. Choose the correct words for the sketch above:

- The net force on the crate is (zero) (greater than zero).
- Evidence for this is that there is no (velocity) (acceleration).

2. A small pull P is exerted on the crate, but not enough to move it.

- A force of friction acts, which is (less than *P*) (equal to *P*) (greater than *P*).
- Net force on the crate is (zero) (greater than zero).

3. Pull P is increased until the crate begins to slide. It is pulled so that it moves at constant velocity across the floor.

- Friction force F_f is (less than *P*) (equal to *P*) (greater than *P*).
- Constant velocity means acceleration is (zero) (greater than zero).
- Net force on the crate is (less than zero) (zero) (greater than zero).

4. Pull P is further increased and is now greater than friction F_f

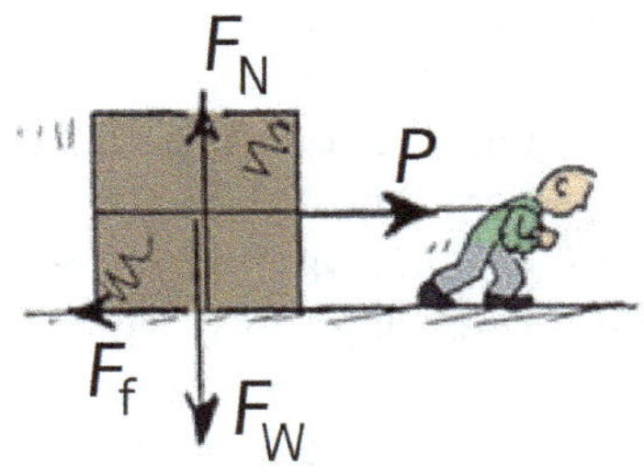

• Net force on the crate is (zero) (greater than zero).
• The net force acts to the right, so acceleration acts toward the (left) (right).

We see here that the friction force varies with different pulling strengths. When the pull is small enough not enough to move the crate, friction equals the pull, and net force is zero. As the pull is increased, friction increases the same, and net force remains zero. When the pulling force is greater than the friction, the net force is greater than zero and the crate accelerates. Once the crate is moving, the pull can be reduced until it matches the friction and the crate moves steadily. Then acceleration is zero and the net force is zero.

Discuss this with your classmates until you understand it. When you do, you're understanding good physics!

CONCEPT CHECK

1. Two forces act on a bowl resting on a table: the bowl's weight and the support force from the table. Does a force of friction also act on the bowl?

2. Suppose that a high-flying jumbo jet flies at constant velocity when the thrust of its engines is a constant 80,000 N. What is the acceleration of the jet? What is the force of air drag acting on the jet?

CHECK YOUR ANSWERS

1. No, not unless the bowl tends to slide or does slide across the table. For example, if you push it toward the left, then friction between the bowl and table will act toward the right. Friction forces occur only when an object tends to slide or is sliding.

2. The acceleration is zero because the velocity is constant (not changing). Since the acceleration is zero, it follows from $a = F/m$ that the net force is zero. This says that the force of air drag must be equal to the thrusting force of 80,000 N and act in the opposite direction. So, the air drag is 80,000 N.

Quick Activity

Feeling the Physics

We've just investigated the steps taken by a guy who encounters friction when pulling on a crate of peaches. Instead of thinking about this, actually DO the same by *pushing* on a piece of furniture. When you push slightly, you can FEEL the friction force that holds it stationary. Push a bit harder so it slides and you'll FEEL how pushing is easier. You'll discover that sliding friction is typically less than friction at rest. Friction truly is a force that tends to resist motion. How nice to feel the physics that you're learning!

2.7 Free Fall

READING CHECK
How is "free fall" defined?

When Galileo developed the concept of acceleration on inclined planes, he was interested in falling objects. Because there were no suitable timing devices at that time, he used inclined planes to effectively slow down acceleration. He found that as the planes were tipped at greater angles, the acceleration of the balls was greater. When tipped all the way vertical, acceleration was that of free fall. We define free fall as falling only under the influence of gravity, where other forces such as air drag can be neglected.

Galileo further discovered that the acceleration didn't depend on mass. On any incline all sliding objects have the same acceleration. Likewise, the acceleration of free fall doesn't depend on mass.

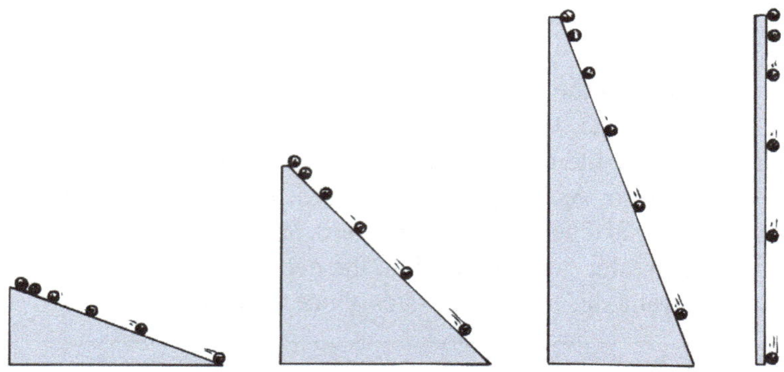

Figure 2.15
The greater the slope of the incline, the greater the acceleration of the ball. What is the acceleration when the incline is vertical?

As discussed earlier, Galileo discovered that heavy and light objects dropped from the Leaning Tower of Pisa fell with equal accelerations. This experiment destroyed the Aristotelian idea that heavy objects always fall faster than lighter ones in the same time. Galileo's experiment and many others demonstrated the same result. But Galileo couldn't say why the accelerations were equal. The explanation comes from Newton's second law.

A falling 10-kg boulder "feels" 10 times the force of gravity (weight) as a 1-kg stone. Followers of Aristotle believed the boulder should therefore fall ten times as fast as the stone—because they considered only the greater weight. But Newton's second law tells us to also consider the mass. As shown in **Figure 2.16**, can you see that ten times as much force acting on ten times as much mass produces the same acceleration as the smaller force acting on the smaller mass? In symbolic notation:

Figure 2.16
The ratio of weight (F) to mass (m) is the same for the 10-kg cannonball and the 1-kg stone.

$$\frac{F}{m} = \frac{\mathbf{F}}{\mathbf{m}}$$

where F stands for the force (weight) acting on the boulder, and m stands for its correspondingly large mass. The small F and m stand for the smaller weight and mass of the stone. We see that the ratio of weight to mass is the same for these or any objects. All freely falling objects have the same force/mass ratio and undergo the same acceleration at the same location. The acceleration due to gravity is represented by the symbol g.

It's of historical interest to note that although Galileo spoke about force, and was the first to propose the concepts of acceleration and inertia, he did not make the connection between these three concepts. It took the brilliance of Isaac Newton to show the connection—namely, $a = F/m$. This rule of mechanics is one of the most profound in physics. With it, scientists and engineers have been able to put people on the Moon.

FOR YOUR INFORMATION

Newton's second law $a = F/m$ can be re-expressed $F = ma$. Letting F be W, and a be g, the acceleration due to gravity, we can define weight as: $W = mg$. This means your weight is always your mass multiplied by the acceleration due to gravity. (We assume here that you are being weighed on Earth's surface. Interestingly, weighing in tomorrow's rotating space habitats calls for a different definition of weight.)

Figure 2.17
The ratio of weight (F) to mass (m) is the same for the large rock and the small feather; similarly, the ratio of circumference (C) to diameter (D) is the same for both the large and small circles.

2.8 Falling Through Air

When air drag is negligible on a falling object we say the object is in a state of *free fall*. Most often, air drag is significant for falling objects. In these cases, the acceleration of fall is *less* than g. The object is in non-free fall. Air drag depends on two things: speed and surface area. When a sky diver steps from a high-flying plane, air drag builds up as falling speed increases. The result is reduced acceleration. More reduction can occur by increasing surface area. A diver does this by orienting the body so more air is encountered—by spreading out like a flying squirrel. So, air drag depends on speed and the surface area encountered by the air (called the frontal area).

For free fall the downward net force is weight. Only weight! But when air is present, the downward net force = weight – air drag. Can you see that the presence of air drag reduces net force? And that less net force means less acceleration? So, as a diver falls faster and faster the acceleration of fall gets less and less. What happens to the net force if air drag builds up to equal weight? The answer is, net force becomes zero. Here we see $\sum F = 0$ again. Then acceleration becomes zero. Does this mean the diver comes to a stop? No! What it means is the diver no longer picks up speed. Acceleration terminates—it no longer occurs. We say the diver has reached **terminal speed**. If we are concerned with direction, down for falling objects, we say the diver has reached **terminal velocity**.

Terminal speed for a human skydiver commonly varies from about 150 to 200 km/h, depending on weight and orientation of the body. A heavier person falls faster for air drag to balance his or her weight. The greater weight is more effective in "plowing through" air. This means more terminal speed for a heavier person. Increasing frontal area reduces terminal speed. That's where a parachute comes in. A parachute greatly increases air drag, and terminal speed can be reduced to a safe 15 to 25 km/h.

READING CHECK
What two things does air drag depend upon?

Figure 2.18
The flying squirrel increases its frontal area by spreading out. This increases air drag and decreases the speed of its fall.

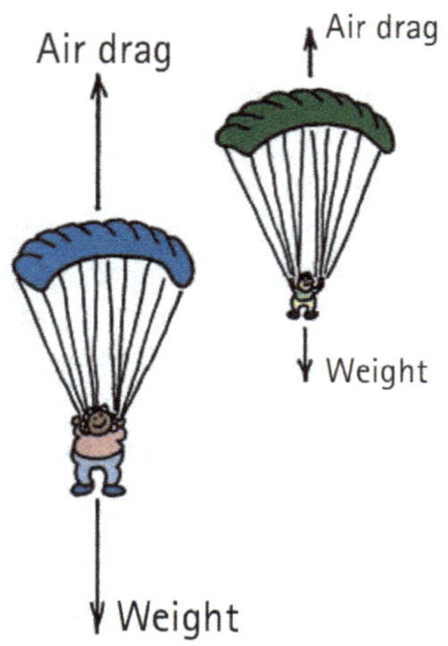

Λ Figure 2.19
The heavier parachutist must fall faster than the lighter parachutist for air drag to cancel her greater weight.

Λ Figure 2.20
A feather and a coin fall at equal accelerations in a vacuum. Try a similar thing by placing a feather on top of a book and then dropping the book.

We all know that a feather falls more slowly than a coin when both are dropped in air. But in a vacuum where air drag is not present, a feather and coin dropped together will fall side by side (**Figure 2.20**). A vacuum pump removes air from the glass tube. There's another way that lets a feather fall at *g*—dropping it atop a heavy object or a book that falls at *g*, as shown in the Introduction chapter. Nicely, the falling feather doesn't encounter the air that is plowed away by the dropped book.

When Galileo reportedly dropped objects of different weights from the Leaning Tower of Pisa, they didn't actually hit at the same time. They almost did, but because of air drag, the heavier one hit a split second before the other. But this contradicted the much longer time difference expected by the followers of Aristotle. The behavior of falling objects was never really understood until Newton announced his second law of motion. Isaac Newton truly changed our way of seeing the world.

CONCEPT CHECK

Consider two parachutists, a heavy person and a light person, who jump from the same altitude with parachutes of the same size.

1. Which person reaches terminal speed first?
2. Which person has the greatest terminal speed?
3. Which person gets to the ground first?
4. What if there were no air drag, like on the Moon?

CHECK YOUR ANSWERS

To answer these questions think of a coin and feather falling in air.

1. Just as a feather reaches terminal speed very quickly, the lighter person reaches terminal speed first.
2. Just as a coin falls faster than a feather through air, the heavy person falls faster and reaches a higher terminal speed.
3. Just like the race between a falling coin and feather in air, the heavier person falls faster and will reach the ground first.
4. If there were no air drag, there would be no terminal speed at all. Both would be in free fall and hit the ground at the same time.

Quick Activity

What Effect Does Air Resistance Have on Falling Objects?

1. Hold a sheet of paper and a book, side by side, over the top of a table. Then drop them at the same time.
2. Place the paper underneath the book and repeat.
3. Place the paper inside the book and repeat.
4. Place the paper on top of the book and repeat. Explain your observations.
5. To be more dramatic, substitute a feather for the sheet of paper!

Chapter 2

Summary of Terms

Acceleration The rate at which velocity changes with time; the change may be in magnitude or direction or both.

Air Drag Frictional resistance due to motion through air.

Free Fall Motion under the influence of gravitational pull only.

Friction The resistive force that opposes the motion or attempted motion of an object past another with which it is in contact, or through a fluid.

Inertia The property of things to resist changes in motion.

Inversely When two values change in opposite directions, so that if one increases and the other decreases by the same amount, they are said to be inversely proportional to each other.

Kilogram The fundamental SI unit of mass.

Newton's Second Law The acceleration produced by a net force on an object is directly proportional to the net force, is in the same direction as the net force, and is inversely proportional to the mass of the object.

Mass The quantity of matter in an object. More specifically, it is the measure of the inertia or sluggishness that an object exhibits in response to any effort made to start it, stop it, deflect it, or change in any way its state of motion.

Terminal Speed The speed at which the acceleration of a falling object terminates because air resistance balances its weight.

Terminal Velocity Terminal speed with direction of motion (down for falling objects).

Volume The quantity of space an object occupies.

Weight The force due to gravity on an object.

Review Questions

1. Distinguish between velocity and acceleration.
2. When are you most aware of motion in a moving vehicle—when it is moving steadily in a straight line or when it is accelerating?
3. What is the acceleration of free fall?
4. Is acceleration proportional to net force or does acceleration equal net force?
5. What relationship does mass have with inertia?
6. What relationship does mass have with weight?
7. Fill in the blanks: Shake something to-and-fro and you're measuring its ________. Lift it against gravity and you're measuring its _______.
8. What is the weight of a 1-kilogram brick?
9. Is acceleration directly proportional to mass, or is it inversely proportional to mass? Give an example.
10. If the net force acting on a sliding block is tripled, by how much does the acceleration increase?
11. If the mass of a sliding block is tripled at the same time the net force on it is tripled, how does the resulting acceleration compare to the original acceleration?
12. Suppose you exert a horizontal push on a crate that rests on a level floor, and it doesn't move. How much friction acts compared with your push?
13. As you increase your push, will friction on the crate increase also?
14. Once the crate is sliding, how hard do you push to keep it moving at constant velocity?

15. Why doesn't a heavy object accelerate more than a light object when both are freely falling?

16. The ratio of circumference/diameter for all circles is π. What is the ratio of force/mass for freely-falling bodies?

17. What two principal factors affect the force of air resistance on a falling object?

18. What is the acceleration of a falling object that has reached its terminal velocity?

19. If two objects of the same size fall through air at different speeds, which encounters the greater air drag?

20. Galileo defined the concept of acceleration. Who discovered its relation to force and mass?

Answers to Odd-Numbered Questions

1. Velocity is speed in a given direction, and acceleration is the time rate of change of velocity. The change can be in speed, direction, or both.

3. The acceleration of free fall is that provided only by gravity. On Earth it has the value 10 m/s^2.

5. Mass is a measure of inertia. The greater the mass, the greater the resistance to changes in motion.

7. Mass; weight.

9. Acceleration is inversely proportional to mass, which means that acceleration and mass change in opposite directions. Large mass means small acceleration, for example.

11. If you triple the force acting on three times the mass, there's no change in acceleration. Acceleration remains the same.

13. Yes, until your push is hard enough to get the crate sliding.

15. Acceleration is the ratio of force/mass. A heavy object has the same ratio as a light object, hence both accelerate equally in free fall.

17. Frontal area of fall and speed.

19. The faster one encounters greater air resistance.

Chapter 3

Newton's Third Law

THE MAIN IDEA

For every force, there is an equal force in the opposite direction.

Can *Conceptual Academy Biology* author Jennifer touch her daughter Io without Io also touching Jennifer? Newton's third law says no—that you can't touch without being touched! When a moving billiard ball hits a stationary billiard ball does the stationary billiard ball also hit the moving billiard ball? Yes! But do they hit each other with the same force? Can the heavyweight boxing champion of the world punch a piece of paper in midair any harder than the paper hits back? What if the paper is taped to a brick wall? In this chapter, we'll see how Newton's third law of motion, often called the law of action and reaction, guides our answers to these and other intriguing questions.

3.1 Interactions

In the previous chapters we've considered force to be simply push or a pull. Looking closer, Newton realized that a force is more than just a single push or pull. A force is part of a mutual action—an **interaction**—between one thing and another. When a truck crashes into a car there is an interaction between the truck and the car. Part of the interaction is the truck exerting a force on the car. The other part is the car exerting a force on the truck. The forces are equal in strength and opposite in direction and they occur at exactly the same time.

In every interaction, forces always occur in pairs. For example, you interact with the floor when you walk on it—you push backward against the floor, and the floor at the same time (simultaneously) pushes forward on you. Likewise, the tires of a car interact with the road—the tires push against the road, and the road pushes back on the tires. In swimming you interact with the water—you push the water backward, and the water pushes you forward. A **force pair** acts in each interaction. The interactions in these examples depend on friction. But a car on slippery ice, by contrast, may not be able to exert a force against the ice to produce the needed opposite force. Then the ice cannot push back and no change in motion occurs. For the car on slippery ice, the wheels will just spin and the car goes nowhere.

READING CHECK

In what specific way do forces occur in every interaction?

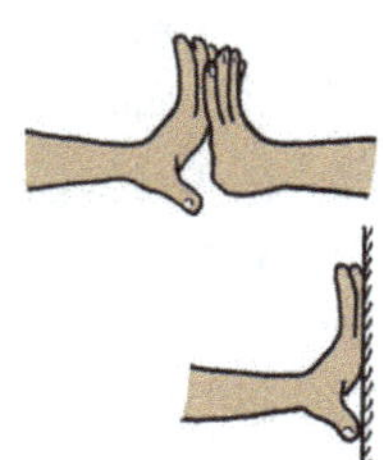

Figure 3.1 > You can feel your fingers being pushed by your friend's fingers. You also feel the same amount of force when you push on a wall and it pushes back on you. As a point of fact, you can't push on the wall unless it pushes back on you!

ʌ Figure 3.2
An autonomous car and truck with no passengers collide. In the interaction between this car and the truck, is the force of impact the same on each? Is the damage the same?

Can things like floors, car tires, and water exert forces? Your friends (not taking this course) may think that only living things like people and animals can exert forces. For example, when you push on a wall, how can the wall push back? It's not alive. It doesn't have muscles. But look at your fingers as you push on a wall. They're bent a little. Something must have pushed on them. Interestingly, the wall exerts the same amount of force on your fingers as your finger exert on the wall.

So, in this chapter we expand our thinking about forces. We see that non-living things can exert them. A force is more than a push or pull. It is part of a mutual interaction between objects.

Does a speeding baseball have force? The answer is no. Force is not something an object possesses, like mass. A speeding baseball exerts a force only when it hits something. Objects don't possess force as a thing in itself. As we will see in the following chapters, a speeding object possesses momentum and kinetic energy—but not force. A force is an *interaction* between one object and another.

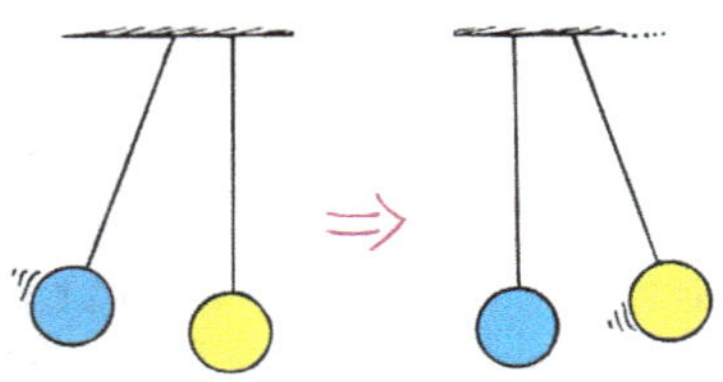

ʌ Figure 3.3
The impact forces between the blue and yellow balls move the yellow ball and stop the blue ball.

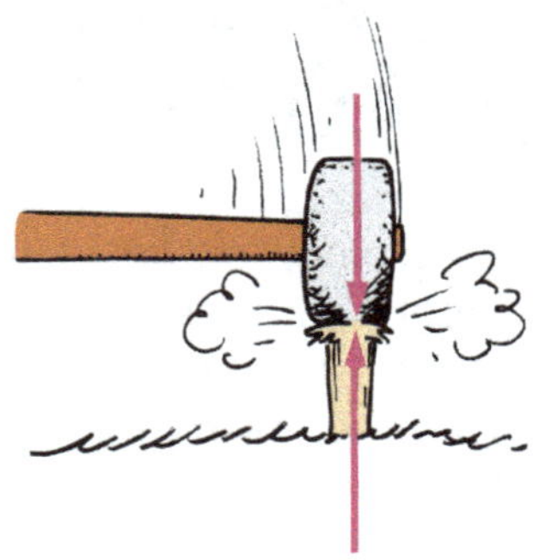

ʌ Figure 3.4
In the interaction between the hammer and the stake, each exerts the same amount of force on the other.

CONCEPT **CHECK**

1. Is it possible to push on something without that the same something pushing back on you? Defend your answer.

2. When tossing a ball, you exert a force on the ball. Does part of that force remain with the ball after it leaves your hand?

3. Does a high-speed baseball possess force?

CHECK YOUR ANSWERS

1. No, because every push is one part of a two-push interaction. You can only push on something if that something pushes back on you. If it doesn't push back, you have not exerted a force on it!

2. No. In tossing a ball your hand presses on the ball while the ball presses on your hand (which is why tossing a ball with a sore hand can be painful). As soon as contact between your hand and the ball ceases, all force by your hand ceases also. From that point on, the ball encounters forces due to gravity and air resistance—but no part of a former force by your hand.

3. No. A speeding baseball may exert a force on something when an interaction occurs, but it does not possess force as an entity in itself. In following chapters we'll learn that a speeding object possesses kinetic energy and momentum—but not force.

Quick Activity

Can An Interaction Consist of Only One Force?

Stretch a rubber band between your thumb and index finger. Ask yourself, which pulls stronger: your thumb or your index finger? Or do they pull the same? If you pull harder with your thumb, how does your index finger respond? Is there a way for one to ever pull harder than the other? Why do we say that forces only occur in pairs?

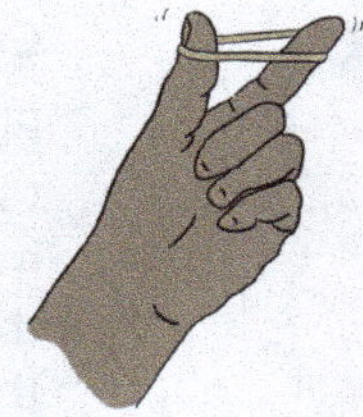

3.2 Action and Reaction

In his investigation of interactions, Newton discovered an underlying principle, called **Newton's third law**:

> **Whenever one object exerts a force on a second object, the second object simultaneously exerts an equal and opposite force on the first.**

We can call one force the action force, and the other the reaction force. Then we can express Newton's third law in the form:

> **To every action there is always an opposed equal reaction.**

It doesn't matter which force we call action and which we call reaction. The important thing is that they are co-parts of a single simultaneous interaction and that neither force exists without the other. Action and reaction forces are equal in strength and opposite in direction.

READING CHECK
Compare the relative strengths and directions of action and reaction forces.

Figure 3.5
When you lean against a wall, you exert a force on the wall. At the same time the wall exerts an equal and opposite force on you. That's why you don't topple over.

Quick Activity

Can There Be Only One Force in An Interaction?

Playing with magnets is fun. Applying Newton's third law to magnets is also fun. Hold a toy magnet near another magnet. Notice that when one magnet moves another, it is also moved by the other. For equal mass magnets the effect is most noticeable. That's because the changes in motion (acceleration) are the same for each. For different size magnets, the smaller magnets move more. Can you see how this ties into Newton's second law? (Newton's second law tells us that the acceleration of the magnet depends not only on force, but on mass.)

CONCEPT CHECK

1. Which exerts more force, the Earth pulling on the Moon, or the Moon pulling on the Earth?

2. When a heavy football player and a light one run into each other, is the third law correct in saying the force exerted on both players has the same strength?

3. Is the damage to the heavy player the same as the damage to the light one?

CHECK YOUR ANSWER

1. This is like asking which is greater, the distance between New York and San Francisco, or the distance between San Francisco and New York. Both distances are the same, but in opposite directions. Likewise, for the pulls between the Earth and the Moon.

2. Yes. In the interaction between the two players, the forces each exert on the other have equal strengths.

3. No. Although the forces are the same on each, the effects are quite unequal, particularly because they have different masses and, hence, will experience different accelerations. The low-mass player may be knocked unconscious while the heavier one may be completely unharmed. There is a difference between the force and the acceleration caused by the force.

A Simple Rule Helps Identify Action and Reaction

Here's a simple rule for identifying action and reaction forces. First, identify the interaction: One thing, say object A, interacts with another, say object B. Then action and reaction forces can be stated in the form:

Action: Object A exerts a force on object B.
Reaction: Object B exerts a force on object A.

We see that A and B are simply switched around. Consider the case of your hand pushing on the wall. The interaction is between your hand and the wall. We'll say the action is your hand (object A) exerting a force on the wall (object B). Then the reaction is the wall exerting a force on your hand. Look carefully to **Figures 3.6** and **3.7** for many examples.

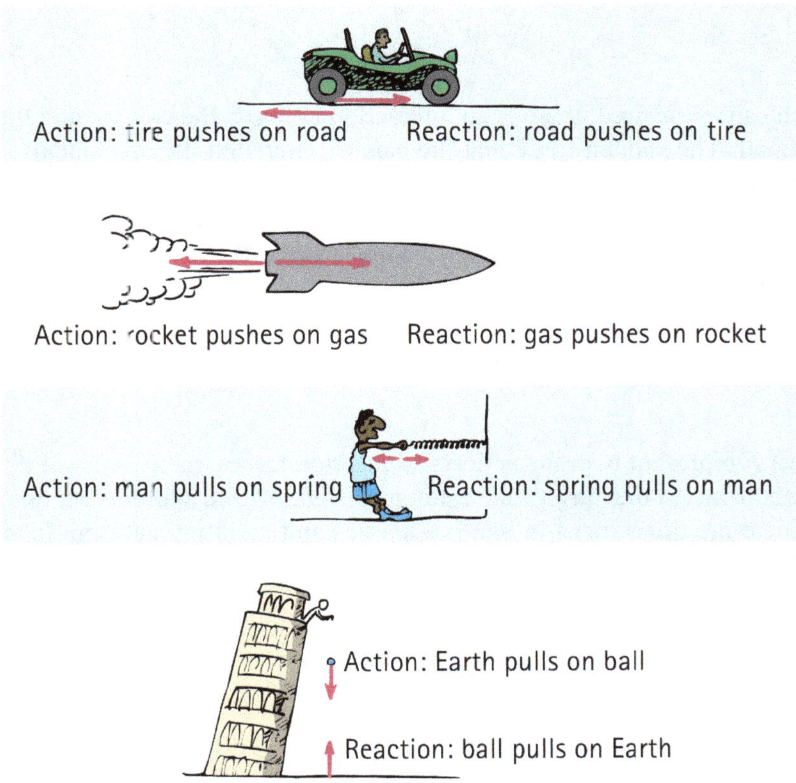

< Figure 3.6
Note that when action is "A exerts force on B," the reaction is simply "B exerts force on A."

Practicing Vectors

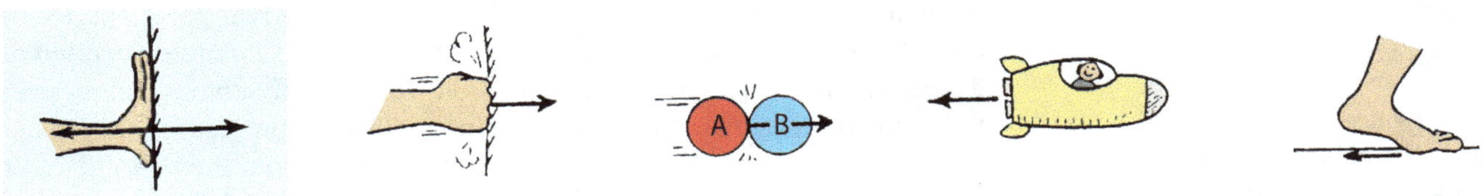

Λ Figure 3.7
The first sketch (boxed in blue) shows the pair of vectors that occur for your hand pushing a wall; your push on the wall to the right—the wall's push on your hand to the left. We see that action and reaction are equal and opposite. Only one force is shown for the other sketches, say, the action force. Draw vectors showing the reaction forces, or verbalize the action-reaction forces in each case.

CONCEPT **CHECK**
Can you identify the action and reaction forces in the case of an object that is falling in a vacuum? (A vacuum is a region of space that is completely empty—no air.)

CHECK YOUR ANSWER
To identify a pair of action-reaction forces in any situation, first identify the pair of interacting objects. In this case the Earth is interacting with the falling object through the force of gravity. So, the Earth pulls the falling object downward (call it action). Then reaction is the falling object pulling the Earth upward. (Hmm... you say, the falling object pulls the Earth upward? Yes, this may be hard to imagine at first but it is true. You don't notice the Earth being pulled upward by a falling object because of the Earth's large mass. More on this in Section 3.3.)

3.3 Different Masses

Λ Figure 3.8
The force exerted against the recoiling cannon is just as great as the force that drives the cannonball along the barrel. Why, then, does the cannonball undergo more acceleration than the cannon? Answer: It has less mass, hence, less inertia.

When a cannon is fired, there is an interaction between the cannon and the cannonball. The sudden force that the cannon exerts on the cannonball is exactly equal and opposite to the force the cannonball exerts on the cannon. This is why the cannon recoils (kicks). But the effects of these equal forces are very different. This is because the forces act on different masses. Recall Newton's second law:

$$a = \frac{F}{m}$$

Let F represent both the action and reaction forces, m the mass of the cannonball, and m the mass of the cannon. Different sized symbols are used to indicate the differences in relative masses and resulting accelerations. Then the acceleration of the cannonball and cannon are:

$$\textbf{Cannonball: } \frac{F}{m} = \boldsymbol{a}$$

$$\textbf{Cannon: } \frac{F}{\boldsymbol{m}} = a$$

READING CHECK
What is the relationship between force and mass when acceleration is produced?

Do you see why the change in velocity of the cannonball is so large compared to the change in velocity of the cannon? A given force exerted on a small mass produces a large acceleration, while the same force exerted on a large mass produces a small acceleration. To work effectively, canons need to be much more massive than the cannonballs they project.

We can extend the idea of a cannon recoiling from the ball it fires, to understanding rocket propulsion. Consider an inflated balloon recoiling when air is expelled. If the air is expelled downward, the balloon accelerates upward. A rocket accelerates the same way. It continually "recoils" from the ejected exhaust gas. Each molecule of exhaust gas is like a tiny cannonball shot from the rocket back end, as shown in **Figure 3.10**.

A common misconception is that a rocket is propelled by the impact of exhaust gases against the atmosphere. In fact, before the advent of rockets, it was commonly thought that sending a rocket to the Moon was impossible. Why? Because there is no air above the Earth's atmosphere for the rocket to push against. But this is like saying a cannon wouldn't recoil unless the cannonball had air to push against. Not true! Both the rocket and recoiling cannon accelerate because of the reaction forces by the material they fire—not because of any pushes on the air. In fact, a rocket works better above the atmosphere where there is no air drag.

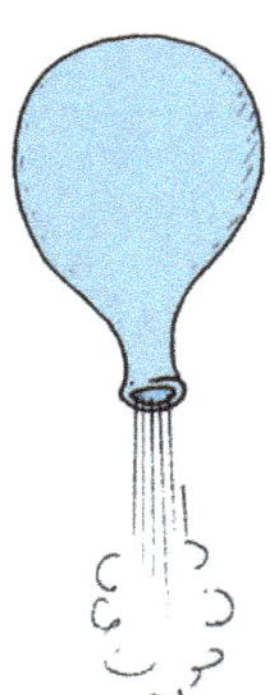

Λ Figure 3.9
The balloon recoils from the escaping air and climbs upward.

CONCEPT **CHECK**

A high-speed bus and an innocent bug have a head-on collision. The force of the bus on the bug splatters the poor bug all over the windshield. Is the corresponding force of the bug on the bus greater, less, or the same? Is the resulting deceleration of the bus greater than, less than, or the same as that of the bug?

CHECK YOUR ANSWERS

The magnitudes of both forces are the same, for they make up an action-reaction force pair in the bus-bug interaction. The accelerations, however, are very different because of the different masses! The bug undergoes an enormous and lethal deceleration, while the bus undergoes a very tiny deceleration—so tiny that the very slight slowing of the bus is unnoticed by its passengers. But if the bug were more massive, as massive as another bus, for example, the slowing down would be quite evident.

∧ Figure 3.10
The rocket recoils and rises from the "molecular cannonballs" it fires.

A note to the teacher about the above Concept Check. Be ready for students who will say "no way is the force on the bus as great as the force on the bug. After all, the bug is splattered and the bus moves as if no bug impact occurred." Stress the difference between a force itself and the *effect* of that force. Although the forces have equal magnitudes, the effects are enormously different, particularly for the bug. Force is one thing; effect is another. Force and its effect are entirely different.]

3.4 Different Objects

Since action and reaction forces are equal and opposite, why don't they cancel to zero? They don't cancel out because they act on *different* objects. Consider kicking a football (**Figure 3.11**). Call the force exerted by your foot action. That's the only horizontal force on the football, so the football accelerates. Reaction is the football exerting a force on your foot, which tends to slow your foot down a bit. You can't cancel the force on the football with a force on your foot. Forces cancel only when they act on the same object.

Now, what would happen if two players kicked the same football with opposite and equal forces at the same time, as shown in **Figure 3.12**? In this case, two interactions occur. Two different forces act on the football and these forces together cancel to zero. Is there a reaction force to each of these two forces that act on the football? The answer is yes, in accordance with Newton's third law.

∧ Figure 3.11
A acts on B, and B accelerates.

READING CHECK
When can forces cancel each other?

CONCEPT **CHECK**

An action has the same force as a reaction but in the opposite direction. Why don't they cancel?

CHECK YOUR ANSWER

The action and reaction forces are applied to different objects. Consider a foot kicking a ball. If the action is the foot applying a force to the ball, then the reaction is the ball applying a force to the foot. What happens to the ball in no way will cancel what happens to the foot. Different objects can impact each other through a single interaction.

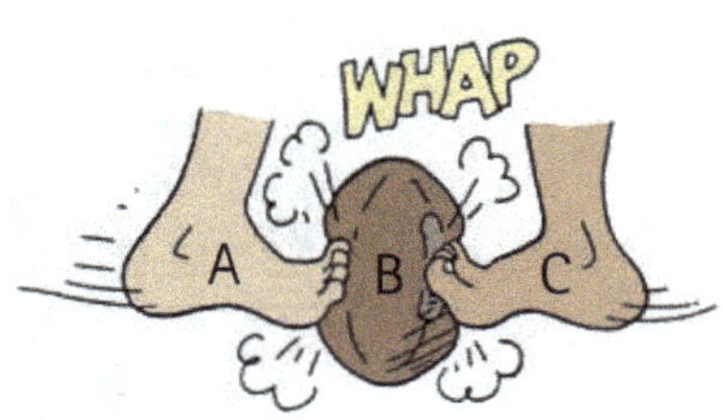

∧ Figure 3.12
Both A and C act on B. They can cancel each other so B does not accelerate.

3.5 Horse-Cart Problem

A situation similar to the kicked football is shown in the comic strip "Horse Sense". Here we pretend that the horse believes its pull on the cart will be canceled by the opposite and equal pull by the cart on the horse, making acceleration impossible. This is the classic horse-cart problem that is a stumper for many students at the university level. By thinking carefully, you can understand it here.

The horse-cart problem can be looked at from different points of view. One is the farmer's point of view where his only concern is getting his cart (the cart system) to market. Then, there is the point of view of the horse (the horse system). Finally, there is the point of view of the horse and cart together (the horse-cart system).

First look at the farmer's point of view—the cart system. The net force on the cart, divided by the mass of the cart, will produce an acceleration. The farmer doesn't care about the reaction on the horse.

Now look at the horse's point of view—the horse system. It's true that the opposite reaction force by the cart on the horse restrains the horse. This force tends to hold the horse back. Without this force the horse could freely gallop to the market. So how does the horse move forward? By interacting with the ground. At the same time the horse pushes backward against the ground, the ground pushes forward on the horse. If the horse pushes the ground with a greater force than its pull on the cart, then there will be a net force on the horse. Acceleration occurs. When the cart is up to speed, the horse needs only to push against the ground with enough force to offset the friction between the cart's wheels and the ground.

Finally, look at the horse-cart system as a whole. From this viewpoint, the pull of the horse on the cart and the reaction of the cart on the horse are internal forces—forces that act and react within the system. They contribute nothing to the acceleration of the horse-cart system. The forces cancel and can be neglected. To move across the ground, there must be an interaction between the horse-cart system and the ground. This is similar to trying to move a car while you're sitting in it. You can push and push against the dashboard, but no acceleration occurs. To get a car moving you must get outside and make the ground push you and the car. The horse-cart system is similar. It is the outside reaction force by the ground that pushes the system.

READING CHECK

Specifically, what force pushes a system across the ground?

< Figure 3.13

All the pairs of forces that act on the horse and cart are shown: (1) the pull P of the horse and the cart on each other; (2) the push F of the horse and the ground on each other; and (3) the friction f between the cart wheels and the ground. Can you see that the acceleration is due to the net force $F - f$ on the horse-cart system (while opposite $F - f$ acts on the ground)?

CONCEPT **CHECK**

1. What is the net force that acts on the cart in **Figure 3.13**? On the horse? On the ground?
2. Once the horse gets the cart moving at the desired speed, must the horse continue to exert a force on the cart?

CHECK YOUR ANSWERS

1. The net force on the cart is P-f; on the horse, F-P; on the ground, F-f.
2. Yes, but only enough to counteract wheel friction and air resistance. Interestingly, air resistance would be absent if there were a wind blowing in the same direction and just as fast as the horse and cart. If the wind blows fast enough to provide a force to counteract friction, the horse could wear rollerblades and simply coast with the cart all the way to the market.

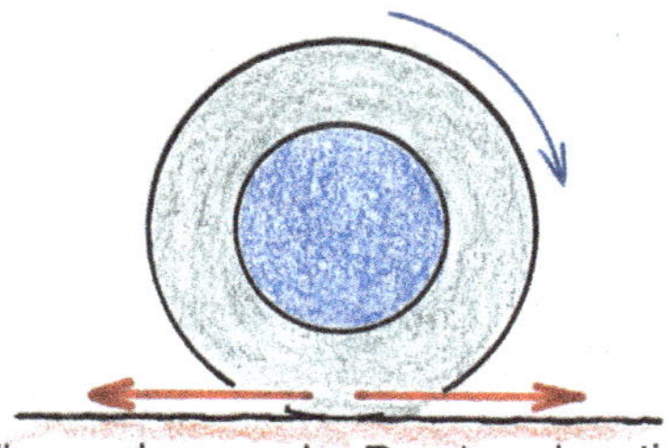

∧ Figure 3.14

Action: The tire of the rotating wheel pushes on the road. Reaction: The road pushes in the opposite direction on the tire (which is the *only* external horizontal force exerted on the car).

Figure 3.14 shows that the tires on the wheels of a car make contact with a road surface. Rotation of the tire produces a backward force of friction on the road (action). The road simultaneously produces a forward friction force on the tire (reaction), which propels the car forward. The originally straight-lined bricks shown in **Figure 3.15** have been pushed backward when cars accelerate forward. Bricks are pushed backward by the tires (action); and tires are pushed forward by the bricks (reaction). Bring your friends into a discussion of what actually pushes vehicle along roadways. You may enjoy explaining that it really is the road that pushes vehicles along a road!

Figure 3.15 >

The bricks were originally straight. Do you see their crookedness as evidence the road pushing back on tires as they accelerate from a stop?

3.6 Action Equals Reaction

When you tie a rope to a wall and pull on it, you produce a tension in the rope. Your pull on the rope and the pull by the supporting wall are equal and opposite. Otherwise there would be a net force on the rope and it would accelerate. The same is true if a friend holds one end of the rope and you have a tug-of-war. Rope tension when pulled at opposite ends is the same as the force provided by each end. Both pulls are the same in magnitude. This leads to a fascinating discovery for people who play tug-of-war. The team to win is not the team to exert the greatest force on the rope, but the greatest force against the ground!

READING CHECK

What must a team do to win in a tug-of-war?

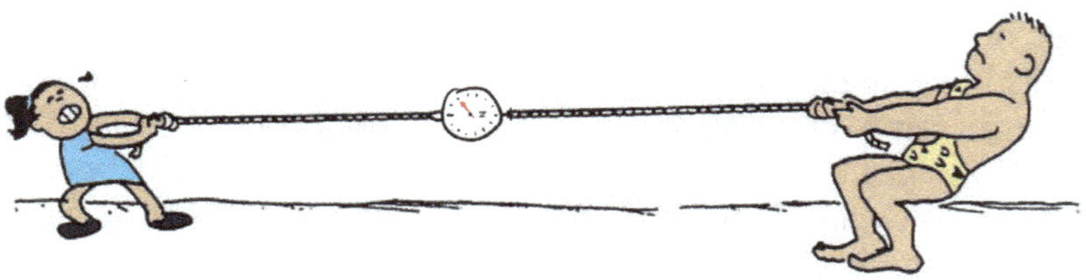

Figure 3.16 >

Arnold and Suzie pull on opposite ends of the rope. The spring scale within the rope measures the tension. Can Arnold pull any harder on the rope than Suzie pulls on it? If Suzie lets go, could Arnold provide tension in the rope?

< Figure 3.17
When a bird pushes air downward with its wings, air at the tips of its wings swirls upward, creating an updraft that is strongest off to its side. A trailing bird gets added lift by positioning itself in this updraft. This bird pushes air down and creates an updraft for the next bird, and so on. The result is a flock flying in a V formation.

Newton's third law tells us how a helicopter gets its lifting force. The whirling blades are shaped to force air particles down (action), and the air forces the blades up (reaction). This upward reaction force is called lift. When lift equals the weight of the craft, the helicopter hovers in midair. When lift is greater, the helicopter rises.

The same is true for airplanes and birds. Birds fly by pushing air downward. The air simultaneously pushes the bird upward. Interestingly, when some birds deflect the air downward, the moving air meets air below and is swirled upward, mostly near the edges of the wings. Off to the side another bird can position itself to get added lift from this updraft. This bird, in turn, creates an updraft for a following bird and so on. This is the physics of why geese and ducks fly in a V formation!

Λ Figure 3.18
The pair of vectors represents the force each wing exerts on the air. What forces act on the bird?

According to Newton's third law, you can't hit a piece of paper any harder than the paper can hit you back. Hold a sheet of paper in midair and tell your friends that nobody can hit the paper with a force of 200 N (45 lbs). You're correct even if the heavyweight boxing champion of the world hits the paper. The reason is that a 200-N interaction between the champ's fist and the sheet of paper in midair isn't possible—the paper is not capable of exerting a reaction force of 200 N. You cannot have an action force without its reaction force. Now, if you hold the paper against the wall, that is a different story. The wall will easily assist the paper in providing 200 N of reaction force, and more if needed.

CONCEPT **CHECK**

1. We said earlier that a car accelerates along a road because the road pushes it. Can we say that a team wins in a tug-of-war when the ground pushes harder on them than on the other team?
2. Does the scale read 100 N, 200 N, or zero?

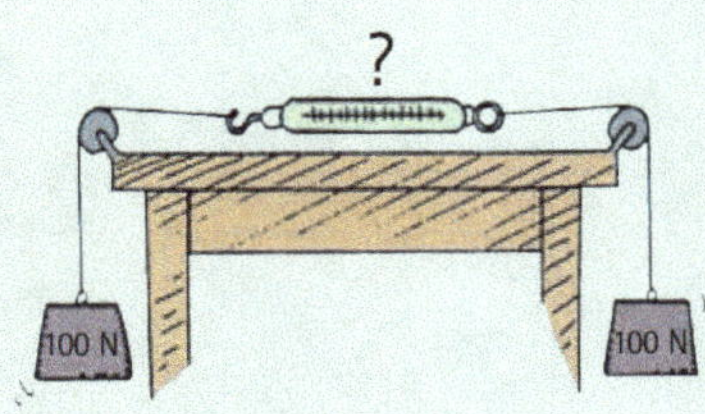

CHECK YOUR ANSWER

1. Yes!
2. Although the net force on the system is zero (as evidenced by no acceleration), the scale reading is 100 N, the tension in the string. Note that the string tension is 100 N in all the positions shown.

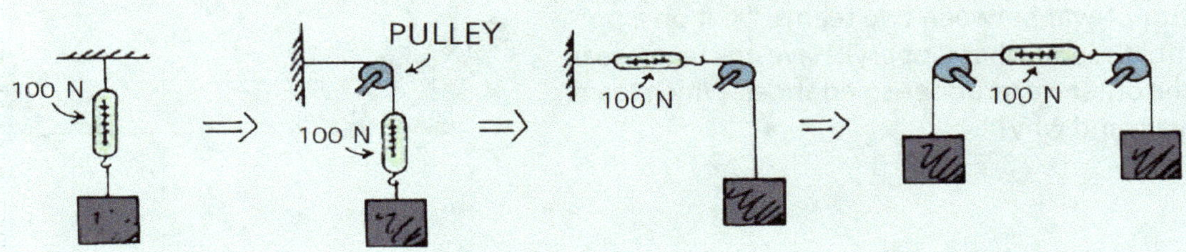

Λ Figure 3.19
He can hit the massive bag with considerable force. But with the same punch he can exert only a tiny force on the tissue paper in midair.

For every interaction between things, there is always a pair of oppositely-directed forces that are equal in strength. If you push hard on the world, for example, the world pushes hard on you. If you touch the world gently, the world touches you gently in return. The way you touch others is the way others touch you.

Summary of Newton's Three Laws

Newton's first law is the law of inertia: An object at rest tends to remain at rest; an object in motion tends to remain in motion at constant speed along a straight-line path. This property of objects to resist change in motion is called *inertia*. Mass is a measure of inertia. Objects will undergo changes in motion only when acted on by a net force.

Newton's second law is the law of acceleration: When a net force acts on an object, the object will accelerate. The acceleration is directly proportional to the net force and inversely proportional to the mass. Symbolically, $a = F/m$. Acceleration is always in the direction of the net force. When objects fall in a vacuum, the net force is simply the weight, and the acceleration is g (the symbol g denotes that acceleration is due to gravity alone.) When objects fall in air, the net force is equal to the weight minus the force of air drag, and the acceleration is less than g. If and when the force of air drag equals the weight of a falling object, acceleration terminates, and the object falls at constant speed (called terminal speed).

Newton's third law, the law of action-reaction: Whenever one object exerts a force on a second object, the second object exerts an equal and opposite force on the first. Forces come in pairs, one action and the other reaction, both of which occur simultaneously and comprise the single interaction between one object and the other. Action and reaction always act on different objects. Neither force exists without the other.

There has been a lot of new and exciting physics since the time of Isaac Newton. Nevertheless, and quite interestingly, it was primarily Newton's laws that got us to the Moon.

Quick Activity

Slip Slide Tug-of-War

Perform a tug-of-war between two teams. Do it on a polished floor (that's somewhat slippery). Have one team wear socks and the other wear rubber-soled shoes. Which team will surely win, and why?

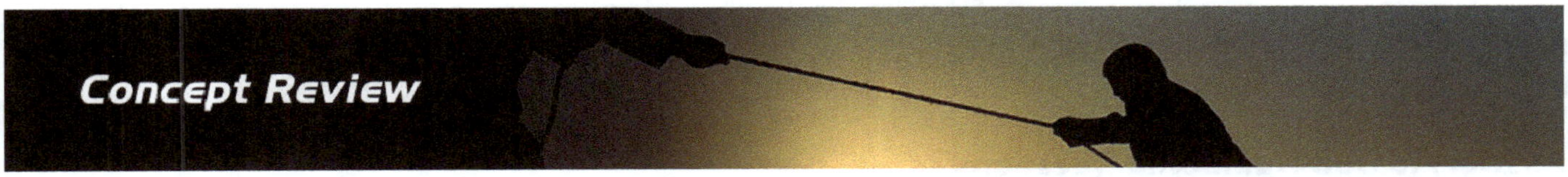

Chapter 3

Summary of Terms

Force pair The action and reaction pair of forces that occur in an interaction.

Interaction Mutual action between objects where each object exerts an equal and opposite force on the other.

Newton's third law Whenever one object exerts a force on a second object, the second object simultaneously exerts an equal and opposite force on the first. Or put another way, "To every action there is always an opposed equal reaction."

Review Questions

1. In the simplest sense, a force is a push or a pull. In a deeper sense, what is a force?
2. How many forces are required for an interaction?
3. When you push against a wall with your fingers, they bend because they experience a force. Identify this force.
4. Why do we say a speeding object doesn't have force?
5. Does Newton's third law of motion only apply when the forces are simultaneous?
6. Consider hitting a baseball with a bat. If we call the force on the bat against the ball the action force, identify the reaction force.
7. If a bat hits a ball with 1000 N of force, how much force does the ball hit back on the bat?
8. If the world gravitationally pulls you downward, what is the reaction force?
9. If the forces that act on a cannonball and the recoiling cannon from which it is fired are equal in magnitude, why do the cannonball and cannon have very different accelerations?
10. Identify the force that propels a rocket.
11. What would have to occur for the net force on a ball to be zero when you kick it?
12. Why does a push on the dashboard of a stalled car not accelerate the car?
13. Referring to Figure 3.13, how many forces are exerted on the cart? What is the horizontal net force on the cart?
14. How many forces are exerted on the horse? What is the net force on the horse?
15. How many forces are exerted on the horse-cart system? What is the net force on this system?
16. Who wins in a tug-of-war, those pulling harder on the rope, or pushing harder against the floor?
17. How does a helicopter get its lifting force?
18. A boxer can hit a heavy bag with great force. Why can't he hit a piece of tissue paper in midair with the same amount of force?
19. Can you physically touch another person without that person touching you with the same magnitude of force?
20. Fill in the blanks: Newton's first law is often called the law of ________; Newton's second law highlights the concept of _______________; and Newton's third law is the law of ______ and _________.

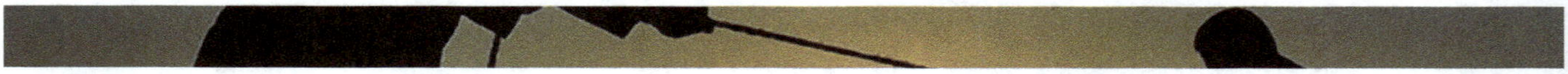

Answers to Odd-Numbered Questions

1. A force is an interaction between one thing and another.

3. When you push on the wall, the wall pushes on you. It is the force of the wall on your fingers that bends them.

5. Yes! A pair of forces act at the same time. If not, Newton's third law wouldn't apply.

7. 1000 N. Action and reaction forces are always equal in magnitude.

9. Equal forces on unequal masses produce unequal accelerations, in accord with Newton's second law.

11. Only if there is another external force on the ball, like someone kicking it in the opposite direction.

13. Two, P and f. The net force is $P - f$.

15. Two, F and f. The net force is $F - f$.

17. The helicopter blades force air downward, and in doing so the air forces the blades upward. The upward force provides lift.

19. No. Touching requires a two-ness: the toucher and the touchee.

Chapter 4

Momentum

THE MAIN IDEA

Inertia in motion is conserved

Drones are an extension of helicopters. Both are able to fly via a spinning rotor or more that continuously push air downward. Drones increase lift with multiple rotors. Newton's third law tells us a rotor cannot push air downward unless the air pushes the rotor upward. In this chapter we extend Newton's third law to the concept of *momentum*—which we think of as inertia in motion. We will see that when air is forced to move and thereby gain momentum, a corresponding change in momentum occurs for the mover of that air. Welcome to the concept of momentum.

4.1 Inertia in Motion

We know that a massive truck is harder to stop than a small car moving at the same speed. We say the truck has more momentum than the car. By momentum we mean "inertia in motion." We define **momentum** as:

$$\text{momentum} = \text{mass} \times \text{velocity}$$

or, in shorthand notation:

$$\text{momentum} = mv$$

When direction is not an important factor, we can say:

$$\text{momentum} = \text{mass} \times \text{speed}$$

which we still abbreviate mv.

READING CHECK
How does momentum relate to inertia?

From the definition we see that a moving object can have a large *momentum* if its mass is large, its speed is large, or if both its mass and speed are large. A huge ship moving at a low speed has a large momentum because its mass is large, and a fast-moving small bullet has a large momentum

Figure 4.1
The boulder, unfortunately, has more momentum than the runner.

Figure 4.2 >
Why are the engines of a supertanker normally cut off 25 km from port? And why do they need tug boats to safely navigate into the port?

because of its high speed. A massive truck with no brakes rolling down a steep hill has a large momentum, whereas the same truck at rest has no momentum at all—because the v part of mv is zero.

CONCEPT CHECK
Can you think of a case where a car and a truck with twice the car's mass would have the same momentum?

CHECK YOUR ANSWER
They'd have the same momentum if the car were traveling twice as fast as the truck. Then ($m \times 2v$) for the car would equal ($2m \times v$) for the truck. Get it? And if they were both at rest, they'd certainly have the same momentum—zero.

4.2 Impulse

When the momentum of an object changes, either its mass changes, its velocity changes, or both change. If its mass doesn't change, as is most often the case, then the object's velocity changes. So, the object accelerates. And what produces an acceleration? The answer is a net force. The greater the force exerted on an object, the greater will be the change in velocity, and hence, the change in momentum.

But something else is important also: *time*—how long the force acts. When you briefly push a stalled automobile, you'll change its momentum only a little. If you exert the same force over a longer period of time, a greater change in momentum results. A long-sustained force produces more change in momentum than the same force applied briefly. So, when changing the momentum of an object, both force and time are important.

READING CHECK
What is the impulse-momentum relationship?

The quantity "force × time interval" is called **impulse**. The greater the impulse exerted on something, the greater will be the change in momentum of that something. This is known as the **impulse-momentum relationship**. Mathematically, the exact relationship is:

$$\text{Impulse} = \text{change in momentum}$$

or

$$Ft = \text{change in } mv$$

or in Δ delta notation for "change in"

$$Ft = \Delta mv$$

which reads, "force multiplied by the time-during-which-it-acts equals change in momentum." Interestingly, this equation is another form of Newton's second law, $a = F/m$, where in place of "a" you write what acceleration is, which is $\Delta v/t$. This gives you $\Delta v/t = F/m$, which can be algebraically rewritten as $Ft = \Delta mv$.

4.3 Changing Momentum

The impulse-momentum relationship helps us analyze a variety of examples of changing momentum. Let's consider familiar examples of impulse for the cases of increasing momentum, and decreasing momentum.

Case 1: Increasing Momentum — Increase Force, Time, or Both

If you want to get a maximum increase in momentum, you not only apply the greatest force, you also extend the *time* of application for as long as possible. For example, why does a long-range cannon have a long barrel? Answer: The longer the barrel, the longer the amount of *time* that exploding gunpowder gases push on the projectile. This also explains why a rifle shoots bullets faster than a shorter barrel handgun. Strictly speaking, it's not the longer barrel that causes the projectile to shoot out faster. Rather, it's that a longer barrel allows for the projectile to be pushed for a longer period of *time*.

Figure 4.3
The expert golfer knows that a follow-through swing extends the time of impact between the club and the ball. What effect does this longer duration of impact have on the ball?

Case 2: Decreasing Momentum in a Long Time Means Less Force

Imagine you are in a fast-moving truck and the brakes fail, and you have a choice of stopping by either hitting a concrete wall or a haystack. You don't have to be a rocket scientist to make a correct decision. Of course, you'd hit the haystack, because it would be a "softer" hit. The reason hitting something soft is quite different from hitting something hard has everything to do with the time of contact.

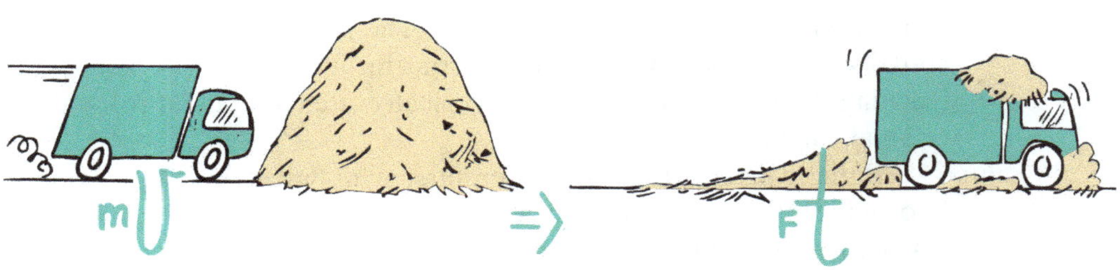

Figure 4.4
A large change in momentum in a long time requires a small force.

Figure 4.5 >
A large change in momentum in a short time requires a large force.

Whether you hit the wall or the haystack, your momentum decreases the same—from something large to zero. According to the impulse-momentum relationship, the same change in momentum means the same impulse. The same impulse, however, doesn't mean the same force, nor the same time. Rather, it means the same *product* of force and time. By hitting the haystack instead of the wall, you extend the contact *time*—the time during which your momentum is brought to zero. A longer time reduces the force and so the deceleration you experience is less. For example, if the time of contact is extended 10 times, the force of contact is reduced 10 times. Whenever you wish the force of contact to be small, extend the time of contact.

READING CHECK
What of the deceleration you experience when your momentum is stopped over a longer period of time?

Someone tosses you a raw egg from across the room. How do you catch that egg without breaking the shell? Answer: You extend your arms to meet the egg. Then as you catch the egg you ride with it to slow it down slowly. In this instance you are bringing the egg to zero momentum over a long period of time so that the force you apply can be as small as possible. Bring it to a quick stop and the force will necessarily be greater, which means the egg may break.

A wrestler thrown to the floor tries to extend his time of arrival on the floor by relaxing his muscles and spreading the crash into a series of impacts as foot, knee, hip, ribs, and shoulder fold onto the floor in turn. The increased time of contact reduces the overall force. Of course, falling on a mat is better than falling on a solid floor, because the force experienced is further decreased by increased time of contact.

When you jump from an elevated position to a floor below, you bend your knees when you make contact. This extends the time during which your momentum is being reduced. Suppose that you extend your time of landing by 5 times that of a stiff-legged, abrupt landing. Then you'll reduce the forces experienced by your bones by 5 times.

Bungee jumping puts the impulse-momentum relationship to a thrilling test (**Figure 4.6**). The momentum gained during fall must be decreased to zero by an impulse of the same amount. The long stretching time of the cord results in a small average force to bring the jumper to a safe halt before reaching the ground. Bungee cords typically stretch to about twice their original length during the fall.

Ballet dancers prefer a wooden floor with "give" to a hard cement floor with little or no "give." The wooden floor allows a longer time of contact whenever the dancer lands, reducing the force of contact and possible injury. A safety net used by acrobats is an obvious example of small contact force over a long time.

< Figure 4.6
A large change in momentum over a long time requires a safely small average force.

Case 3: Decreasing Momentum Over a Short Time Means More Force

When boxing, if you move into a punch instead of away, you're in trouble. This is similar to catching a high-speed baseball while your hand moves toward the ball instead of away upon contact. For short contact times, the forces are large.

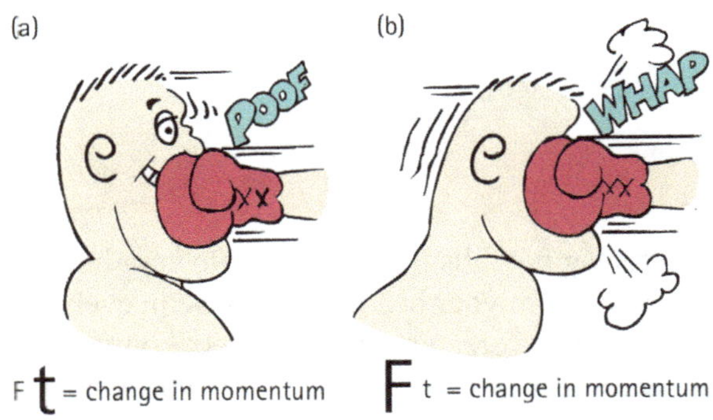

< Figure 4.7
In both cases the boxer's jaw provides an impulse that reduces the momentum of the punch. (a) The boxer is moving away when the glove hits, thereby extending the time of contact. This means the force is less than if the boxer had not moved. (b) The boxer is moving into the glove, thereby lessening the time of contact. This means that the force is greater than if the boxer had not moved.

The idea of short contact time explains how a karate expert can sever a stack of bricks with the blow of her bare hand (**Figure 4.8**). She brings her arm and hand swiftly against the bricks with a lot of momentum. This momentum is quickly reduced when she hits the bricks. The impulse is the

< Figure 4.8
Cassy imparts a large impulse to the bricks in a short time and produces a considerable force.

force of her hand against the bricks multiplied by the time her hand makes contact with the bricks. By swift execution she makes the time of contact very short, which makes the force huge. If her hand bounces upon contact, the force is even greater.

CONCEPT **CHECK**

1. If the boxer in Figure 4.7 makes the time of contact 3 times as long by riding with the punch, by how much is the force reduced?
2. If the boxer instead moves into the punch and shortens the contact time by half, by how much is the force increased?
3. A boxer being hit with a punch tries to extend time for best results, whereas a karate expert delivers a force in a short time for best results. Isn't there a contradiction here?

CHECK YOUR ANSWERS

1. The force will be three times less than if he didn't pull back.
2. The force will be two times greater than if he held his head still. Forces of this kind account for many knockouts.
3. There is no contradiction because the best results for each are quite different. The best result for the boxer is reduced force, accomplished by maximizing time. The best result for the karate expert is increased force delivered in minimum time.

4.4 Bouncing

READING CHECK

Compare the impulses in catching a ball with catching it and tossing it back to the thrower?

You know that if a flower pot falls from a shelf onto your head, you may be in trouble. If it bounces from your head, you may be in even more trouble. Why? Because impulses are greater when an object bounces. The impulse required to bring an object to a stop and then to "throw it back again" is greater than the impulse required merely to bring it to a stop. Suppose, for example, that you catch the falling pot with your hands. You provide an impulse to reduce its momentum to zero. If you throw the pot upward again, you have to provide additional impulse. It takes more impulse to catch it and throw it back up than merely to catch it. This increased amount of impulse is supplied by your head if the pot bounces upward from it.

Figure 4.9 >
Teacher Howie Brand shows that the block topples when the swinging dart bounces from it. When he changes the head of the dart so it doesn't bounce when it hits the block, no tipping occurs.

An interesting application of the greater impulse that occurs for bouncing was employed with great success in California during the gold rush days of the 19th century. The water wheels used in gold-mining operations were not very effective. A man named Lester A. Pelton saw that the problem had to do with their flat paddles. He designed curved-shape paddles that would cause the incident water to make a U-turn—to "bounce." In this way the impulse exerted on the water wheels was increased. Pelton patented his idea and made more money from his invention, the Pelton wheel, than most gold miners made from gold.

A flowerpot dropped onto your head bounces quickly. Ouch! If bouncing took a longer time, as with a safety net, then the force of the bounce would be much smaller.

CONCEPT **CHECK**

1. Refer to Figure 4.8. How does the force that Cassy exerts on the bricks compare with the force exerted on her hand?
2. How will the impulse differ if her hand bounces back when striking the bricks?

CHECK YOUR ANSWER

1. In accord with Newton's third law, the forces are equal. Only the resilience of the human hand and the training she has undergone to toughen her hand allow her to do this without breaking bones.
2. The impulse will be greater if her hand bounces from the bricks. If the time of contact is not increased, a greater force is then exerted on the bricks (and her hand!).

Λ Figure 4.10
The Pelton wheel. The curved blades cause water to bounce and make a U-turn that produces a greater impulse to turn the wheel.

4.5 Momentum Conservation

According to Newton's third law, a force is an interaction: If object A pushes on object B, then object B pushes on object A. Because the force is *one interaction*, the magnitude of A pushing on B is necessarily the same as B pushing on A, though each object is pushed in exactly the opposite direction. Furthermore, the duration of that push is the same on each other. Think carefully about this and you'll come to realize that the impulse (Ft) on each object is exactly equal and opposite. This means that the change in momentum of A will be exactly equal and opposite to the change in momentum of B. If you were to combine the change in momentum for A moving in one direction and B, now moving in the opposite direction, you'd come up with zero!

Thought Activity

Skateboard Throw

Consider standing at rest on a skateboard and throwing a massive object in front of you. You would recoil in the opposite direction! This makes sense if you understand momentum. The net momentum before the throw was zero. The net momentum just after is also zero—because your recoil momentum is equal and opposite to the momentum of the tossed object. You would see that momentum is conserved. Now consider repeating, but don't let go when you "throw" the object. Do you still recoil when you go through the motions of throwing, but don't really release the object? Defend your answer with your classmates.

In this sense, momentum is *conserved*, which means that within a given system there is no net change in the momentum. If you start of out with a certain amount of momentum, then you'll always end up with that same amount of momentum. Provided there are no forces applied from outside the system, then the total momentum within the system remains unchanged.

Consider a cannon firing a cannonball. The expanding gunpowder applies a force to both the cannon and the cannonball. And, as we described earlier (Newton's second law), the cannon accelerates backward just a little bit because of its greater mass. The cannonball, by contrast, accelerates forward quickly because it has much less mass (inertia). Though the accelerations are different, the change in momentum (ΔMv) of the cannon in one direction, will be equal to the change in momentum (ΔmV) of the cannonball in the opposite direction. Therefore, the total momentum of

Figure 4.11 >
These cannons are now antiques. But in their day they were cutting-edge technology, exerting impulses on cannonballs to simultaneously change their momenta.

< Figure 4.12
The momentum before firing is zero. After firing, the net momentum is still zero, because the momentum of the cannon is equal and opposite to the momentum of the cannonball.

this system before the kaboom will be the same as the total momentum after the kaboom. The momentum within the cannon-cannonball system remains the same regardless of what happens within that system. We say the momentum is conserved.

Two important ideas are to be learned from the canon-and-cannonball example. The first is that momentum, like velocity, is a vector quantity that is described by both *magnitude* and *direction*; we measure both "how much" and "which direction." Therefore, like velocity, when momenta act in the same direction, they are simply added. When they act in opposite directions, they are subtracted.

The second important idea is conservation. For the cannon-cannonball system, no momentum was gained; none was lost. When a physical quantity remains unchanged during a process, we say that quantity is *conserved*. We say momentum is conserved.

The concept that momentum is conserved when no external force acts is so important it is considered a *law* of mechanics. It is called the **law of conservation of momentum**:

In the absence of an external force, the momentum of a system remains unchanged.

When a system undergoes changes in which all forces are internal, the net momentum remains unchanged. This happens when cannons fire cannonballs, cars collide, or stars explode. In such cases, the net momentum of the system before and after the event remains the same.

READING CHECK

What does it mean to say that a physical quantity is conserved?

CONCEPT **CHECK**

A high-speed bus and an innocent bug have a head-on collision. The sudden change of momentum for the bug spatters it all over the windshield. Is the change in momentum of the bus greater, less, or the same as the change in momentum of the unfortunate bug?

CHECK YOUR ANSWER

The momentum of both bug and bus change by the same amount because both the force and the time, and therefore the amount of impulse, is the same on each. Momentum is conserved. Speed is another story. Because of the huge mass of the bus, its reduction of speed is very tiny—too small for the passengers to notice.

4.6 Collisions

Momentum is conserved in collisions because the forces that act are internal forces—acting and reacting within the system itself. There is only a redistribution or sharing of whatever momentum exists before the collision.

In any collision, we can say:

net momentum before collision = net momentum after collision

Elastic Collisions

When a moving billiard ball hits another ball at rest head on, the first ball comes to rest and the second ball moves away with a velocity equal to the initial velocity of the first ball. We see that momentum is transferred from the first ball to the second ball. When objects collide without being permanently deformed and without generating heat, we say the collision is an **elastic collision**. Colliding objects bounce perfectly in perfectly elastic collisions (**Figure 4.13**).

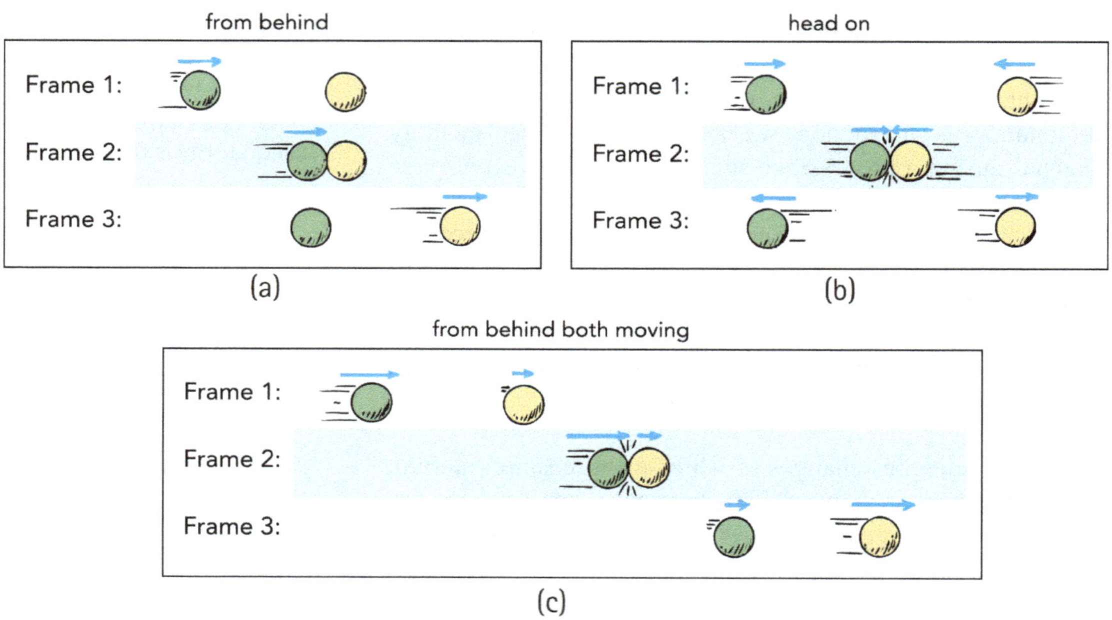

Figure 4.13
Elastic collisions of equally massive balls. (a) A green ball strikes a yellow ball at rest. (b) A head-on collision. (c) A collision of balls moving in the same direction. In each case, momentum is transferred from one ball to the other.

Inelastic Collisions

Momentum is conserved even when the colliding objects become distorted and generate heat during the collision. Whenever colliding objects become tangled or coupled together, we have an **inelastic collision**. In a perfectly inelastic collision, both objects stick together. Consider, for example, the case of a freight car moving along a track and colliding with another freight car at rest (**Figure 4.14**). If the freight cars are of equal mass and are coupled by the collision, can we predict the velocity of the coupled cars after impact?

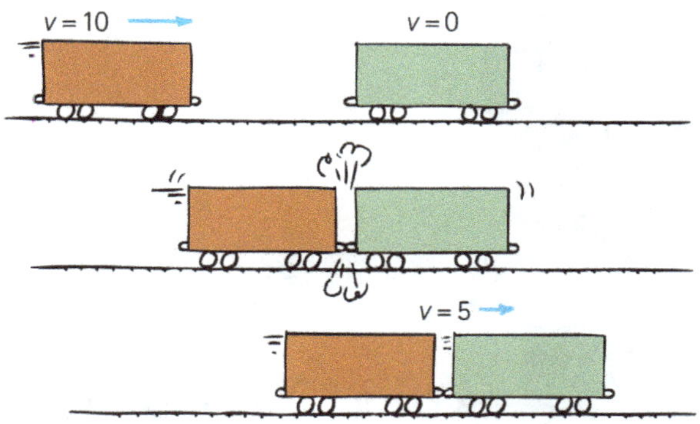

< Figure 4.14
Inelastic collision. The momentum of the freight car on the left is shared with the freight car on the right after collision.

Suppose the single car is moving at 10 meters per second, and we consider the mass of each car to be m. Then, from the conservation of momentum:

$$\text{Net } mv_{\text{before}} = \text{net } mv_{\text{after}}$$

$$(m \times 10 \text{ m/s})_{\text{before}} = (2m \times v)_{\text{after}}$$

By simple algebra, $v = 5$ m/s. This makes sense, for twice as much mass moves after the collision, with half as much as the velocity as before collision. Then both sides of the equation are equal.

Note the inelastic collisions in **Figure 4.15**. Can you see the net momentum after collisions is the same? We see that when external forces don't play a role, momentum is conserved for both elastic and inelastic collisions.

READING CHECK

Is momentum conserved in elastic collision, inelastic collisions, or both?

< Figure 4.15
Inelastic collisions. The net momentum of the trucks before and after collision is the same.

If your teacher has an air track similar to that shown in **Figure 4.16**, you may be treated to fascinating demonstrations of momentum conservation. The air that spurts from the tiny holes in the track lets you see an almost friction-free performance. In the everyday world, friction usually shows itself. Ideally, the net momentum of a couple of freight cars that collide is the same before and just after collision. But as the combined cars move along the track, friction provides an impulse to decrease momentum.

Another thing: Perfectly elastic collisions are not common in the everyday world. Usually some heat is generated in collisions. Drop a ball, and after it bounces from the floor, both the ball and the floor are a bit warmer. Even a dropped "superball" will not bounce to its initial height. At the microscopic level, however, perfectly elastic collisions are

Figure 4.16
An air track. Blasts of air from tiny holes provide a friction-free surface for the carts to glide upon.

commonplace. For example, gas molecules can bounce off one another without generating heat.

Conservation of momentum and, as the next chapter will discuss, conservation of energy are the two most powerful tools of mechanics. These laws let us understand the details of interactions among subatomic particles to the details of entire galaxies.

Quick Activity

Colliding Marbles

Place four or five marbles, all identical in size and shape, in the center groove of a ruler. Launch another identical marble toward the stationary ones. Note the changes in motion of the marbles. Repeat, but with two launched marbles at about the same speed. Note any changes in motion of the marbles. Remove all but one marble. Launch an identical marble toward it. Note how the first marble stops and the one at rest moves. What would occur if three marbles rolling to the right hit two marbles rolling to the left? Is a general rule at play here?

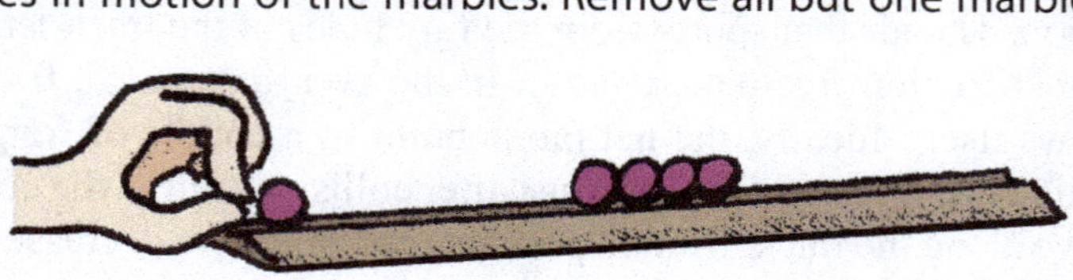

CONCEPT **CHECK**

Refer to the gliders on the air track in Figure 4.16 to answer these questions.

1. Suppose both gliders have the same mass. They move toward each other at the same speed and experience an elastic collision. Describe their motion after the collision.

2. Suppose both gliders have the same mass and move toward each other at equal speed. This time they stick together when they collide. Describe their motion after the collision.

3. Suppose one of the gliders is at rest and is loaded so that it has twice the mass of the moving glider. Again, the gliders stick together when they collide. Describe their motion after the collision.

CHECK YOUR ANSWER

1. Since the collision is elastic, the gliders reverse directions upon colliding and move away from each other at the same speed as before.

2. Before the collision, the gliders had equal and opposite momenta, since their equal masses were moving in opposite directions at the same speed. The net momentum of both was zero. Since momentum is conserved, their net momentum after they stick must also be zero. They slam to a dead halt.

3. Before collision, the net momentum equals the momentum of the unloaded, moving glider. After the collision, the net momentum is the same as before, but now the gliders are stuck together and moving as a single unit. The mass of the stuck-together gliders is 3 times that of the unloaded glider. Thus, the velocity is 1/3 of the unloaded glider's velocity before collision and in the same direction as before, since the direction as well as the amount momentum is conserved.

CONSERVATION LAWS

A conservation law tells us that certain quantities in a system remain precisely constant, regardless of what changes may occur in that system. A system may be as simple as a bat hitting a ball or a solar system orbiting around the center of a galaxy. A conservation law is one of constancy during change—that "something" doesn't change while other parts may change. In this chapter, we saw that the total momentum is unchanged in a collision so long as no outside force acts during the collision. We say momentum is *conserved*. In the next chapter, we'll learn that energy is conserved as it transforms—the amount of energy in light, for example, transforms completely to thermal energy when light is absorbed. When we study electricity we'll learn that electric charge is conserved, which means it can be neither created or destroyed. A study of nuclear physics shows that conservation laws rule in the submicroscopic world. Conservation laws are a source of deep insights into the simple regularity of nature and are often considered the most fundamental physical laws. Can you think of things in your own life that remain constant as other things change?

Calculation Corner

For a numerical example of momentum conservation, consider a pair of carts, A and B, on the air track shown in Figure 4.16. Suppose that cart B is at rest and that A slides against it and they stick together. Let cart A have a mass of 5 kg and move 1 m/s toward cart B, which has a mass of 1 kg. What will be the velocity of both carts when they link together?

$$
\begin{aligned}
\text{Net } mv_{\text{before}} &= \text{net } mv_{\text{after}} \\
[m_A v_A + m_B v_B]_{\text{before}} &= [(m_A + m_B)v]_{\text{after}} \\
(5\text{ kg})(1\text{ m/s}) + (1\text{ kg})(0\text{ m/s}) &= (5\text{ kg} + 1\text{ kg})\, v \\
5\text{ kg m/s} &= (6\text{ kg})\, v \\
v &= 5/6\text{ m/s}
\end{aligned}
$$

Here we see that the small cart has no momentum before collision because its velocity is zero. After collision, the combined mass of both carts moves at velocity v, which by simple algebra is seen to be 5/6 m/s. This velocity is in the same direction as that of cart A.

Now suppose cart B is not at rest, but moves toward the left at a velocity of 4 m/s. The carts are moving toward each other ready for a head-on collision. Let the direction of cart A be +, and cart B – (to indicate the opposite directions). Then we see that:

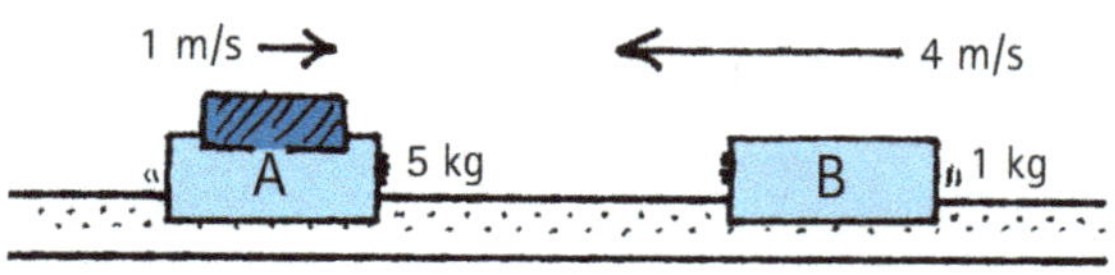

$$
\begin{aligned}
\text{Net } mv_{\text{before}} &= \text{net } mv_{\text{after}} \\
[m_A v_A + m_B v_B]_{\text{before}} &= [(m_A + m_B)v]_{\text{after}} \\
(5\text{ kg})(1\text{ m/s}) + (1\text{ kg})(-4\text{ m/s}) &= (5\text{ kg} + 1\text{ kg})\, v \\
(5\text{ kg m/s}) + (-4\text{ kg m/s}) &= (6\text{ kg})\, v \\
1\text{ kg m/s} &= 6\text{ kg}\, v \\
v &= 1/6\text{ m/s}
\end{aligned}
$$

Note that the negative momentum of cart B before collision has more effect in slowing cart A after collision. If cart B were moving twice as fast, then

$$
\begin{aligned}
\text{Net } mv_{\text{before}} &= \text{net } mv_{\text{after}} \\
[m_A v_A + m_B v_B]_{\text{before}} &= [(m_A + m_B)v]_{\text{after}} \\
(5\text{ kg})(1\text{ m/s}) + (1\text{ kg})(-8\text{ m/s}) &= (5\text{ kg} + 1\text{ kg})\, v \\
(5\text{ kg m/s}) + (-8\text{ kg m/s}) &= (6\text{ kg})\, v \\
-3\text{ kg m/s} &= 6\text{ kg}\, v \\
v &= -1/2\text{ m/s}
\end{aligned}
$$

Here we see the final velocity is –1/2 m/s. What is the significance of the minus sign? It tells us that the final velocity of the two-cart system is opposite to the initial velocity of cart A. After collision the two-cart system moves toward the left.

We have discussed carts on an air track. Our example could well have been football players or even swimming fish, which we explore in the video of this chapter section.

Concept Review

Chapter 4

Summary of Terms

Elastic collision A collision in which colliding objects rebound without lasting deformation or the generation of heat.

Impulse The product of the force acting on an object and the time during which it acts. Impulses are equal and opposite in an interaction.

Impulse-momentum relationship Impulse is equal to the change in the momentum of the object that the impulse acts on. In symbol notation, $Ft = \Delta mv$.

Inelastic collision A collision in which the colliding objects become distorted, generate heat, and possibly join together.

Law of conservation of momentum When no external net force acts on an object or a system of objects, no change of momentum takes place. Hence, the momentum before an event involving only internal forces is equal to the momentum after the event: mv(before event) = mv(after event)

Momentum The product of the mass of an object and its velocity.

Review Questions

1. Which has a greater momentum, a heavy truck at rest or a moving automobile?
2. How can a supertanker have a huge momentum when it moves relatively slowly?
3. Why is it incorrect to say that impulse equals momentum?
4. To impart the greatest momentum to an object, what should you do in addition to exerting the largest force possible?
5. For the same force, which cannon imparts the greater speed to a cannonball—a long cannon or a short one? Explain.
6. If you're in a car with faulty brakes and you have to hit something to stop, the momentum will change to zero whether you hit a brick wall or a haystack. So why is hitting a haystack a safer bet?
7. Why is it less damaging if you fall on a mat than if you fall on a solid floor?
8. Why is it a good idea to extend your hand forward when catching a fast-moving baseball with your bare hand?
9. In boxing, why is it advantageous to roll with the punch?
10. In karate why is a short time of the applied force advantageous?
11. Which is the greater change in momentum, stopping something dead in its tracks, or stopping it and then reversing its direction?
12. Which requires the greater impulse, stopping something dead in its tracks, or stopping it and then reversing its direction?
13. When can the momentum of two moving objects be cancelled?
14. What does it mean to say that momentum (or any quantity) is conserved?

15. When a cannonball is fired, its momentum does change! Is momentum conserved for the cannonball?

16. When a cannonball is fired, the cannon recoils. Is momentum conserved for the cannon?

17. When a cannonball is fired, is momentum conserved for the cannon-cannonball system as a whole? (Why is your answer different than in the previous two questions?)

18. Distinguish between an elastic collision and an inelastic collision. For which type of collision is momentum conserved?

19. Railroad car A rolls at a certain speed and makes a perfectly elastic collision with car B of the same mass. After the collision, car A is observed to be at rest. How does the speed of car B compare with the initial speed of car A?

20. If the equally massive cars of the previous question stick together after colliding inelastically, how does their speed after the collision compare with the initial speed of car A?

Answers to Odd-Numbered Questions

1. The truck at rest has no speed, hence no momentum. So, the moving automobile has greater momentum.

3. Impulse equals a change in momentum, not momentum itself.

5. A cannonball will have more momentum coming from the long cannon due to the force acting for more time.

7. The time to come to a stop will be longer on the mat. This means a longer time to stop and a smaller force.

9. The force of impact will be less if momentum changes over a long time. By making t long, F will be smaller.

11. Stopping and then reversing direction is a greater change in momentum.

13. When their momentum is equal and opposite and they comprise the same system.

15. No, momentum is not conserved, for an impulse acts on the cannonball.

17. Momentum is conserved for the cannon-cannonball system, where the forces are internal to the system. Both forces act within the system, whereas if the system in question is only the cannon (or cannonball), force is from outside the system.

19. Momentum is completely transferred from car A to car B. Since the masses are the same, the speed of car B is equal to the initial speed of car A.

Chapter 5

Energy

THE MAIN IDEA

Energy can change form without a net loss or gain.

At an amusement park, why is the first summit on a roller coaster always the highest one? And why are following summits lower in height? In ancient times, people had great difficulty lifting huge loads, yet today simple pulleys or pistons can be employed to lift heavy loads unimaginable in past times. How do pulleys and pistons so greatly increase output force? The answers involve the concept of *energy*—what this chapter is about.

Energy is a most central concept, not only to science but to society. Energy comes from the Sun in the form of sunlight, which is most abundant. Solar energy is more and more being tapped for our energy needs. We find energy in the foods we eat, which sustains life. Our study of energy begins with a related concept: *work*.

5.1 Work and Power

Newton's second law tells us that a force is required to change the motion of an object. If the object is at rest and you want it moving, you need to apply a force. If the object is already moving and you want it to speed up or slow down, again, you need to apply a force. This should be intuitive from your everyday experiences—to get things moving, you apply a force.

A force may be applied, but here's a key question: For "how long" is that force being applied? In the previous chapter we talked about how long in terms of *time*, which gives what we learned to be *impulse*:

$$\text{Impulse} = \text{force} \times \text{time}$$

As we saw in the previous chapter, an impulse results in a change in the object's momentum. For this chapter we explore a different term for "how long"—not time, but *distance*. Applying a force on a load over a particular distance leads to a quantity known as **work**:

$$\text{Work} = \text{force} \times \text{distance}$$

Λ Figure 5.1
Compared with the work done in lifting a load one story high, twice as much work is required to lift the same load two stories high. That's because the distance is twice as much.

∧ **Figure 5.2**
When twice the load is lifted to the same height, twice as much work is done because the force needed to lift it is twice as much.

READING CHECK
What two things does the amount of work done depend upon?

Compared with the work done in lifting a load one story high, twice as much work is done in lifting the same load two stories high. Twice the work is done because the distance is twice as much.

We do work when we apply a force to lift a load against Earth's gravity. The heavier the load or the higher we lift it, the more work we do. The amount of work done on an object depends on (1) how much force is applied and (2) the distance moved by the force. Note: An assumption here is that the force and distance are in the same direction. For example, you push upward on the load and it is lifted upward in the same direction.

When a weight lifter raises a heavy barbell, he does work on the barbell. Interestingly, when a weight lifter simply holds a barbell overhead, he does no work on it. He may get tired holding the barbell still, but if the barbell is not moved by the force he exerts, he does no work on the barbell. Work may be done on his muscles as they stretch and contract, which is force × distance on a biological scale. But this work is not done on the barbell. Lifting the barbell is different than holding the barbell.

∧ **Figure 5.3**
Work is done in lifting the barbell. Lifting it twice as high requires twice as much work.

The unit of work combines the unit of force (N) with the unit of distance (m), the newton-meter (Nm). We call a newton-meter the joule (J) (rhymes with cool). One joule of work is done when a force of 1 newton is exerted over a distance of 1 meter, as in lifting an apple over your head. For larger values we can speak of kilojoules (kJ), thousands of joules, or megajoules (MJ), millions of joules. The weight lifter in **Figure 5.3** does work in kilojoules. The work done to vertically raise a heavily loaded truck can be in megajoules.

∧ **Figure 5.4**
He may expend energy when he pushes on the wall, but if it doesn't move, no work is done on the wall.

CONCEPT **CHECK**

1. How much work is needed to lift an object that weighs 500 N to a height of 4 m?
2. How much work is needed to lift it twice as high?
3. How much work is needed to lift a 1000 N to a height of 8 m?

CHECK YOUR ANSWER

1. $W = F \times d = 500\text{ N} \times 4\text{ m} = 2000\text{ J}$.
2. Twice the height means twice the work. That is, $W = F \times d = 500\text{ N} \times 8\text{ m} = 4000\text{ J}$.
3. Lifting twice the load twice as high requires four times the work. That is, $F \times d = 1000\text{ N} \times 8\text{ m} = 8000\text{ J}$.

Lifting a load quickly is more of a challenge than lifting the same load slowly. If equal loads are lifted to the same height, one quickly and the other slowly, the work done is the same. What's different is the *power*. **Power** is equal to the amount of work done per the time it takes to do it:

$$\text{Power} = \frac{\textbf{work done}}{\textbf{time interval}}$$

READING CHECK

What do work and time have to do with power?

By analogy, if you think of work as money, then power is the rate at which you spend that money. Consider two cars, one with an engine twice as powerful. Both cars can reach a speed of 50 mph. However, the car with the doubly powerful engine can reach that speed in half the time. Car racing fans are impressed with the "power" of the race car engines. Likewise, fans of the 100-meter dash are impressed by the "power" of the sprinter's muscles. In each case, greater power means the same amount of work can be performed more quickly.

The unit of power is the joule per second, called the **watt**. This is in honor of James Watt, the eighteenth-century developer of the steam engine. One watt (W) of power is consumed when one joule of work is done in one second. One kilowatt (kW) equals 1000 watts. One megawatt (MW) equals one million watts.

< Figure 5.5

A space rocket burns fuel at an enormous rate to generate the power needed to lift it against gravity.

CONCEPT **CHECK**

1. You do work when you do push-ups. If you do the same number of push-ups in half the time, how does your power output compare?

2. How many watts of power are needed when a force of 1 N moves a box 2 m in a time of 1 s?

CHECK YOUR ANSWER

1. Your power output is twice as much.

2. The power expended is 2 watts

$$P = \frac{W}{t} = \frac{F \times d}{t} = \frac{1\ \text{N} \times 2\ \text{m}}{1\ \text{s}} = 2\ \text{W}$$

5.2 Mechanical Energy

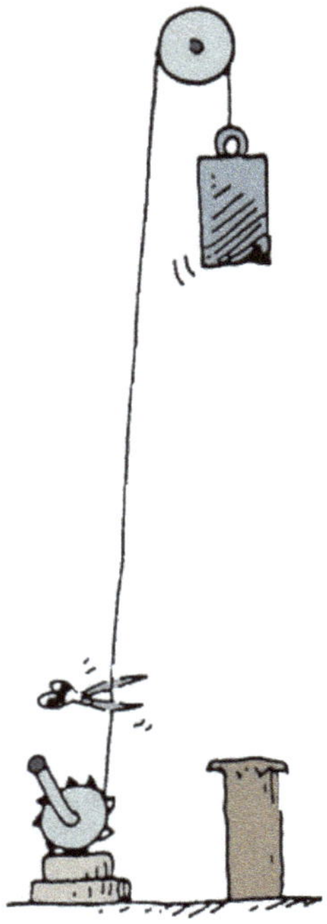

Λ Figure 5.6
Work is required to lift the massive ram of the pile driver.

Work is done in lifting the heavy ram of a pile driver (**Figure 5.6**). When raised, the ram then has the ability to do work on a piling beneath it when it falls. When work is done by an archer in drawing a bow, the bent bow has the ability to do work on the arrow. When work is done to wind a spring mechanism, the spring then has the ability to do work on various gears to run a clock, ring a bell, or sound an alarm.

In each case, the ability to do work has been acquired by an object. This ability to do work is **energy**. Like work, energy is measured in joules.

Energy appears in many forms, such as heat, light, sound, electricity, and radioactivity. As a society, we are very keen on sources of energy. Having energy allows us to do work, such as when we drive a car, build a treehouse, or mow the lawn. On a biological level, it's the energy from our food that allows us to go for a walk, breathe, and even just to think, which all involve doing work on a cellular scale.

In this chapter, we keep it simple as we focus on only two forms of energy: potential energy and kinetic energy. **Potential energy** is energy that arises because of an object's position. **Kinetic energy** is energy possessed by an object in motion. Potential and kinetic energy are both considered to be kinds of mechanical energy. So mechanical energy may be in the form of either potential energy, kinetic energy, or both.

READING CHECK
What are the two major forms of mechanical energy?

CONCEPT CHECK

1. She pushes the block five times farther up the incline than the man lifts it to the same height. How much more force does the man exert when he lifts the ice?
2. Who does more work on the ice?
3. If both jobs take the same time, who expends more power?

CHECK YOUR ANSWER

1. The man exerts 5 times as much force as the woman exerts.
2. Although he exerts more force, both do the same amount of work on the ice.
3. They both do the same amount of work in the same time, so both expend the same power.

5.3 Potential Energy

There are times when an object is able to do work merely by virtue of its position. Consider a brick sitting on the edge of a kitchen counter. On the floor below is a nail poised to be pounded into a piece of wood. If the brick were to fall precisely right, it could hit the nail thus applying a *force* to the nail and driving it a certain *distance* into the wood. This is to say that the brick, by virtue of its position, has the potential of doing work. This is a form of energy we call *potential energy*, which is the energy due to position. Potential energy (PE) is stored-up and held in readiness. For example, when an archer draws an arrow with a bow, energy is stored in the bow. When released, energy is transferred to the arrow.

There are various kinds of potential energy. The potential energy easiest to visualize is in an object elevated against Earth's gravity. The potential energy due to an elevated position is called *gravitational potential energy*. The elevated ram of a pile driver and of water in an elevated reservoir both have gravitational potential energy.

The amount of gravitational potential energy that an elevated object has is equal to the work done to lift it—the force required to move it upward multiplied by the vertical distance moved ($W = F \times d$). Once upward motion begins, the upward force to keep an object moving at constant speed equals its weight. So, the work done in lifting an object is its weight × height. We say:

$$\text{Gravitational potential energy} = \text{weight} \times \text{height}$$

An object's weight is its mass m multiplied by the acceleration of gravity g. We write weight as mg. So, the work done in lifting mg through a height h is equal to its gain in gravitational potential energy (PE):

$$\text{PE} = mgh$$

Note that the height h is the distance above some base level, such as the ground or the floor of a building. The potential energy is relative to that level and depends only on weight and height h. You can see in **Figure 5.8** that the potential energy of the ball at the top of the structure depends on height only, and not on the path taken to get it there.

Figure 5.7
The potential energy of Tenny's drawn bow equals the work (average force × distance) she did in drawing the arrow into position.

READING CHECK
How does the gravitational potential energy of an object relate to the work required to elevate it?

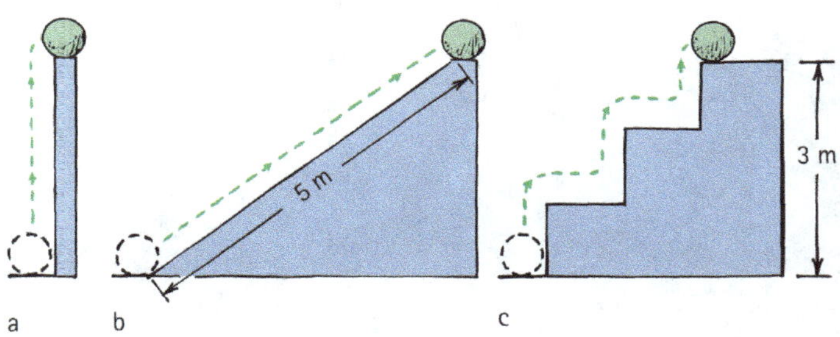

Figure 5.8
The PE of the 10-N ball is the same (30 J) in all three cases. That's because the work done in elevating it 3 m is the same whether it is (a) lifted with 10 N of force, (b) pushed with 6 N of force up the 5-m incline, or (c) lifted with 10 N up each 1-m step. No work is done in moving it horizontally (neglecting friction).

Figure 5.9
The man raises a block of ice by lifting it vertically. The woman pushes an identical block of ice up the ramp. When both blocks are raised to the same height, both have the same potential energy.

CONCEPT CHECK

1. How much work is done in lifting the 200-N block of ice shown in Figure 5.9 a vertical distance of 2.5 m?
2. How much work is done in pushing the same block of ice up the 5-m long ramp? The force needed is only 100 N (which is why inclines are used).
3. What is the increase in the block's potential energy in each case?

CHECK YOUR ANSWER

1. 500 J. (We get this either by *Fd* or *mgh*.)
2. 500 J. (She pushes with half the force over twice the distance.)
3. Either way increases the block's potential energy by 500 J. The ramp simply makes this work easier to perform.

5.4 Kinetic Energy

READING CHECK
How does the kinetic energy of an object relate to its mass and speed?

When you push on an object, you can make it move. We find a moving object is capable of doing work. A moving car, for example, can apply a *force* to a guard rail, which dents by a given *distance* in stopping the car. A moving object has energy of motion, or kinetic energy (KE). The kinetic energy of an object depends on its mass and speed. Specifically, kinetic energy is equal to the mass of an object multiplied by the square of its speed, multiplied by the constant 1/2:

$$\text{Kinetic energy} = 1/2 \text{ mass} \times \text{speed}^2$$

$$\text{KE} = 1/2\, mv^2$$

Because kinetic energy depends on mass, heavy objects have more kinetic energy than light ones moving at the same speed. For example, a car moving along the road has a certain amount of kinetic energy. A twice-as-massive car moving at the same speed has twice as much kinetic energy.

Figure 5.10 >
The downhill "fall" of the roller coaster results in its roaring speed in the dip, and this kinetic energy sends it up the steep track to the next summit.

Kinetic energy also depends on speed. In fact, kinetic energy depends on speed more than it depends on mass. Why? Look at the equation. Kinetic energy depends on speed multiplied by speed, or speed squared. So, if a car moving along the road has a certain amount of kinetic energy, a twice-as-fast car (with the same mass) has 2^2 or four times as much kinetic energy! The same car moving with three times the speed has 3^2 or nine times as much kinetic energy. So, we see that small changes in speed produce large changes in kinetic energy.

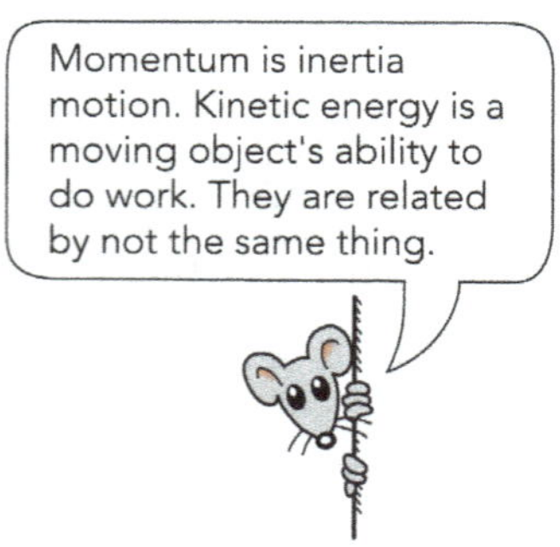

Kinetic energy appears in many forms. For example, when fast-moving air molecules bump against the surface of your skin, they transfer kinetic energy to you similar to the way colliding billiard balls transfer energy. From this, you'll sense the warmth of the air. A vibrating tuning fork pushes rhythmically against surrounding air molecules. Now vibrating with kinetic energy from the tuning fork, those air molecules can collide with your ear allowing you to sense the sound of the tuning fork. Light itself is a form of kinetic energy arising from the rapid vibrations of electrons. We see that kinetic energy has many applications in our lives, and is far-reaching.

CONCEPT CHECK

1. A car travels at 30 km/h and has kinetic energy of 1 MJ. If it travels twice as fast, 60 km/h, how much kinetic energy will it have? (Note that 1 MJ = 1,000,000 J)
2. If it travels three times as fast, at 90 km/h, what will be its kinetic energy?
3. If it travels four times as fast, at 120 km/h, what will be its kinetic energy?

CHECK YOUR ANSWER

1. Twice as fast means (2^2) four times the kinetic energy, or 4 MJ.
2. Three times as fast means (3^2) nine times the kinetic energy, or 9 MJ.
3. Four times as fast means (4^2) sixteen times the kinetic energy, or 16 MJ.

5.5 Work-Energy Theorem

Consider an object moving along with a certain amount of kinetic energy. By applying a force you can speed it up or slow it down. Either way, you'll be changing the object's speed, hence its kinetic energy. In fact, the only way to change an object's kinetic energy is to apply a force over some distance, which is to do work. This important relationship between work and changes in energy is called the work-energy theorem. We abbreviate "change in" with the delta symbol, Δ, and say

READING CHECK
What is the work-energy theorem?

$$\text{Work} = \Delta E$$

When the energy that is changed is kinetic energy, we say

$$\text{Work} = \Delta KE$$

Work equals change in kinetic energy. The work in this equation is the net work—that is, the work done by the net force.

Figure 5.11
When the car goes twice as fast, it has four times the kinetic energy (and will need four times the stopping distance when braking).

The work-energy theorem emphasizes the role of change. If there is no change in an object's energy, then we know no work was done on it. This theorem applies to changes in potential energy also. Recall our previous example of the weight lifter raising the barbell. When work was being done on the barbell, its potential energy was being changed. But when it was held stationary, no further work was being done *on the barbell*—as evidenced by no further change in its energy.

Similarly, push against a box on a floor. If it doesn't slide, then you are not doing work on the box. Put the box on a very slippery floor and push again. If it slides, then you're doing work on it. When the amount of work done to overcome friction is small, the amount of work done on the sliding box is practically equal to its gain in kinetic energy.

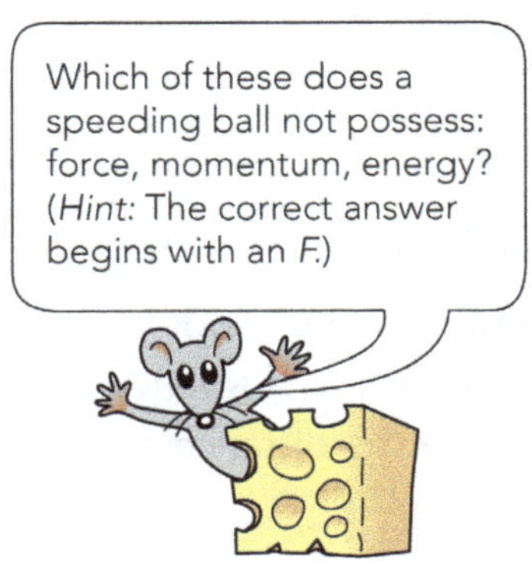

The work-energy theorem applies to decreasing speed as well. The more kinetic energy something has, the more work is required to stop it. Twice as much kinetic energy means twice as much work. When we apply the brakes to slow a car, we do work on it. This work is the friction force supplied by the brakes, multiplied by the distance over which the friction force acts.

Interestingly, the maximum friction that the brakes can supply is nearly the same whether the car moves slowly or quickly. In a panic stop with antilock brakes, the only way for the brakes to do more work is to act over a longer distance. A car moving at twice the speed of another has four times ($2^2 = 4$) as much kinetic energy, and will require four times as much work to stop. Therefore, it takes four times as much distance to stop. Accident investigators are well aware that an automobile going 100 kilometers per hour has four times the kinetic energy as it would have at 50 kilometers per hour. So, a car going 100 kilometers per hour will skid four times as far when its brakes are applied as it would going 50 kilometers per hour. Kinetic energy depends on speed squared.

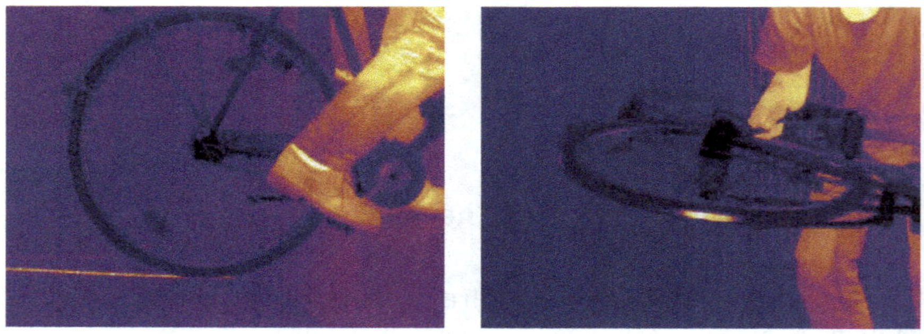

Figure 5.12 >
Due to friction, energy is transferred both into the floor and into the tire when the bicycle skids to a stop. (a) An infrared camera reveals the heated tire track on the floor. (b) The warmth of the tire is also revealed.

Courtesy of Michael Vollmer

∧ Figure 5.13
The potential energy of the elevated ram is converted to kinetic energy when released.

CONCEPT CHECK

1. When the brakes of a car are locked, the car skids to a stop. How much farther will the car skid if it's moving 3 times as fast?
2. Can an object have energy?
3. Can an object have work?

CHECK YOUR ANSWER

1. Nine times farther. The car has nine times as much energy when it travels three times as fast: $1/2\, m(3v)^2 = 1/2\, m9v^2 = 9(1/2\, mv^2)$. The friction force will ordinarily be the same in either case. Therefore, to do nine times the work requires nine times as much sliding distance.
2. Yes, but only in a relative sense. For example, an elevated object may possess PE relative to the ground, but none relative to a point at the same elevation. Similarly, the kinetic energy of an object is relative to a frame of reference, usually taken to be ground level at Earth's surface.
3. No, unlike energy, work is not something an object has. Work is something an object does to some other object. However, an object can do work only if it has energy.

Quick Activity

Hands-On Exploration

Rub your hands briskly together. The friction between them multiplied by the distance of rubbing produces work that becomes heat. Note how quickly your palms are warmed.

5.6 Conservation of Energy

By studying how energy changes from one form to another, scientists have developed one of the greatest generalizations in science—the law of **conservation of energy**:

Energy cannot be created or destroyed; it may be transformed from one form into another or transferred from one object to another, but the total amount of energy never changes.

For any system, whether as simple as a swinging pendulum or as complex as an exploding star, energy remains the same. Energy may change form or may be transferred from one place to another, but the total energy score stays the same.

Consider the circus diver of **Figure 5.14**. At the top of the pole he has a maximum amount of potential energy, which is 10,000 J. Sitting at the top of the pole he is not moving, hence is kinetic energy is 0 J. Upon jumping his potential energy decreases. Why? Because his height is also decreasing. Meanwhile, he's speeding up, which means his kinetic energy is increasing. The law of energy conservation tells us that the total energy of the system remains the same. Therefore, as much potential energy that he loses is exactly as much kinetic energy that he gains. By the time he reaches the pool below, his potential energy is 0 J while his kinetic energy is 10,000 J. Energy can change from one form to another, but the total amount of energy remains the same.

Consider also the swinging pendulum of **Figure 5.15**. At its highest point, the bob has its greatest amount of potential energy. At this highest point, however, the bob is also momentarily at rest, which means it has a minimum amount of kinetic energy. At its lowest point, the bob has its least amount of potential energy, which is where it will have its maximum amount of kinetic energy. So, we see that as the pendulum swings there is a perpetual conversion between potential and kinetic energy. As one goes up, the other necessarily goes down. The total amount of energy remains the same.

Most all the energy available to us here on Earth has its origins within the core of the Sun, where hydrogen atoms are combined to form helium atoms. This happens through a process known as *thermonuclear fusion*, which we explore in a later chapter. The energy from thermonuclear fusion within the Sun reaches Earth in the form of sunshine, which provides photosynthetic plants the energy they need to grow. A portion of this energy is then transferred to animals who eat those plants, and then again to animals who eat the animals who ate those plants. Also, under proper geologic

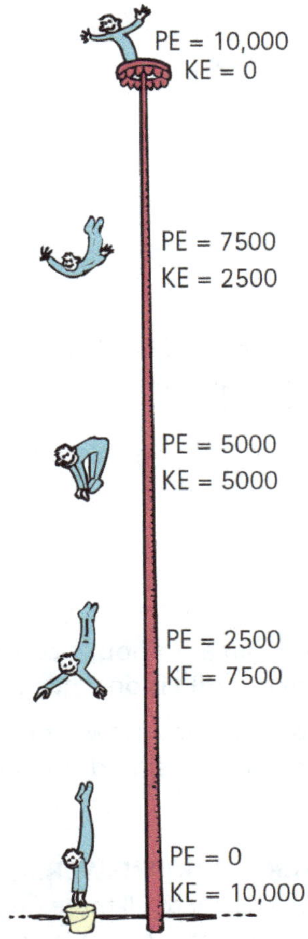

Figure 5.14
A circus diver at the top of a pole has a potential energy of 10,000 J. As he dives, his potential energy converts to kinetic energy. Note that at successive positions one-fourth, one-half, three-fourths, and all the way down, the total energy is constant.

READING CHECK
What remains the same as systems change form or transfer energy from one place to another?

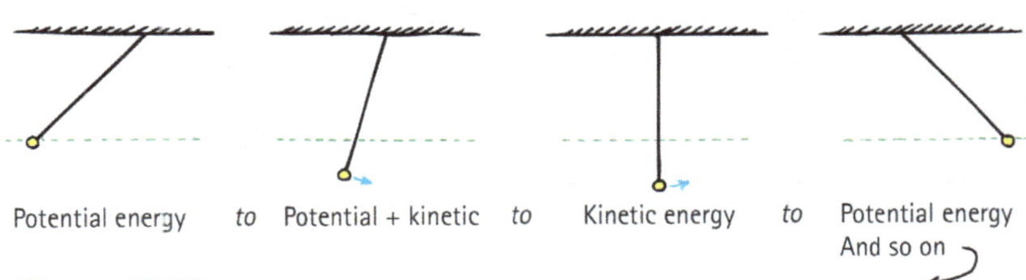

Figure 5.15
Energy transitions in a pendulum. PE is relative to the lowest point of the pendulum when it is vertical.

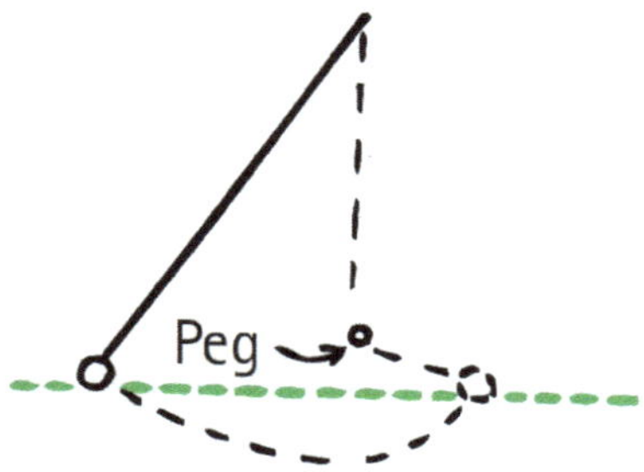

∧ Figure 5.16
The pendulum bob will swing to its original height whether or not the peg is present.

conditions, ancient plant matter can be concentrated into this gooey fluid known as *petroleum*, which can be converted to the fossil fuels we use, such as gasoline. From these examples, we see how, in the natural world, energy readily changes from one form to another.

Consider also how sunshine evaporates water from the ocean, and part of this returns to Earth's surface as rain that may be trapped behind a dam. The potential energy of the dammed water may be used to power a generating plant below, where it will be transformed to electric energy. So, the energy generated at dams began in the Sun. And this energy travels through wires to homes, where it is used for lighting, heating, cooking, and operating electric gadgets. How wonderful that energy changes from one form to another! But as we'll also be discussing later, all energy ultimately ends up dispersed as heat. Energy always tends to flow from where it is concentrated, such as within the Sun's core, to where it is dilute, such as within the vastness of outer space. As it moves in this direction we are able to tap into it while still concentrated enough to be useful.

CONCEPT **CHECK**

1. Does an automobile consume more energy when its air conditioner is turned on? When its lights are on? When its radio is on while the auto is at rest in the parking lot?

2. Rows of wind-powered generators are used in various windy locations to generate electric power. Does the power generated affect the speed of the wind? Would locations behind the "windmills" be windier if they weren't there?

CHECK YOUR ANSWERS

1. The answer to all three questions is yes, whether the energy source is fuel or an electric battery. Even energy from the battery must be given back to the battery by the alternator, which is turned by the engine. All used energy must come from some source. There's no free lunch!

2. Windmills generate power by taking kinetic energy from the wind, so the wind is slowed by interaction with the windmill blades. So yes, it would be windier behind the windmills if they weren't there.

Calculation Corner

The values of kinetic energy and potential energy for the block freely sliding down a ramp are shown only at the bottom of the ramp. Fill in the missing values.

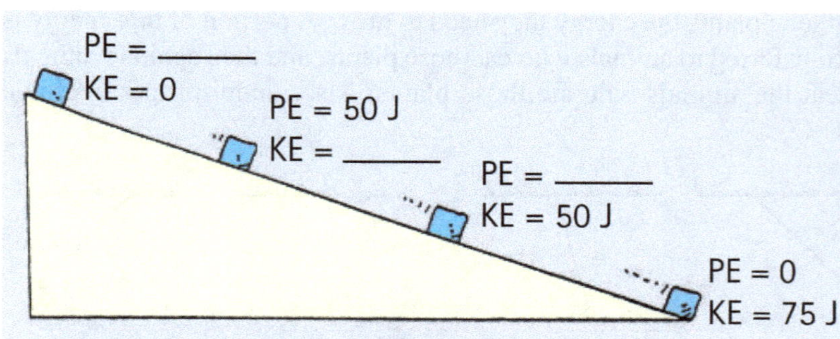

5.7 Machines

A machine is a device for multiplying forces or simply changing the direction of forces. All machines employ the conservation of energy. Machines may be mechanical that deal with levers and pulleys, or hydraulic that employ pistons. We first examine mechanical machines, the most simple being the lever (**Figure 5.17**). When you do work by pushing one end of the lever down, work is done at the other end. The work done on the output side of the lever raises a load. If heat from friction forces is small enough to neglect, the work input is equal to the work output:

$$\text{Work input} = \text{work output}$$

Since work equals force multiplied by distance, we can say that input force × input distance = output force × output distance.

$$(\text{force} \times \text{distance})_{\text{input}} = (\text{force} \times \text{distance})_{\text{output}}$$

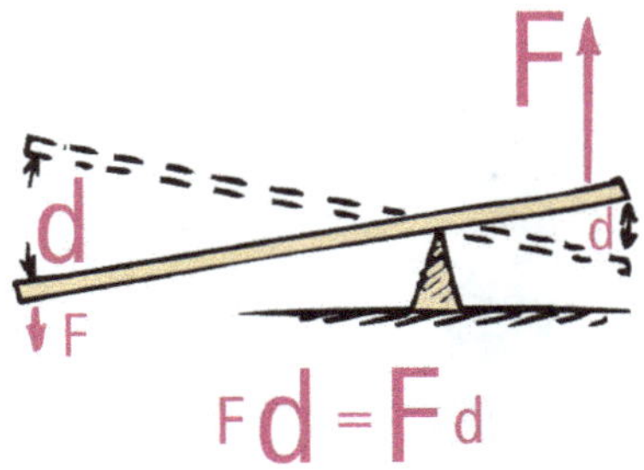

Λ Figure 5.17
A simple lever.

The point of support on which a lever rotates is called a fulcrum. When the fulcrum is close to the load, a large output force is produced by a small input force. This is because the input force is exerted through a large distance and the load is moved over a short distance. In this way, a lever can multiply force. But no lever or machine has been found that can multiply work or energy. Our understanding of energy suggests that none ever will be found. We are so confident of this that we say energy is never created or destroyed.

The principle of the lever was understood by the Greek scientist Archimedes in the 3rd century BC. He said that he could move the whole world if he had a long enough lever and a place to put the fulcrum. Some good science has been around for a long time!

Another simple machine is a pulley. Can you see that it is a lever "in disguise"? When used as in **Figure 5.20**, it changes only the direction of the force. But when used as in **Figure 5.21**, the output force is doubled. Force is increased and distance moved is decreased.

Forces can be nicely multiplied with a system of pulleys. Such pulley arrangements are common in gyms, athletic training centers, and automobile

Λ Figure 5.18
Work done on one end equals the work done on a load at the other end.

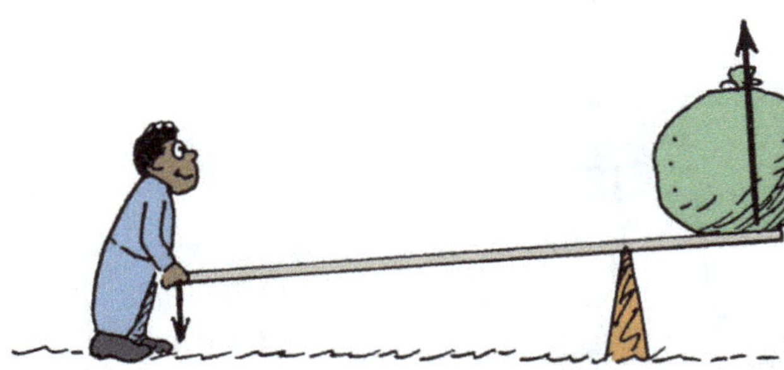

Λ Figure 5.19
Force is multiplied. Note that a small input force × large distance = large output force small distance.

Figure 5.20
This pulley acts like a lever. It changes only the direction of the input force.

Figure 5.21
In this arrangement, a load can be lifted with half the input force.

service centers or machine shops wherever heavy loads are lifted. An ideal pulley system is shown in **Figure 5.22**. The man pulls 7 meters of rope with a force of 50 newtons and lifts 500 newtons through a vertical distance of 0.7 meter. The work the man does when pulling the rope is numerically equal to the increased potential energy of the 500-N bag.

A common alternative to mechanical machines are hydraulic machines that use pistons rather than levers and pulleys to perform work. As with cables that transmit forces, pistons do the same through fluids (usually a mineral oil or a gas). As we'll be discussing in Chapter 7, a principle of hydraulics is that pressure applied to a confined fluid is transmitted undiminished to all points in the fluid.

Figure 5.22
Input force × input distance = output force × output distance. Note the load is supported by 7 strands of rope. Each strand supports 1/7 the load. The tension in the rope pulled by the man is likewise 1/7 the load.

Pressure and force are closely related but different concepts: by definition pressure is force per surface area. A given force exerted over a small area produces greater pressure than when exerted over a larger area. That's why the push of a pin hurts more than an equal push by a finger—more about pressure in Chapter 7. For now let's briefly investigate how forces on pistons are similar to forces on levers and pulleys.

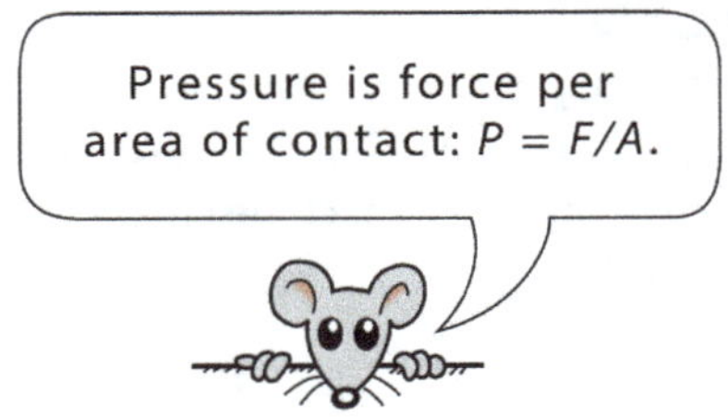

< **Figure 5.23**
Gears, pulleys, and cables have yielded to hydraulic pistons in almost all construction equipment.

Fill a U-tube with a liquid and place pistons at each end (**Figure 5.24**). Pressure exerted against the left piston will be transmitted throughout the liquid and up against the bottom of the right piston. The pistons are simply "plugs" that can slide freely but snugly inside the tube. Doing this with same-size pistons is no big deal, but using a wider piston on the right is more impressive (**Figure 5.25**). As with a lever, force is multiplied at the expense of distance. A unique aspect of hydraulics is that the energy is transmitted using a fluid within a sealed pipe or channel.

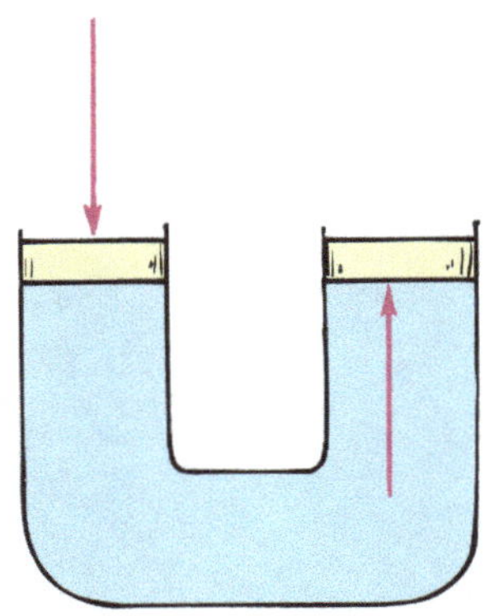

∧ **Figure 5.24**
The downward force exerted on the left piston is transmitted up against the right piston.

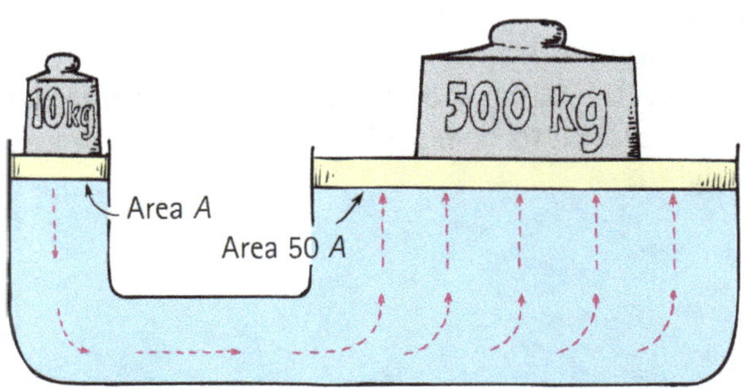

∧ **Figure 5.25**
A 10-kg load on the left piston will support 500 kg on the right piston.

READING CHECK
A machine can transfer or transform energy, but can a machine create energy?

Hydraulics is employed by modern devices ranging from very small to enormous. Their operation follows the same rules that underlie mechanical machines—the conservation of energy.

Any machine, mechanical or hydaulic, that multiplies force does so at the expense of distance. Likewise, any machine that multiplies distance does so at the expense of force. No machine or device can put out more energy than is put into it. No machine can create energy; it can only transfer it or transform it from one form to another.

Figure 5.26
Hydraulic pistons are common in earth movers.

CONCEPT **CHECK**
If a lever is arranged so that input distance is twice output distance, can we predict that energy output will be doubled?

CHECK YOUR ANSWER
No, no, a thousand times no! We can predict output force will be doubled, but never energy. Work and energy stay the same, which means force × distance stays the same. Shorter distance means greater force, and vice versa. Be careful to distinguish between the concepts of force and energy!

5.8 Efficiency

The previous examples of machines were considered to be ideal. All the work input was transferred to work output. An ideal machine would have 100% efficiency. No real machine can be 100% efficient. In any machine, some energy is transformed into atomic or molecular kinetic energy—making the machine warmer. We say this wasted energy is dissipated as thermal energy, or roughly speaking, as heat. Efficiency can be expressed by the ratio:

READING CHECK
Why can't a machine be 100% efficient?

$$\textbf{Efficiency} = \frac{\textbf{work done}}{\textbf{energy used}}$$

Even a lever converts a small fraction of input energy into heat when it rotates about its fulcrum. We may do 100 joules of work but get out 98 joules. The lever is then 98 percent efficient, and we waste 2 joules of work input on heat. In a pulley system, a larger fraction of input energy goes into heat. If we do 100 joules of work, the forces of friction acting through the distances through which the pulleys turn and rub about their axles may dissipate 60 joules of energy as heat. So, the work output is only 40 joules, and the pulley system has an efficiency of 40 percent. The lower the efficiency of a machine, the greater the amount of energy wasted as heat.

A traditional automobile engine is a machine that transforms chemical energy stored in gasoline into mechanical energy. But only a fraction of the energy in the gas is used by the car to move forward. Some of the fuel energy in the gas goes out in the hot exhaust gases and is wasted. Also, nearly half of the energy stored in the gas is wasted in the friction of the moving engine parts. In addition to these inefficiencies, some of the gas doesn't even burn completely. So, the energy in the unburned gasoline also goes unused.

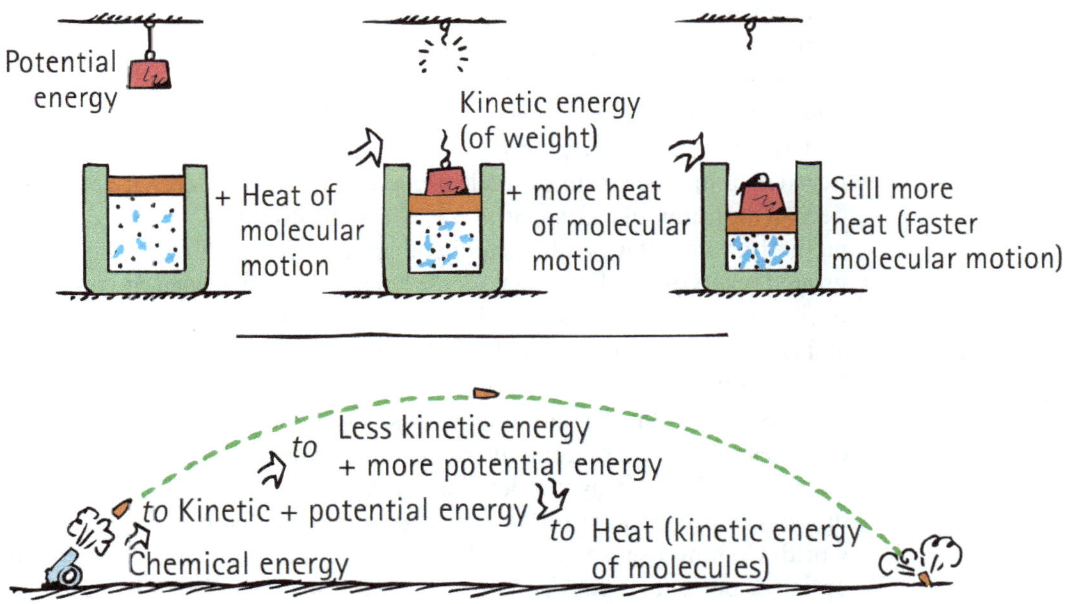

Figure 5.27
Energy transitions. The graveyard of kinetic energy is thermal energy.

CONCEPT **CHECK**

Consider an imaginary miracle car that has a 100 percent efficient engine and burns fuel that has an energy content of 120 megajoules per kg. If the air drag plus frictional forces on the car traveling at highway speed is 500 N, what is the maximum distance the car can go on one liter of fuel?

CHECK YOUR ANSWER

From the definition work = force × distance, simple rearrangement gives distance = work/force. If all 120 million J of energy in 1 kg is used to do the work of overcoming the air drag and frictional forces, the distance covered is 240 kilometers:

$$\text{Distance} = \frac{\text{work}}{\text{force}} = \frac{\mathbf{120{,}000{,}000\ J}}{\mathbf{500\ N}} = \mathbf{240{,}000\ m = 240\ km}$$

The important point to learn is that even with a perfect engine, there is still an upper limit of fuel economy dictated by the conservation of energy.

5.9 Energy Sources

READING CHECK

From where did fossil fuels originally acquire their energy?

Our greatest energy source is the Sun. Even the energy we obtain from petroleum, coal, natural gas, and wood comes from the Sun. That's because these fuels are created by photosynthesis, the process by which plants trap solar energy and store it as plant tissue.

Sunlight is directly transformed into electricity by photovoltaic cells. We use the energy in sunlight to generate electricity indirectly as well. Sunlight evaporates water, which later falls as rain. Rainwater flows into rivers and turns water wheels, or it flows into modern generator turbines as it returns to the sea.

Wind, caused by unequal warming of the Earth's surface, is another form of solar energy. The energy of wind can be used to turn generator turbines within specially equipped windmills. Wind farms spread out in windy regions feed electrical energy to power grids that service consumers.

We find there is also appreciable energy stored in the elements uranium, plutonium, and thorium. These elements can be processed into nuclear fuels that we discuss in Chapter 15. Interestingly, Earth's interior is kept hot because of another form of nuclear energy called *radioactivity*, which has been within this planet since it was formed billions of years ago. Earth's naturally warm interior allows for geothermal energy, which is held in underground reservoirs of hot water. Geothermal energy has been predominantly limited to areas of volcanic activity, such as Iceland, New Zealand, Japan, and Hawaii. These and dry-rock geothermal plants are tapping Earth's natural heat to provide steam for running turbo-generators.

As the world population increases, so does our need for energy. With the rules of physics to guide them, technologists are presently researching newer and cleaner ways to develop energy sources. But they race to keep up with world population growth and greater demand in the developing world. Common sense dictates that in the meantime we should continue to optimize present sources and use what we consume efficiently and wisely.

Concept Review

Chapter 5

Summary of Terms

Conservation Of Energy Energy cannot be created or destroyed; it may be transformed from one form into another, but the total amount of energy never changes. In an ideal machine, where no energy is transformed into heat, $work_{input} = work_{output}$ and $(Fd)_{input} = (Fd)_{output}$.

Efficiency The percent of the work put into a machine that is converted into useful work output.

Energy The property of a system that enables it to do work.

Kinetic Energy Energy of motion, described by the relationship: Kinetic energy = $1/2\ mv^2$.

Machine A device that increases (or decreases) a force or simply changes the direction of a force.

Potential Energy The stored energy that a body possesses due to its position.

Power The time rate of doing work: Power = work/time.

Watt The unit of power, the joule per second.

Work The product of the force and the distance through which the force moves: $W = Fd$.

Work-Energy Theorem The work done on an object is equal to the energy gained by the object. Work = ΔE. The energy change can be PE or KE.

Review Questions

1. A force sets an object in motion. When the force is multiplied by the time of its application, we call the quantity impulse, which changes the momentum of that object. What do we call the quantity force × distance?
2. Cite an example where a force is exerted on an object without doing work on the object.
3. Which requires more work—lifting a 50-kg sack a vertical distance of 2 m or lifting a 25-kg sack a vertical distance of 4 m?
4. If both sacks in the preceding question are lifted their respective distances in the same time, how does the power required for each compare? How about for the case where the lighter sack is moved its distance in half the time?
5. What are the two main forms of mechanical energy?
6. Exactly what is it that a body having energy is capable of doing?
7. A car is lifted a certain distance in a service station and therefore has potential energy relative to the floor. If it were lifted twice as high, how much potential energy would it have?
8. Two cars are lifted to the same elevation in a service station. If one car is twice as massive as the other, how do their potential energies compare?
9. How many joules of potential energy does a 1-N book gain when it is elevated 4 m? When it is elevated 8 m?
10. A moving car has kinetic energy. If it speeds up until it is going four times as fast, how much kinetic energy does it have in comparison?
11. Compared to some original speed, how much work must the brakes of a car supply to stop a four-times-as-fast car? How will the stopping distance compare?
12. What will be the kinetic energy of pile driver ram when it undergoes a 10 kJ decrease in potential energy? (Assume no energy goes to heat.)

13. Can a machine multiply input force? Input distance? Input energy? (If your three answers are the same, seek help, for the last question is especially important.)

14. If a machine multiplies force by a factor of four, what other quantity is diminished, and how much?

15. If the man in Figure 5.22 pulls 1 m of rope downward with a force of 100 N, and the load rises 1/7 as high (about 14 cm), what is the maximum load that can be lifted?

16. What is the efficiency of a machine that miraculously converts all the input energy to useful output energy?

17. Is a machine physically possible that has an efficiency greater than 100 percent? Discuss.

18. What is the ultimate source of energies for the burning of fossil fuels, dams, and windmills?

19. What is the ultimate source of geothermal energy?

20. The energy we require for existence comes from the chemically stored potential energy in food, which is transformed into other forms when it is metabolized. What happens to a person whose work output is less than the energy he or she consumes? Whose work output is greater than the energy he or she consumes? Can an undernourished person perform extra work without extra food? Briefly discuss.

Answers to Odd-Numbered Questions

1. Work.

3. Both require the same work because the force × distance is the same for each.

5. Potential energy and kinetic energy.

7. Twice.

9. PE = 1 N at 4 m = 40 J. At 8 m the PE = 80 J.

11. A four-times as fast car has 16 times as much KE and will require 16 times as much work to stop, and 16 times as much stopping distance.

13. Yes; yes; NO! (Machines are good at multiplying force or distance, but no machine can multiply energy—a conservation of energy no no!

15. Seven times as much, 700 N.

17. Not in the world as we know it. Such a machine would clearly violate the conservation of energy, a law that has never been violated.

19. Nuclear energy from radioactive decay.

Chapter 6

Gravity

THE MAIN IDEA

Satellites are freely-falling projectiles in the grip of gravity

Is an astronaut taking a "space walk" in the grips of Earth gravity? Was gravity discovered by Isaac Newton? Or was gravity discovered by earlier people who fell from trees? If Newton didn't discover gravity, what did he discover about gravity? Does gravity reach to the Moon? Does it reach to the planets? To the stars? How far does gravity reach? How does gravity affect the motion of projectiles? Of satellites? Is it true that satellites, including planets, are freely-falling projectiles? We'll learn the answers to these questions in this chapter.

6.1 Newton's Insights

Legend tells us that when Newton was a young man sitting under an apple tree, he made a connection that changed the way we see the universe. He saw an apple fall. Perhaps he looked up through the tree branches toward the origin of the falling apple and noticed the Moon. In any event, Newton had the insight to realize that the force pulling on a falling apple is the same force that pulls on the Moon. Newton realized that Earth gravity reaches to the Moon.

A force by Earth on the Moon would explain why the Moon does not follow a straight line path, but instead circles about Earth. Circular motion is accelerated motion, which requires a force. Newton reasoned that the Moon is falling for the same reason the apple falls—they are both pulled by Earth's gravity.

Newton developed this idea further. How can the Moon fall, and not hit Earth? He realized that the Moon is also moving

READING CHECK

What similarity did Newton see between a falling apple and the Moon?

Figure 6.1

Newton realizes that Earth's gravity affects both the apple and the Moon.

sideways causing the Moon to fall not into Earth but around it instead. So, the Moon is actually falling in the sense that it falls beneath the tangential straight line it would follow if no force acted on it. He hypothesized that the Moon was a freely-falling object under the attraction of gravity. But because it also moved sideways, the net result was it circling around us. Newton further realized that the Moon's path around the Earth is similar to the paths of the planets around the Sun.

READING CHECK

In what sense is the Moon in a continual state of free fall?

This concept is illustrated in an original drawing by Newton, shown in **Figure 6.3**. Newton compared motion of the Moon with a cannonball fired from the top of a high mountain. He imagined that the mountain top was above Earth's atmosphere, so that air resistance would not slow the motion of the cannonball. If the cannonball were fired with a small horizontal speed, it would follow a curved path and soon hit the ground below. If it were fired faster, its path would be less curved and it would hit the ground farther away. If the cannonball were fired fast enough, Newton reasoned, its curvature of fall would match the curvature of Earth and the cannonball would circle indefinitely. It would be in orbit. These ideas have very much changed the way people think.

FOR YOUR INFORMATION

Newton believed gravity acted instantaneously. Einstein showed that gravity travels at the speed of light. Earth responds to the Sun's mass as it was about 8 minutes in the past. Why? Because, at the speed of light, it takes about 8 minutes for the effect of gravity to travel from the Sun to Earth. Like Newton before him, Einstein also changed how we think. Very much so.

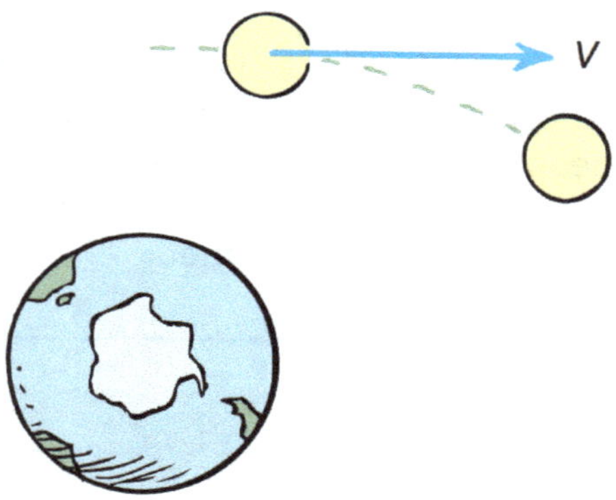

Figure 6.2

The sideways velocity of the Moon allows it to fall around the Earth rather than directly into it.

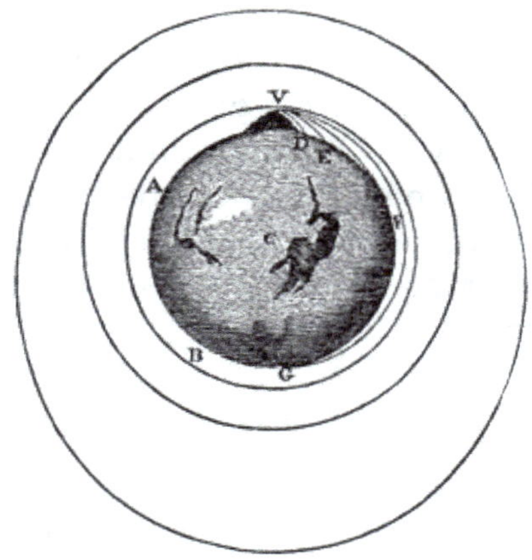

Figure 6.3

This original drawing by Isaac Newton shows how an object projected fast enough would fall around Earth and become an Earth Satellite.

CONCEPT **CHECK**

We see that the Moon falls around the Earth rather than straight into it. If the sideways tangential velocity were zero, how would the Moon move?

CHECK YOUR ANSWER

If the Moon's sideway tangential velocity were zero, it would fall straight down and crash into the Earth!

6.2 Inverse-Square Law

Newton further realized that everything pulls on everything else. He discovered that a force of gravity acts between all things in a beautifully simple way—involving only mass and distance. According to Newton, every mass attracts every other mass with a force that is directly proportional to the product of the two interacting masses. This statement is known as the **law of universal gravitation**. The force is inversely proportional to the square of the distance separating them.

READING CHECK
State Newton's law of gravity in words.

$$\text{Force} \sim \frac{\text{mass}_1 \times \text{mass}_2}{\text{distance}^2}$$

Expressed in symbol shorthand:

$$F = G\frac{m_1 m_2}{r^2}$$

where m_1 and m_2 are the masses, and r is the distance between their centers (the r stands for "*radial* distance"). Thus, the greater the masses m_1 and m_2, the greater the force of attraction between them. The greater the distance of separation, r, the weaker is the force of attraction—weaker as the inverse square of the distance between their centers.

CONCEPT **CHECK**

1. According to the equation for gravity, what happens to the force between two bodies if the mass of one body is doubled?
2. What happens if instead the mass of the other body is doubled?
3. What happens if the masses of both bodies are doubled?
4. What happens if the mass of one body is doubled, and the other tripled?

CHECK YOUR ANSWERS

1. When one mass is doubled, the force between them doubles.
2. The force is still doubled, for it doesn't make any difference which mass doubles. (2 × 1 = 1 × 2; same product either way!)
3. The force is four times as much.
4. Double × triple = six. So, the force is six times as much. (If you don't see why, discuss this with a friend before going further.)

Gravity gets weaker with distance the same way a light gets dimmer as you move farther away from it. Consider the candle flame in **Figure 6.4**. Light from the flame travels in all directions in straight lines. As the light spreads out, its brightness decreases. When you're twice as far away, it appears 1/4 as bright. Why? Because the same amount of light is now spread out over 4 times the area. When you're 3 times as far away, the light is spread over 9 times the area, which makes the light appear 1/9 as bright at this distance. There is a rule here: The intensity of the light gets less as the inverse square of the distance. This is the **inverse-square law**.

The inverse-square law also applies to a paint sprayer. Pretend you hold a spray paint can at the center of a sphere with a radius of 1 meter (**Figure 6.5**). Suppose that a burst of paint produces a square patch of paint 1 millimeter thick. How thick would the patch be if the experiment were done in a

Figure 6.4 >

Light from the flame spreads in all directions. At twice the distance, the same light is spread over 4 times the area; at 3 times the distance it is spread over 9 times the area.

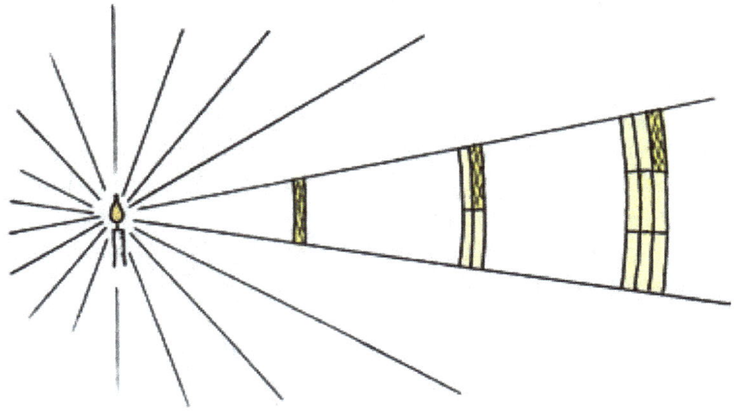

sphere with twice the radius—that is, with the spray gun twice as far away? The answer is not half as thick, because the paint would spread to a patch twice as tall and twice as wide. It would spread over an area four times as big, and its thickness would be only 1/4 millimeter. Can you see that for a sphere of radius 3 meters the thickness of the paint patch would be only 1/9 millimeter? Do you see that the thickness of paint quickly decreases as the square of the distance? The inverse-square law holds for light, for paint spray, and for gravity. It holds for all phenomena where something from a localized source spreads uniformly throughout the surrounding space. We'll see this to be true of the electric field about an electron, light from a match, radiation from a piece of uranium, and sound from a cricket.

So, we see that gravity decreases according to the inverse-square law. The force of gravity weakens as the square of the distance, which means it decreases fairly rapidly with increasing distance. In using Newton's equation for gravity, the distance term r is the distance between the *centers* of the masses of objects attracted to each other. Note in **Figure 6.6** that the girl at the top of the ladder weighs only 1/4 as much as she weighs at the Earth's surface. That's because she is twice the distance from the Earth's *center*. This corresponds to a ladder of about 6400 km tall. Interestingly, space craft usually orbit about 300 km in altitude where the force of gravity is practically the same as it is on the surface.

READING CHECK

How does the force of gravity change with distance?

Figure 6.5 >

The inverse-square law. Paint spray travels in straight lines away from the nozzle of the can. Like gravity, the "strength" of the spray obeys the inverse-square law. Fill in the blanks.

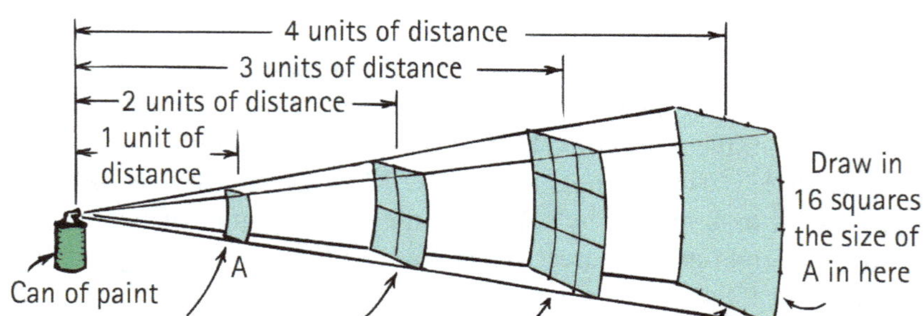

	1 area unit	4 area units	( ) area units	( ) area units
Paint spray	1 layer thick	¼ layer thick	( ) layer thick	( ) layer thick

CONCEPT **CHECK**

1. How much does the force of gravity change between the Earth and a receding rocket when the distance between them is doubled? Tripled? Ten times as much?

2. Consider an apple at the top of a tree. The apple is pulled by Earth's gravity with a force of 1 N. If the tree were twice as tall, would the force of gravity be only 1/4 as strong? Defend your answer.

CHECK YOUR ANSWERS

1. When the distance is doubled, the force is 1/4 as much. When tripled, 1/9 as much. When 10 times, 1/100 as much.

2. No, because the twice-as-tall apple tree is not twice as far from the Earth's center. The taller tree would have to be 6,370 km tall (the Earth's radius) for the apple's weight to reduce to 1/4 N. For a decrease in weight by 1 percent, an object must be raised 32 km—nearly four times the height of Mt. Everest. So, as a practical matter we disregard the effects of everyday changes in elevation for gravity. The apple has practically the same weight at the top of the tree as at the bottom.

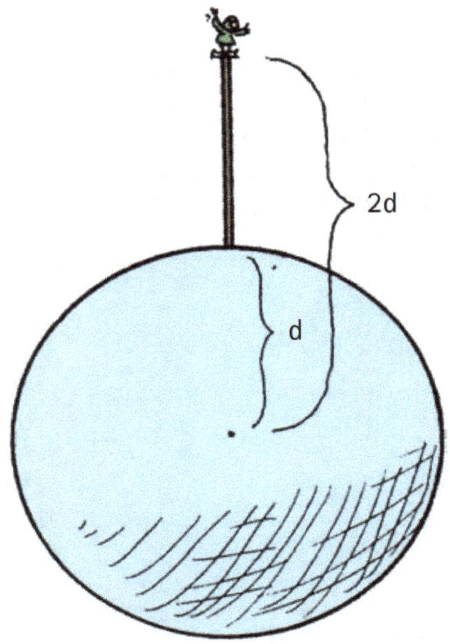

∧ Figure 6.6

At the top of the ladder she is twice as far from Earth's center, and weighs only 1/4 as much as at the bottom of the ladder. She's in outer space. If she stepped off the ladder, she would fall downward because gravity is still acting on her, albeit not as strongly.

Earth's gravity weakens with distance, but no matter how far away, Earth's gravity approaches, but never reaches zero. Even if you traveled to the far reaches of the universe, the gravitational influence of home would still be with you. It may be overwhelmed by the gravitational influences of nearer and/or more massive bodies, but it is there. The gravitational influence of every material object, however small or far, is exerted through all of space.

< Figure 6.7

As the rocket gets farther from Earth, gravitation between the rocket and the Earth decreases, but never falls to zero.

CONCEPT **CHECK**

1. Light from the Sun, like gravity, obeys the inverse-square law. If you were on a planet twice as far from the Sun, how bright would the Sun appear?

2. How bright would the Sun appear if you were on a planet twice as close to the Sun?

CHECK YOUR ANSWERS

1. One-quarter as bright.

2. Four times brighter.

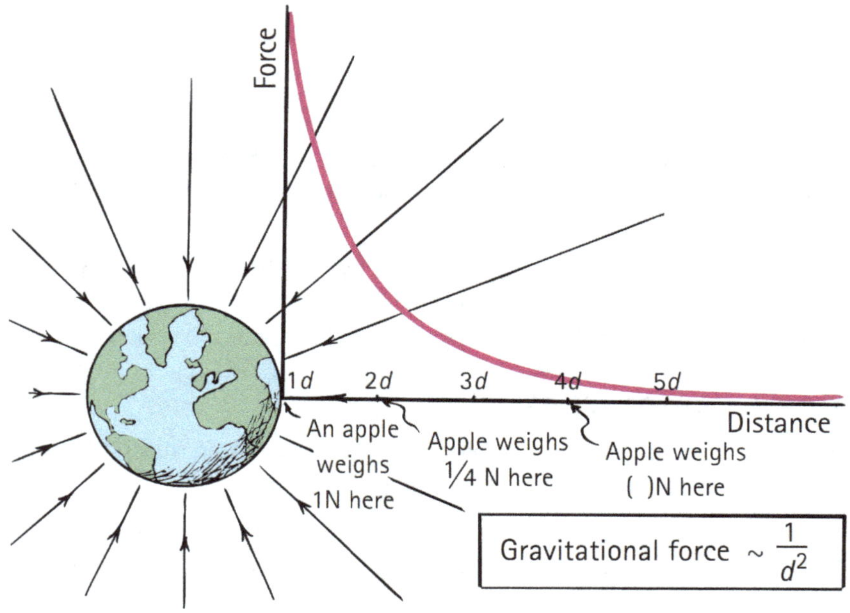

Figure 6.8 >

If an apple weighs 1 N at the Earth's surface, it weighs only 1/4 N twice as far from Earth's center. At three times the distance, it weighs only 1/9 N. What would it weigh at four times the distance? Five times?

Quick Activity

Distance Illusion

Hold your hands outstretched with one hand twice as far from your eyes as the other. Make a casual judgment about which hand looks bigger. Try it again, only this time overlap your hands slightly and view them with one eye closed.

Many people see their hands in this position as the same size, while many see the nearer hand as slightly bigger. Very few people see the nearer hand as four times as big. But by the inverse-square law, the nearer hand actually appears twice as tall and twice as wide. Twice times twice means four times as big!

A person's belief that their hands are the same size is so strong that they likely overrule what is seen This raises an interesting question: What other illusions do we have that are not so easily checked?

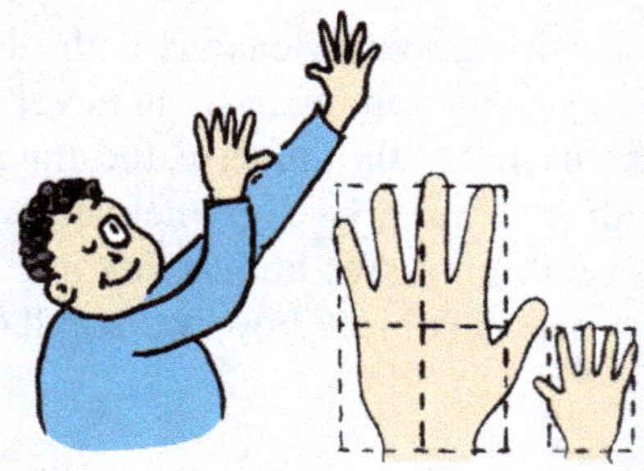

6.3 The Mass of Earth

READING CHECK

How do the units for G affect the unit of force?

The universal law of gravitation can be written as an exact equation when the **universal constant of gravitation,** *G*, is used. Then we have:

$$F = G\frac{m_1 m_2}{r^2}$$

The units of *G* make the force come out in newtons. The magnitude of *G* is the same as the gravitational force between two 1-kilogram masses that are 1 meter apart: 0.0000000000667 newton.

$$G = 6.67 \times 10^{-11}\ \mathrm{Nm^2/kg^2}$$

This is an extremely small number. It shows that, relatively speaking, gravity is a very weak force—two 1 kg masses 1 meter apart barely affect each other. The large net gravitational force we feel as weight is because of the enormity of planet Earth pulling on us.

The value of G wasn't measured until a century after the publication of Newton's theory of universal gravitation. One method of measuring it, though not the first, is shown in Figure 6.9. Once we knew the value of G, we had enough information to calculate the mass of the Earth! Here's how: The force that the Earth exerts on a 1-kilogram mass at its surface is 10 newtons (more precisely, 9.8 newtons). The distance between the 1-kilogram mass and the center of the Earth is the Earth's radius, 6.4×10^6 meters. Using

$$F = G\frac{m_1 m_2}{r^2}$$

and using F as 9.8 N, m_1 is the mass of the 1-kilogram mass, and m_2 is the mass of the Earth:

$$9.8\ \text{N} = 6.67 \times 10^{-11}\,\frac{\text{N}\cdot\text{m}^2}{\text{kg}^2} \times \frac{1\ \text{kg} \times m_2}{(6.4 \times 10^6\ \text{m})^2}$$

only unknown

For this equation there is only one unknown quantity, m_2, the mass of the Earth. Solving, we find $m_2 = 6 \times 10^{24}$ kilograms. This is a huge number, which makes sense, given that, relative to us, the Earth is exceedingly massive.

When G was first measured in the eighteenth century, people all over the world were excited about it. That's because newspapers everywhere announced the discovery as one that measured the mass of planet Earth. How exciting that Newton's formula gives the mass of the entire world, with all its oceans, mountains, and inner parts yet to be discovered. G and the mass of the Earth were measured when much of the Earth's surface was still undiscovered.

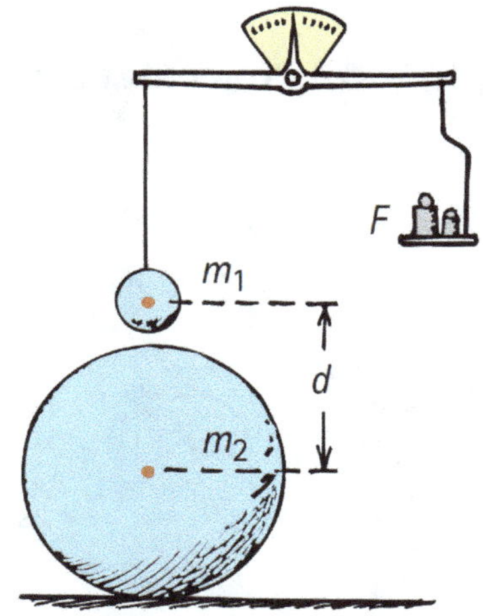

Figure 6.9

Philipp von Jolly's method of measuring G. A small ball of known mass is attracted to a 6-ton sphere rolled beneath it. The force of attraction can be measured by the weights needed to restore balance.

READING CHECK

Why were people all over the world excited about the measurement of G?

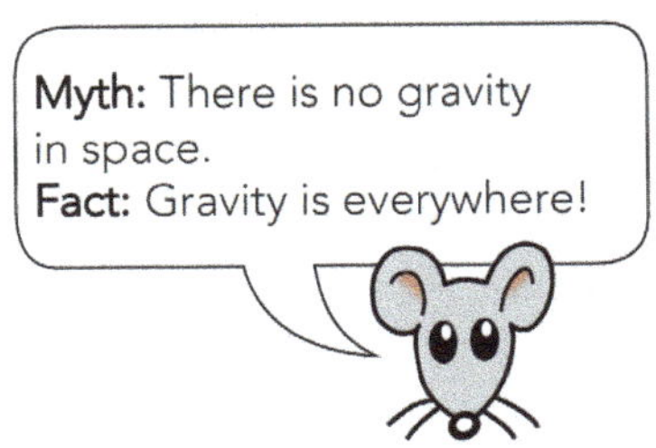

CONCEPT CHECK

1. What value will result if you let your mass be m_1, the mass of the Earth m_2, and d the Earth's radius, in the equation for gravity?
2. If your mass increases, does your weight increase also?

CHECK YOUR ANSWERS

1. Your weight.
2. Yes, in direct proportion. That is, if you double your mass, your weight also doubles.

6.4 Projectile Motion

A tossed stone, a cannonball, or any object projected by any means that continues in motion is called a **projectile**. A very simple projectile is a falling stone, **Figure 6.10**. This is a version of an illustration we studied in Chapter 2. The stone gains speed as it falls straight downward, as indicated by a speedometer. Remember that a freely falling object gains 10 m/s during each second of fall—the acceleration due to gravity, 10 m/s^2. If it begins its fall from rest, 0 m/s, then at the end of the first second of fall its speed is 10 m/s. At the end of 2 seconds, its speed is 20 m/s, and at the end of 3 seconds, it is 30 m/s—and so on. It keeps gaining 10 m/s each second it falls.

Although the change in speed is the same each second, the distance of fall keeps increasing. That's because the average speed of fall increases each second. Let's apply this to a new situation—throwing the stone horizontally from a high cliff.

First, imagine that gravity doesn't act on the stone (of course it does act, but we're just pretending for now). In **Figure 6.11** we see the positions that a horizontally thrown stone would have with no gravity. Note that the positions each second are the same distance apart. That's because there is no force acting on it. The motion of the stone is like the motion of a bowling ball rolling along a bowling lane. Horizontal motion is constant because no horizontal force acts. Both the stone and the ball move without accelerating. They move at constant velocities, covering equal distances in equal times.

In the real world there is gravity, and the thrown stone falls beneath the straight line it would follow with no gravity. Three different directions of throw are shown in **Figure 6.13**. Each shows the straight-down vertical motion of the stone would fall if dropped from rest (vertical dashed line), and the direction thrown (top dashed line). Both of these directions occur

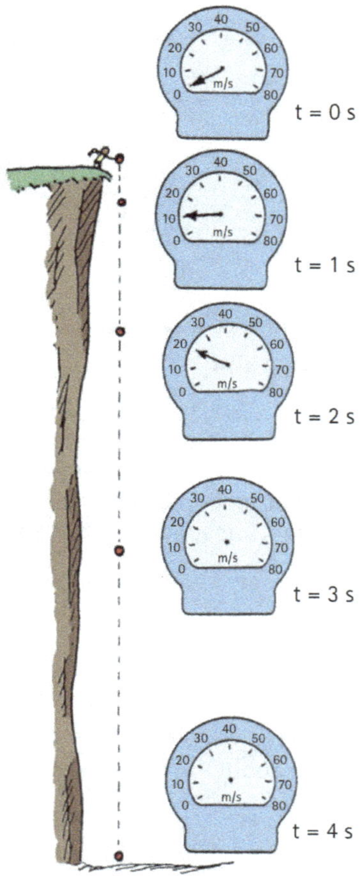

Figure 6.10

The falling stone gains a speed of 10 m/s each second. What might be the speedometer readings for the times 3 and 4 seconds?

Figure 6.11

If there were no gravity, a stone thrown horizontally would move in a straight-line path and cover equal distances in equal time intervals.

< **Figure 6.12**

A bowling ball rolling along a lane similarly covers equal distances in equal times. It rolls at constant velocity.

simultaneously to produce a different curved paths for each. As a tossed stone moves in the direction thrown, it also falls straight downward—beneath the place it would be if there were no gravity.

The curved path of a projectile is the result of constant motion horizontally and accelerated motion vertically under the influence of gravity. This curve is a **parabola**.

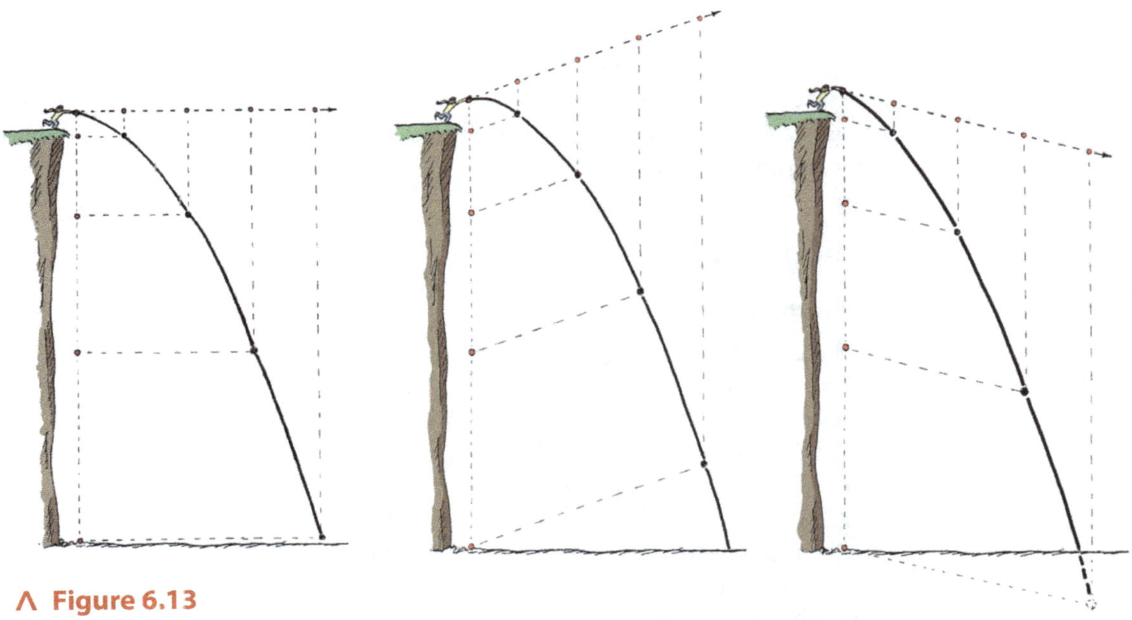

∧ **Figure 6.13**

A stone thrown horizontally (left), at an angle upward (center), and at an angle downward (right). The resulting parabolas differ in shape.

CONCEPT **CHECK**

At the instant a horizontal cannon fires a cannonball from atop a high cliff, another cannonball is simply dropped from the same height. Which hits the ground below first, the one fired downrange, or the one that drops straight down?

CHECK YOUR ANSWER

Both cannonballs hit the ground at the same time, for both fall the same vertical distance. We can reason this another way by asking which one would hit the ground first if the cannon were pointed at an upward angle. Then the dropped cannonball would hit first, while the fired ball is still in the air. Now consider the cannon pointing downward. In this case the fired ball hits first. So, projected upward, the dropped one hits first; downward, the fired one hits first. Is there some angle at which there is a dead heat—where both hit at the same time? Can you see that this occurs when the cannon is horizontal?

If you toss a ball vertically it will lose speed going up and gain the same speed in coming down, if air drag is negligible. Without air drag time going up is the same as time coming down (**Figure 6.14**). For fast-moving balls, air drag is usually not negligible.

Air drag greatly affects the range of balls batted and thrown in baseball games. Without air drag, a ball normally batted to the middle of center field would be a home run. If baseball were played on the Moon (not scheduled in the near future!), the range of balls would be considerably farther—about six times ideal range on Earth. This is because there is no atmosphere on the Moon, so air drag on the Moon is completely absent. Second, gravity is one-sixth as strong on the Moon, which allows higher and longer paths.

Back here on Earth, baseball games normally take place on reasonably level ground. Baseballs curve over a flat playing field. The speeds of baseballs are not great enough for the Earth's curvature to affect the ball's path. For very long-range projectiles, however, the curvature of the Earth's surface must be taken into account. As we will now see, when an object is projected fast enough, it can fall all the way around the Earth to become an orbiting *satellite*.

READING CHECK

What does a projectile on the Moon not encounter?

Quick Activity

Projectile Model

Make your own model of projectile paths. On a ruler or a stick, at position 1, hang a bead from a string 1-cm long as shown. At position 2 hang a bead from a string 4-cm long. At position 3, do the same with a 9-cm length of string. At position 4, use 16 cm of string, and for position 5, 25 cm of string. Hold the stick horizontally and you have a version of Figure 6.13. Hold it at a slight upward angle to show a version of Figure 6.13 (center). Held at an angle downward, you have Figure 6.13 (right).

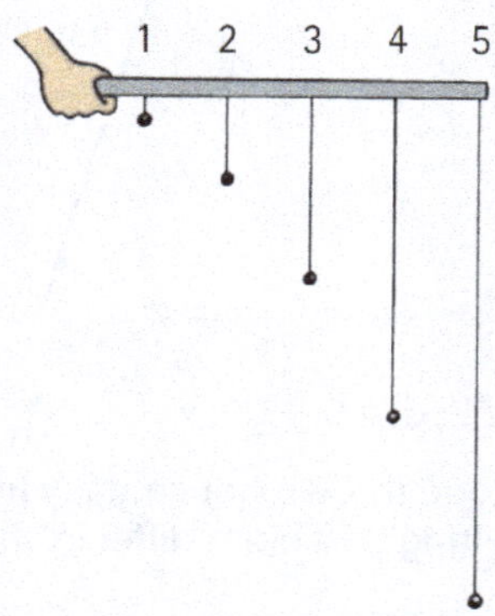

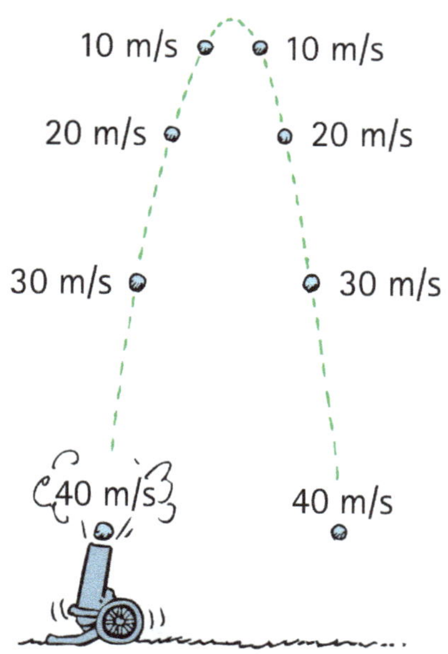

Figure 6.14

Without air resistance, speed lost while going up equals speed gained while coming down: time going up equals time coming down.

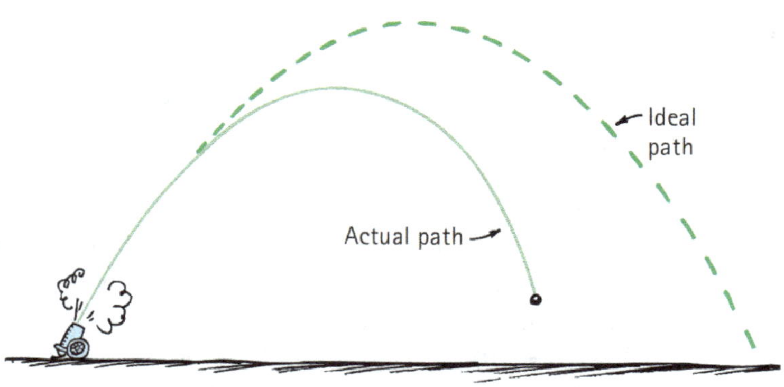

Figure 6.15

In the presence of air resistance, a high-speed projectile falls short of a parabolic path. The dashed line shows an ideal path with no air resistance. The solid line indicates an actual path.

CONCEPT **CHECK**

The boy on the tower throws a ball 20 m downrange. Assuming no air drag, what is his pitching speed?

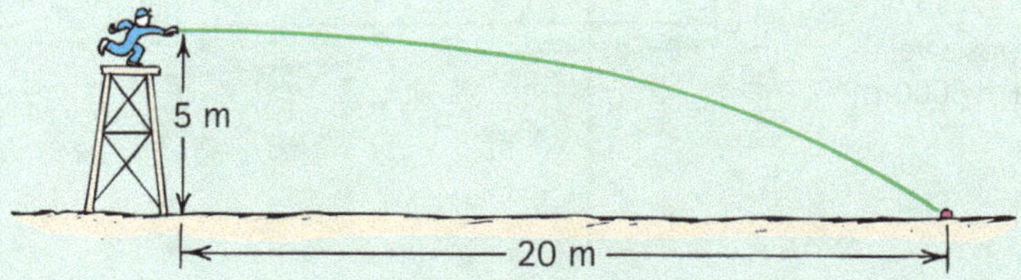

CHECK YOUR ANSWER

The ball is thrown horizontally, so the pitching speed is horizontal distance divided by time. A horizontal distance of 20 m is given, but the time is not stated. However, knowing the vertical drop is 5 m, you remember that a 5-m drop takes 1 s! From the equation for constant speed (which applies to horizontal motion), $v = d/t = (20\text{ m})/(1\text{ s}) = 20\text{ m/s}$. It is interesting to note that the equation for constant speed, $v = d/t$, guides our thinking about the crucial factor in this problem—the time.

6.5 Satellites

Suppose a cannon fires a cannonball so fast that its curved path matches the curvature of the Earth. Then without air drag, it would be an Earth satellite! The same would be true if you could throw a stone fast enough. Any satellite is simply a projectile moving fast enough to fall continually around the Earth.

READING CHECK

Exactly what is a satellite?

In **Figure 6.16**, we see the curved paths of a stone thrown horizontally at different speeds. Whatever the pitching speed, in each case the stone drops the same vertical distance in the same time. Like in the Concept Check above, for a 1-second drop, that distance is 5 meters. So, if you simply drop a stone from rest, it will fall 5 meters in 1 second of fall. Toss the stone sideways, and in 1 second it will be 5 meters below where it would have been without gravity. To be an Earth satellite, the stone's horizontal velocity must be great enough for its falling distance to match the Earth's curvature.

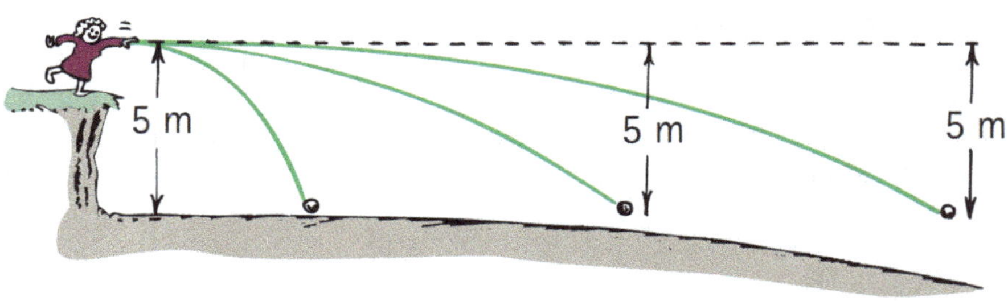

Figure 6.16
Throw a stone at any speed and 1 second later it falls 5 meters below where it would have been if there were no Earth gravity.

It is a geometrical fact that the surface of the Earth drops a vertical distance of 5 meters for every 8000 meters tangent to the surface. A tangent to a circle or to the Earth's surface is a straight line that touches the circle or surface at only one place. We live on a round Earth.

A projectile that moves fast enough to travel a horizontal distance of 8 kilometers during 1 second is an Earth satellite. Neglecting air drag, it would follow the curvature of the Earth. A little thought tells you that this speed is 8 kilometers per second. If this doesn't seem fast, convert it to kilometers per hour and you get an impressive 28,800 kilometers per hour (17,900 mi/h). Fast, indeed!

READING CHECK

What name is given to a projectile that travels a horizontal distance of 8 km in 1 second?

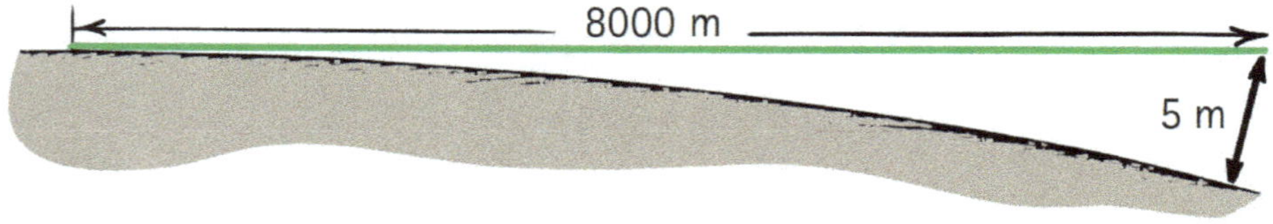

Figure 6.17 >
The Earth's curvature drops a vertical distance of 5 m for each 8000-m tangent (not to scale).

At this speed, atmospheric friction would incinerate the projectile. This happens to grains of sand and other meteorites that graze Earth's atmosphere, burn up, and appear as "falling stars." That is why satellites are launched to altitudes higher than 150 kilometers—to be above the atmosphere. Are they above gravity? No! Remember, gravity extends forever. Furthermore, the force of gravity at an altitude of only 150 km is about the same as it is on Earth's surface. Satellites are launched into space to avoid atmospheric friction.

Newton estimated the speed that a cannonball fired high above the atmosphere would need to circle Earth. Even if his estimate was about 8 km/s, he knew that no cannons could fire cannonballs that fast So, he did not foresee humans launching satellites.

As we have learned, the Moon has a **tangential** ("sideways") velocity, parallel to the Earth's surface. This velocity is enough to ensure motion *around* the Earth rather than *into* it. Without air drag to reduce speed, the Moon or any Earth satellite "falls" around and around the Earth indefinitely. Similarly, with the planets that continually fall around the Sun in closed paths.

Why don't the planets crash into the Sun? They don't because of their tangential velocities. What would happen if their tangential velocities were reduced to zero? The answer is simple enough: Their motion would be straight toward the Sun and they would indeed crash into it. Any objects in the solar system without sufficient tangential velocities have long ago crashed into the Sun. What remains is the harmony we observe.

Satellites are payloads carried above the atmosphere by rockets. Putting a payload into orbit requires control over the speed and direction of the rocket. A rocket initially fired vertically is intentionally tipped from the vertical course as it rises. Then, once above the drag of the atmosphere, it is aimed horizontally, whereupon the payload is given a final thrust to orbital speed (**Figure 6.19**).

∧ Figure 6.18

If the stone is thrown fast enough so that its curve matches the Earth's curvature, it will be a satellite.

CONCEPT **CHECK**
Can we also say a satellite stays in orbit because it is above Earth's main pull of gravity?

CHECK YOUR ANSWER
No, no, no! No satellite is completely "above" Earth's gravity. If the satellite were not in the grip of Earth's gravity it would not orbit and follow instead a straight-line path, which means it would be leaving Earth forever.

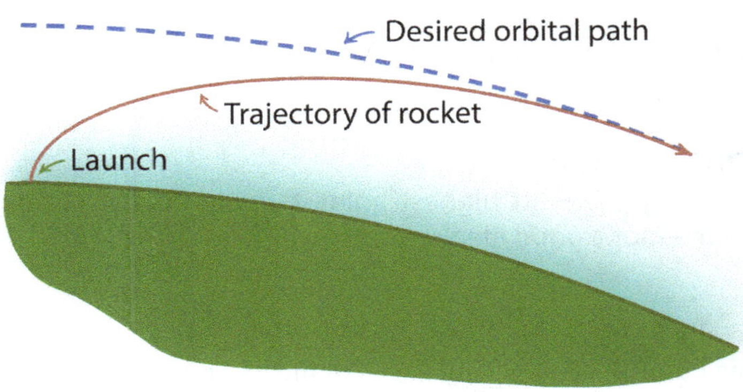

Figure 6.19

The initial thrust of the rocket pushes it above the atmosphere. Another thrust to a tangential speed of at least 8 km/s is needed if it is to fall around rather than into the Earth.

CONCEPT CHECK

1. To review Section 6.5, calculate the speed of a projectile that covers a horizontal distance of 8000 meters in the time it takes for it to fall vertically 5 meters?
2. At this high speed, how would the shape of its trajectory compare with Earth's curvature?
3. What do we call such a high-speed projectile?

CHECK YOUR ANSWER

1. Speed = (distance traveled)/(time to fall 5 m) = (8000 m)/(1 s) = 8000 m/s = 8 km/s.
2. The two curves match.
3. Without air drag or other obstacles, the projectile would be an Earth satellite.

Grains of sand and other meteorites that graze Earth's atmosphere burn up and appear as "falling stars." That's why space vehicles are launched to altitudes above the atmosphere.

Quick Activity

The Water Bucket Swing

Swing a bucket of water in a vertical circle. If you swing it sufficiently fast, the water won't spill. The explanation is similar to why satellites don't "fall" to Earth. Actually, both water in the bucket and satellites are falling. The water doesn't spill at the top of the swing because the bucket swings downward at least as fast as the water falls. Similarly, a satellite doesn't get closer to Earth because it falls a distance that matches the Earth's curvature. Analogies are a good way to understand concepts!

Astronauts inside orbiting space facilities lack a support force. That's why they have no weight. Interestingly, the force of gravity between them and planet Earth is only slightly reduced from what it is at ground level.

CONCEPT **CHECK**

What would be the fate of a rocket launched vertically that remains vertical as it rises?

CHECK YOUR ANSWER

After the rocket reaches its highest point it would fall back to its launching site—not a good idea!

For a satellite close to the Earth, the period (the time for a complete orbit about the Earth) is about 90 minutes. For higher altitudes, gravitation is less and the orbital speed is less—the period is longer. For example, communication satellites located at an altitude of 5.5 Earth radii have a period of 24 hours. This period matches the period of daily Earth rotation. For an orbit around the equator, these satellites stay above the same point on the ground, which is helpful for communication purposes. The Moon is even farther away and has a period of 27.3 days. The higher the orbit of a satellite, the less its speed and the longer its period.

6.6 Elliptical Orbits

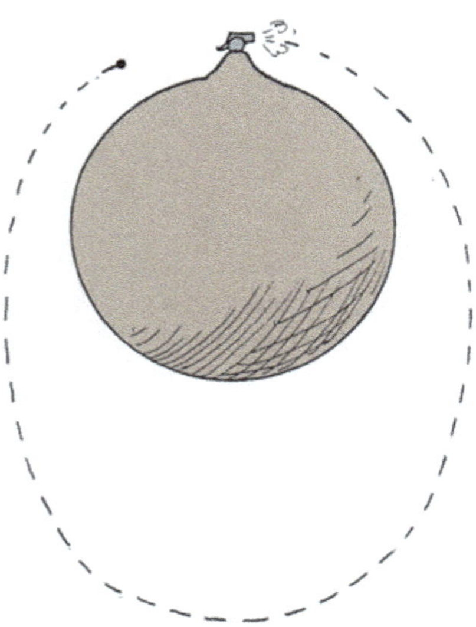

Figure 6.20

A cannonball fired at 9 km/s will overshoot a circular path and follow an ellipse.

If a payload above the drag of the atmosphere is given a horizontal speed somewhat greater than 8 kilometers per second, it will overshoot a circular path and trace an elliptical path, **Figure 6.20**.

An **ellipse** is a specific curve. Technically, it is an oval-like path along which any point has the same sum of distances from two fixed points (called foci). For a satellite orbiting Earth, one focus is at the Earth's center; the other focus could be inside or outside the Earth. An ellipse can be easily constructed by using a pair of tacks, one at each focus, a loop of string, and a pencil. As the foci get farther apart, the path gets more elongated, or "eccentric." The closer the foci are to each other, the closer the ellipse is to a circle. When both foci are *together*, the ellipse is a circle. So, we see that a circle is a special case of an ellipse.

Quick Activity

Build an Ellipse

Construct your own ellipses with string and tacks as shown. Note that the closer your tacks, the more circular your ellipse. When the tacks are farther apart, the ellipse is more eccentric.

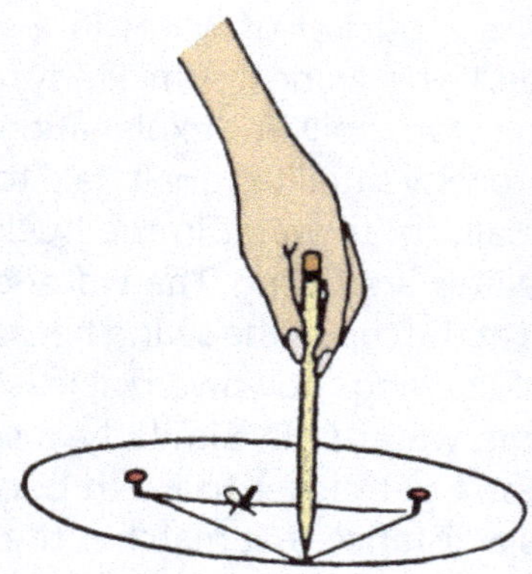

Unlike the constant speed of a satellite in a circular orbit, speed varies in an elliptical orbit.

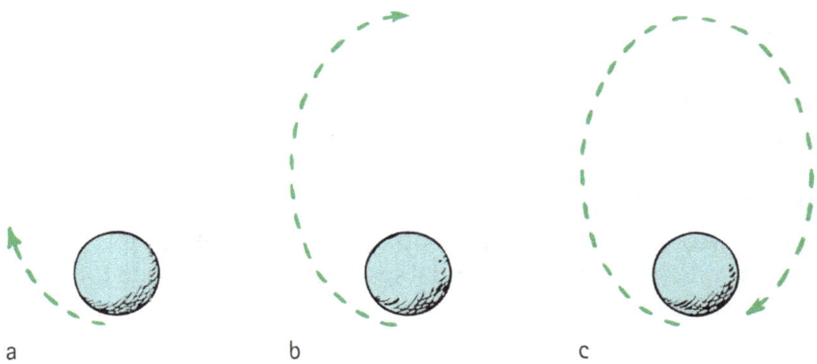

Figure 6.21

The satellite loses speed in receding from the Earth, and regains it when falling back toward the Earth. The cycle is repeated and the satellite remains in an elliptical orbit.

A satellite following an elliptical orbit is different. Half the time the satellite moves away from the Earth, and half the time it falls back toward the Earth. When it moves away, against the force of gravity, it loses speed. Like a stone thrown into the air, it slows to a point where it no longer recedes and then begins to fall back toward the Earth. Then it gains speed. The speed lost in receding is regained as it falls back toward the Earth. Then the satellite rejoins its original path with the same speed it had initially. The procedure repeats over and over, and an ellipse is traced each cycle (**Figure 6.21**).

READING CHECK

What similarity did Newton see between a falling apple and the Moon?

6.7 Escape Speed

We know that a cannonball fired horizontally at 8 kilometers per second from Newton's mountain would be in orbit. But what would happen if the cannonball were instead fired at the same speed vertically? It would rise to some maximum height, reverse direction, and then fall back to Earth. Then the old saying "What goes up must come down" would hold true, just as surely as a stone tossed skyward will be returned by gravity (unless, as we shall see, its speed is great enough).

Today it is more accurate to say, "What goes up *may* come down," because there is a critical speed at which a projectile can outrun gravity and escape Earth. This critical speed is called **escape speed** or, if direction is involved, *escape velocity*. From the surface of Earth, escape speed is 11.2 kilometers per second. A projectile launched at any greater speed will leave Earth, traveling slower and slower due to Earth's gravity, never stopping. Although it never escapes the tug of Earth's gravitation, it escapes Earth itself.

READING CHECK

What is escape speed from planet Earth?

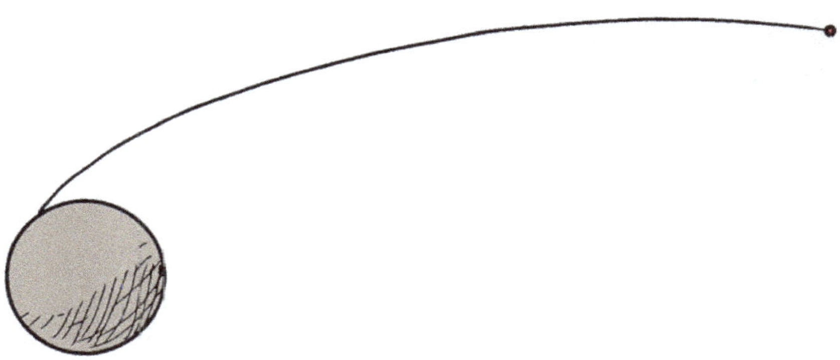

Figure 6.22

Project anything away from the Earth at 11.2 km/s and its trip is a one-way affair. It won't return.

FOR YOUR INFORMATION

Why is Earth and other planets round? Why is the Sun round? They are round because everything attracts everything else, and all parts of these bodies have attracted themselves together as much as they can! Any "corners" of them have been pulled in, making them spheres. Rotational effects make some of them bulge slightly at their equators.

CONCEPT CHECK

If a flight mechanic drops a wrench from a high-flying airplane, it crashes to Earth. But if an astronaut outside an orbiting space station drops a wrench, it doesn't crash to Earth. Explain.

CHECK YOUR ANSWER

If a wrench or anything else is "dropped" from an orbiting space vehicle, it has much the same tangential speed as the vehicle and remains in orbit. If a wrench is dropped from a high-flying airplane, it too has the tangential speed of the airplane. But this speed is too small for the wrench to orbit the Earth. Instead it soon crashes to the Earth's surface..

Several positions of a satellite in an elliptical orbit are shown. At which position does the satellite have the greatest

1. speed?
2. velocity?
3. mass?
4. gravitational attraction to Earth?
5. kinetic energy?
6. potential energy?
7. total energy?
8. acceleration? (Let the equation $a = F/m$ guide you.)

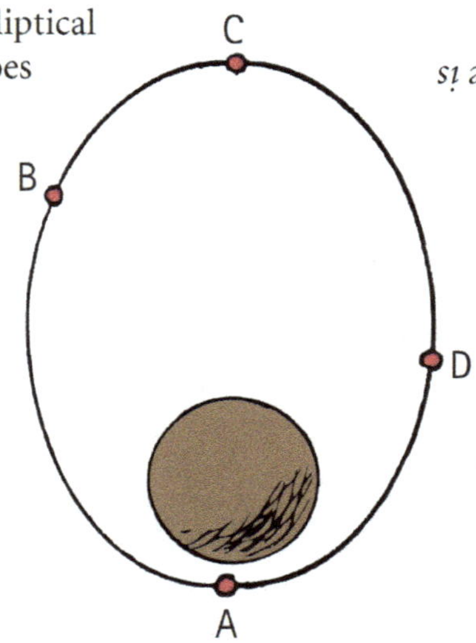

Answers

1. *Greatest speed at A.*
2. *Greatest velocity where greatest speed occurs, A.*
3. *Same mass everywhere. Mass does not depend on location.*
4. *Greatest at A, where it is closest to the Earth.*
5. *Greatest at A, where speed is greatest.*
6. *Greatest at C, where distance is greatest and speed least.*
7. *Same everywhere, in accord with energy conservation.*
8. *Greatest at A, where gravitational force is greatest.*

Concept Review

Chapter 6

Summary of Terms

Ellipse The oval path followed by a satellite. The sum of the distances from any point on the path to two points called foci is a constant. When the foci are together at one point, the ellipse is a circle. As the foci get farther apart, the path gets more "eccentric."

Escape Speed The speed that a projectile, space probe, or similar object must reach to escape the gravitational influence of the Earth or celestial body to which it is attracted.

Law of Universal Gravitation Every body in the universe attracts every other body with a mutually attracting force. For two bodies, this force is directly proportional to the product of their masses and inversely proportional to the square of the distance separating them.

$$F = G\frac{m_1 m_2}{r^2}$$

Inverse-Square Law A law relating the intensity of an effect to the inverse square of the distance from the cause:

$$Intensity \sim \frac{1}{distance^2}$$

Parabola The curved path followed by a projectile near the Earth under the influence of gravity only.

Projectile Any object that moves through the air or through space under the influence of gravity.

Satellite A projectile or small body that orbits a larger body.

Tangential Velocity Velocity that is parallel (tangent) to a curved path.

Universal Constant of Gravitation, *G* The proportionality constant in Newton's law of universal gravitation.

Review Questions

1. What connection did Newton make between a falling apple and the Moon?
2. What does it mean to say something moving in a curve has a tangential velocity?
3. In what sense does the Moon "fall?"
4. State Newton's law of universal gravitation in words. Then do the same with one equation.
5. How does the force of gravity between two bodies change when the distance between them is doubled?
6. How does the thickness of paint sprayed on a surface change when the sprayer is held twice as far away?
7. How does the brightness of light change when a point source of light is brought twice as far away?
8. What is the magnitude of gravitational force between two 1-kilogram bodies that are 1 meter apart?
9. What is the magnitude of the gravitational force between the Earth and a 1-kilogram body?
10. What do we call the gravitational force between the Earth and your body?
11. When G was first measured, the experiment was called the "weighing the Earth experiment." Why?
12. With no gravity, a horizontally-moving projectile follows a straight-line path. With gravity, how far below the straight-line path does a projectile fall compared with the distance of free fall?
13. A ball is batted upward at an angle. What happens to the vertical component of its velocity as it rises? As it falls?
14. With no air drag, what happens to the horizontal component of velocity for the batted baseball?
15. How can a projectile "fall around the Earth"?
16. Are the planets of the solar system simply projectiles falling around and around the Sun?

17. Why does the speed of a satellite undergo change in an elliptical orbit?

18. At what part of an elliptical orbit does a satellite have the greatest speed? The least speed?

19. What is the minimum speed for orbiting Earth in close orbit? The maximum speed? What happens above this speed?

20. What physical concept accounts for a round Earth?

Answers to Odd-Numbered Questions

1. He realized they were both under the influence of Earth gravity.

3 It falls beneath the straight-line path it would follow if there were no gravity.

5 The force is 1/4 as much, in accord with the inverse-square law.

7 Brightness is 1/4 as much, in accord with the inverse-square law.

9. 10 N, or more precisely 9.8 N.

11. Because once *G* was known, it was a simple matter to calculate the mass of the Earth.

13. The vertical component changes the same way a ball tossed straight upward changes, and changes on the way down just as a stone does in free fall.

15. It can fall around Earth if its tangential velocity is sufficient to ensure its curved path matches the curve of Planet Earth.

17. Because half the time is going away from gravity, and the other half going with gravity. When going against gravity, speed decreases. When going with gravity, speed increases.

19. 8 km/s; 11.2 km/s; above this speed the satellite leaves Earth permanently.

Chapter 7

Fluid Mechanics

THE MAIN IDEA

Boats float and birds fly because fluids flow

Liquids and gases have the ability to flow; hence, they are called *fluids*. Because both are fluids we find liquids and gases obey similar mechanical laws. So we wonder, why is it easier to float in salt water than in fresh water? Why does a large rock feel less heavy when underwater? How is it that iron boats don't sink in water and that helium balloons rise to the sky? Why is gas compressible while liquids are not? Why do your ears pop when riding an elevator? How do fish and SCUBA divers control their depth? How is the field of hydraulics changing the way machines operate. To answer these sorts of questions about fluids, it is important to start with two related concepts—*density* and *pressure*.

7.1 Density and Pressure

An important property of materials is the measure of compactness: density. We think of density as the "lightness" or "heaviness" of materials of the same size. Density is a measure of how much mass occupies a given space, whether solid, liquid, or gas. **Density** is the amount of mass per unit volume:

$$\text{Density} = \frac{\text{mass}}{\text{volume}}$$

READING CHECK

How do mass and volume relate to the density of solids, liquids, and gases?

Quick Activity

Egg Float

Place an egg in a small container of water. Note that it rests on the bottom. Gradually pour salt in the water. Let it dissolve. Observe what happens for the egg. How does the density of an egg compare to that of tap water? To salt water?

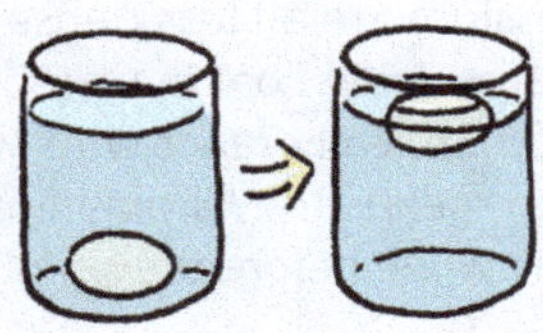

^ Figure 7.1

When the volume of the bread is reduced, its density increases.

The densities of a few materials are listed in **Table 7.1** showing units of grams per cubic centimeter (g/cm^3). Water has a density of about 1 g/cm^3. Mercury's density is 13.6 g/cm^3, which means it has 13.6 times as much mass as an equal volume of water. Osmium, a hard, brittle, silvery-white metal, is the densest substance on Earth's surface with a density of about 22.6 g/cm^3.

A quantity known as **weight density**, commonly used when discussing liquid pressure, is expressed by the amount of weight per unit volume:

$$\text{Weight density} = \frac{\text{weight}}{\text{volume}}$$

Table 7.1 Densities of Some Materials

Material	Grams per Cubic Centimeter (g/cm^3)
Liquids	
Mercury	13.6
Seawater	1.03
Water at 4°C	1.00
Ethyl alcohol	0.81
Solids	
Osmium	22.6
Platinum	21.1
Gold	19.3
Uranium	19.0
Lead	11.3
Silver	10.5
Copper	8.9
Iron	7.8
Aluminum	2.7
Ice	0.92
Gases *(sea level atmospheric pressure) (kg/m^3)*	
Dry air	
0°C	1.29
10°C	1.25
20°C	1.20
Helium	0.178
Hydrogen	0.090

CONCEPT **CHECK**

1. Which has the greater density—1 kg of water or 10 kg of water?
2. Which has the greater density—0.0001 g of lead or 10 kg of aluminum?
3. Which has the greater density—an entire candy bar or half a candy bar?

CHECK YOUR ANSWERS

1. The density of any amount of water (at the same temperature) is the same: 1 g/cm^3.
2. Density is a ratio of mass per volume. The amount of material you have doesn't matter. A smaller amount of lead will always have a greater density than a larger amount of aluminum.
3. Both the entire candy bar and half of it have the same density.

Place a book on a bathroom scale and, whether you place it on its back, on its side, or balanced on a corner, it still exerts the same force, which is its weight. Now balance the book on the palm of your hand and you sense a difference—the pressure of the book depends on the area of contact (**Figure 7.2**). There is a difference between force and pressure. **Pressure** is defined as the force exerted over a unit of area, such as a square meter.

$$\text{Pressure} = \frac{\text{force}}{\text{area}}$$

fyi FOR YOUR INFORMATION

The metals lithium, sodium, and potassium are all less dense than water and will float in water. They also chemically react with water, which means they'll only float for a little while before they explode.

< Figure 7.2

Although the weight of both books is the same, the upright book exerts greater pressure against the table.

The standard international (SI) unit of pressure, the newton per square meter, is called the **pascal** (Pa), after the seventeenth-century theologian and scientist, Blaise Pascal. A pressure of 1 Pa is very small and approximately equals the pressure exerted by a dollar bill resting flat on a table. Science types prefer kilopascals (1 kPa = 1000 Pa).

CONCEPT CHECK

Does a bathroom scale measure weight, pressure, or both?

CHECK YOUR ANSWER

A bathroom scale measures weight, the force that compresses an internal spring or equivalent. The weight reading is the same whether you stand on one or both feet (although the pressure on the scale is twice as much when standing on one foot).

7.2 Pressure in a Fluid

When you swim under water, you can feel the water pressure acting against your eardrums. The deeper you swim, the greater the pressure. What causes this pressure? Pressure in the water is due to the weight of the fluid directly above you—both the weight of water plus the weight of the atmosphere. More water is above you when you swim deeper. That means more pressure. If you swim twice as deep, twice the weight of water pushes on you. So, the water's contribution to the pressure you feel is doubled. Added to the water pressure is the pressure of the atmosphere, which is equivalent to an extra 10.3-meter depth of water. Because atmospheric pressure at Earth's surface is nearly constant, pressure *differences* you feel under water depend only on changes in the depth of the water.

READING CHECK

What causes the pressure you feel when you are beneath the surface of water?

The pressure due to a liquid is precisely equal to the product of weight density and depth:

Liquid pressure = weight density × depth

It is important to note that pressure does not depend on the volume of liquid. You feel the same pressure a meter deep in a small pool as you do a meter deep in the middle of the ocean. This is illustrated by the connecting vases shown in Figure 7.3. Pressure depends on depth, not volume.

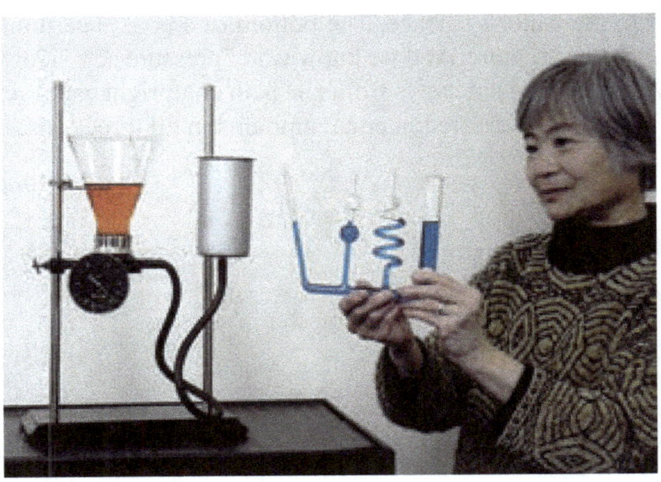

Figure 7.3

Liquid pressure is the same for any given depth below the surface, regardless of the shape of the containing vessel.

Λ Figure 7.4

This water tower does more than store water. The depth of water above ground level assures reliable water pressure to the many homes it serves.

FOR YOUR INFORMATION

Let's see how the equation for liquid pressure is derived:

$$\text{Pressure} = \frac{\text{Force}}{\text{Area}}$$

Recognize that *force* is the *weight* of the fluid and you get:

$$\text{Pressure} = \frac{\text{Weight}}{\text{Area}}$$

The next step is to recognize that weight is equal to weight density × volume and you have:

$$\text{Pressure} = \frac{\text{(Weight Density)(Volume)}}{\text{Area}}$$

But volume is Area × Depth, which gives:

$$\text{Pressure} = \frac{\text{(Weight Density)}\cancel{\text{(Area)}}\text{(Depth)}}{\cancel{\text{Area}}}$$

The units of area cancel out and you end up with:

$$\text{Liquid pressure} = \text{(Weight density)(Depth)}$$

Water seeks its own level. This can be demonstrated by filling a garden hose with water and holding the two ends upright. The water levels will be equal whether the ends are held close together or far apart. If one end is raised higher than the other, water will flow out of the lower end, even if it has to flow "uphill" part of the way. This fact was not fully understood by some of the early Romans, who built elaborate aqueducts with tall arches and roundabout routes to ensure water would always flow slightly downward along a long route. Later pipes were used. When unperturbed, water seeks its own level.

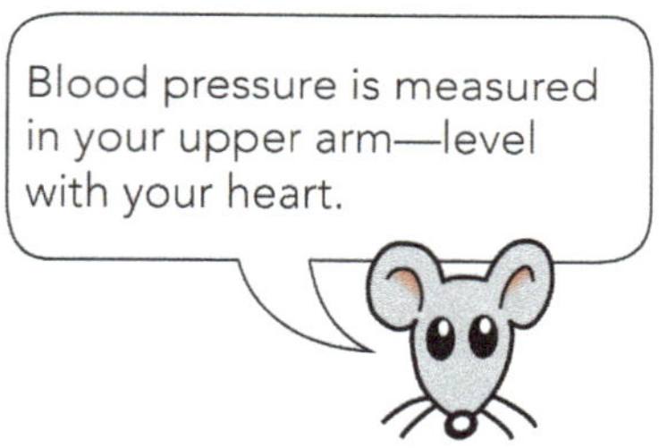

In addition to pressure depending on depth, liquid pressure is exerted equally in all directions. For example, if you are submerged in water, it makes no difference which way you tilt your head—your ears feel the *same* amount of water pressure. Because a liquid can flow, the pressure isn't only downward. We know pressure acts upward when we try to push a beach ball beneath the water's surface. The bottom of a boat is certainly pushed upward by water pressure. And we know water pressure acts sideways when we see water spurting sideways from a leak in an upright can. Pressure in a liquid at any point is exerted in equal amounts in all directions.

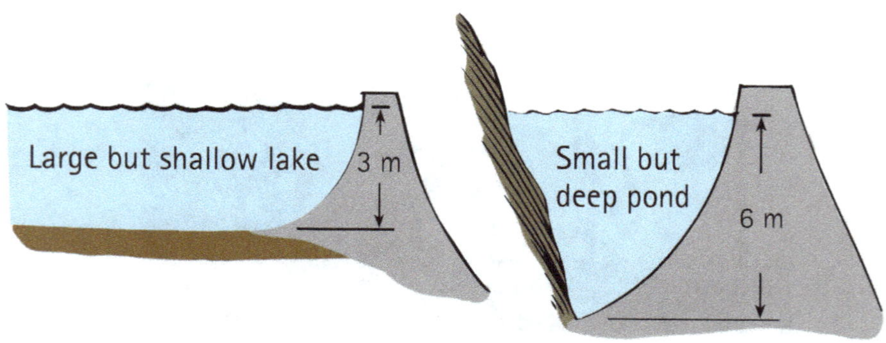

Figure 7.5 >

The average water pressure acting against the dam depends on the average depth of the water—not on the volume of water held back. The large shallow lake exerts only one-half the average pressure that the small deep pond exerts.

When liquid presses against a surface, there is a net force directed perpendicular to the surface (**Figure 7.6**). If there is a hole in the surface, the liquid spurts at right angles to the surface before curving downward due to gravity (**Figure 7.7**). At greater depths the pressure is greater and the speed of the exiting liquid is greater.

Figure 7.6

The forces due to liquid pressure against a surface combine to produce a net force that is perpendicular to the surface.

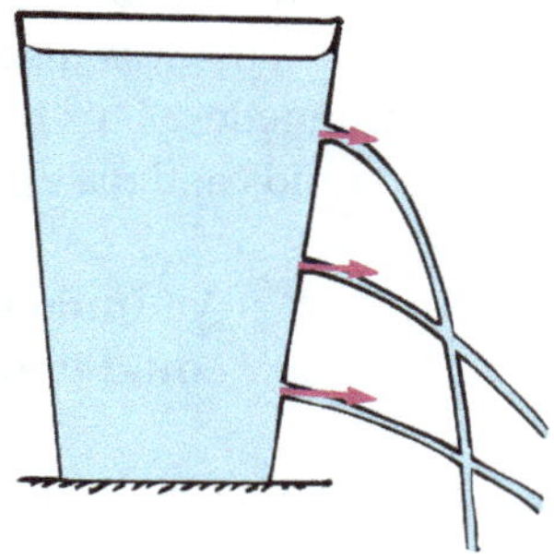

Figure 7.7

The force vectors act in a direction perpendicular to the inner container surface and increase with increasing depth.

7.3 Buoyancy in a Liquid

Lift a large rock off the ground. You'll feel it's quite heavy. Now, lift that same rock while submerged underwater, say, at the bottom of a swimming pool. So long as the rock remains underwater, you'll find the rock is much easier to lift. How can this be? Please understand that the weight of the rock is the same whether out of the water or in the water. When submerged, however, the water itself is helping you to lift the rock. This is how: The water pressure on the bottom of the rock is greater than the water pressure on the top of the rock. The reason is simply because the bottom side of the rock is deeper. Because the upward forces against the bottom are greater than the downward forces against the top, the forces do not cancel, and there is a net force upward. The result is a **buoyant force**—the net upward force that a fluid exerts on an immersed object. The buoyant force makes the rock feel lighter.

Any submerged object will have a buoyant force acting upon it. Whether the object sinks, remains level, or rises depends upon a comparison of the buoyant force to the object's weight. If the weight of the submerged object is greater than the buoyant force, the object will sink. If its weight is equal to the buoyant force, the object will remain at any level, like a fish. If the buoyant force is greater than the weight of the completely submerged object, it will rise to the surface and float.

Consider a container filled to the brim with water. Now place a stone into this water. What happens next is that water will overflow the brim (**Figure 7.9**). Water is *displaced* by the stone. A little thought will tell us that the *volume of the stone*—that is, the amount of space it occupies—is equal to the volume of water displaced. Place any object in a container partially

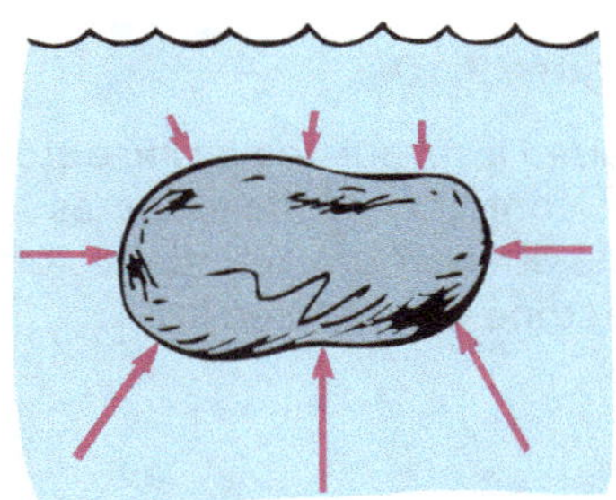

Figure 7.8

The greater pressure against the bottom of a submerged object produces an upward buoyant force.

READING CHECK

How does the comparison of weight and buoyant force determine whether an object will sink, float, or do neither?

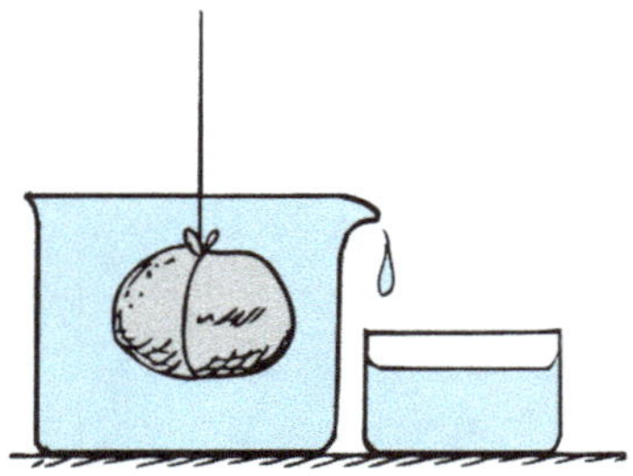

Figure 7.9

When a stone is submerged, it displaces a volume of water equal to the volume of the stone.

filled with water, and the level of the liquid surface rises (**Figure 7.10**). How high? Water will rise to exactly the level that would be reached by pouring in a volume of water equal to the volume of the submerged object. This is a good method for determining the volume of irregularly shaped objects: A completely submerged object always displaces a volume of liquid equal to its own volume.

There is a direct relationship between buoyancy and the displaced liquid. This was discovered in the third century BC by the Greek scientist Archimedes. This relationship is called **Archimedes' principle**. It applies to liquids and gases, which are both fluids. It is stated as follows:

> **An immersed body is buoyed up by a force equal to the weight of the fluid it displaces.**

As an example, consider a sealed 1-liter container. If we immerse this container halfway into water, it will displace 0.5 liters of water. According to Archimedes principle, the buoyant force equals the weight of 0.5 liters of water, which is about 5 newtons. Now, push the container all the way into the water so that it is fully submerged. A total of 1.0 liters of water will be displaced and the buoyant force equals about 10 newtons. What if the container is pushed all the way to the bottom? What will be the buoyant force? To answer this question, just ask yourself how much water is displaced. Assuming the container is not compressed and retains its shape, the amount of water displaced is the same where ever you might place that container underwater, so long as it remains fully submerged. At the bottom, the buoyant force is still about 10 newtons.

The key idea is that buoyant force is equal to the weight of the water displaced, which means it has everything to do with the *volume* of the submerged object. Interestingly, this also means buoyant force has nothing to do with the weight of that submerged object. You could fill the container with sand, lead, helium, gold, little green monkeys, or even water, and this makes no difference. Again, for emphasis: the *weight* of the submerged object makes no difference to the buoyant force. Archimedes' principle tells us that only the *volume* of the submerged object is important.

Figure 7.10

The raised level due to placing a stone in the container is the same as if a volume of water equal to the volume of the stone were poured in.

Figure 7.11

A liter of water occupies a volume of 1000 cm^3, has a mass of 1 kg, and weighs 10 N (more precisely, 9.8 N). Its density may therefore be expressed as 1 kg/L and its weight density as 10 N/L. (Seawater is slightly denser than 10 N/L).

< Figure 7.12

A 3-N block weighs more in air than it does in water. When submerged in water, the block's apparent loss in weight is the buoyant force, which equals the weight of water displaced.

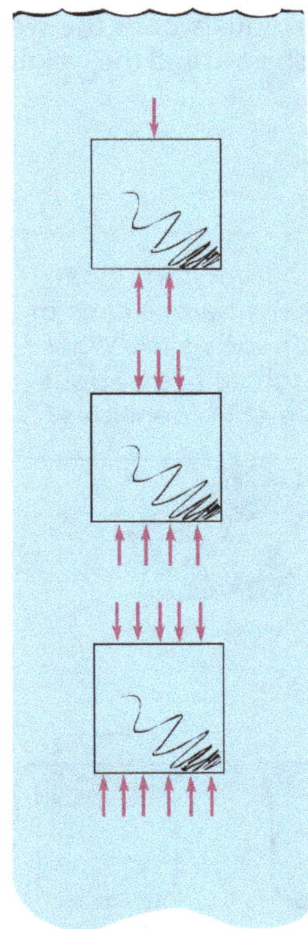

∧ Figure 7.13

The difference in the upward and downward forces acting on the submerged block is the same at any depth.

CONCEPT CHECK

1. Does Archimedes' principle tell us that if an immersed block displaces 10 N of fluid, the buoyant force on the block is 10 N?

2. A 1-liter container completely filled with lead has a weight of 110 N and is submerged in water. What is the buoyant force acting on it?

3. A boulder is thrown into a deep lake. As it sinks deeper and deeper into the water, does the buoyant force on it increase? Decrease?

CHECK YOUR ANSWERS

1. Yes. Looking at it in a Newton's-third-law way, when the immersed block pushes 10 N of fluid aside, the fluid reacts by pushing back on the block with 10 N.

2. The buoyant force is 10 N because 10 N of water is displaced (the weight of 1 liter of water). The 110-N weight of the lead is irrelevant; 1 L of anything submerged in water will displace 1 L and be buoyed upward with a force 10 N. (Get this straight before going further!)

3. Buoyant force remains the same. It doesn't change as the boulder sinks because the boulder displaces the same volume of water at any depth. Because the volume of the boulder remains the same, the buoyant force also remains the same.

FOR YOUR INFORMATION

Archimedes and the Gold Crown

According to legend, Archimedes (287—212 BCE) was given the task of determining whether a crown made for King Hiero II of Syracuse was of pure gold or some less expensive metal such as silver. Archimedes' problem was to determine the density of the crown without destroying it. He could weigh the gold, but determining its volume was a problem. The story tells us that Archimedes came to the solution when he noted the rise in water level while immersing his body in the public baths. Legend reports that he excitedly rushed naked through the streets shouting "Eureka! Eureka!" ("I have found it! I have found it!").

What Archimedes discovered was a simple and accurate way of finding the volume of an irregular object—the displacement method we use today with measuring cups. Once he knew both the weight and volume, he could calculate the density, which could be compared with the density of gold. Archimedes found that the crown indeed was not pure gold.

7.4 Flotation

READING CHECK

What is the special case where buoyant force equals both the weight of water displaced and the weight of the object?

Iron is much more dense than water and therefore sinks. But an iron ship floats. Why is this so? Consider a solid 1-ton block of solid iron. Iron is nearly eight times as dense as water, so when the block is submerged it will displace only 1/8 ton of water—not enough to prevent it from sinking. No mystery there.

But suppose we reshape the same iron block into a bowl, as shown in Figure 7.14. It still weighs 1 ton. However, because of its new shape, when placed in water, the iron displaces a greater volume of water than before. The deeper it is immersed, the more water it displaces and the greater the buoyant force acting on it. When the buoyant force equals 1 ton, it will sink no further. Only in the special case of floating does the buoyant force acting on an object equal the object's weight.

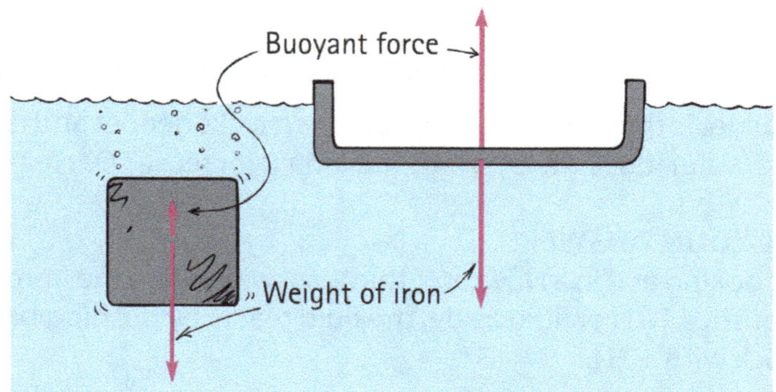

Λ Figure 7.14

An iron block sinks, while the same quantity of iron shaped like a bowl floats.

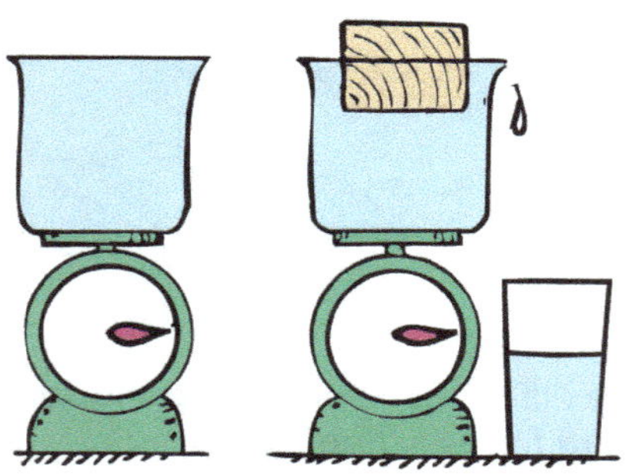

Λ Figure 7.15

A floating object displaces a weight of fluid equal to its own weight.

When the iron boat displaces a weight of water equal to its own weight, it floats. This is called the **principle of flotation**, which states:

A floating object displaces a weight of fluid equal to its own weight.

Every ship, submarine, or dirigible airship must be designed to displace a weight of fluid equal to its own weight. Thus, a 10,000-ton ship must be built wide enough to displace 10,000 tons of water before it immerses too deep in the water. The same applies to vessels in air. A dirigible or huge balloon that weighs 100 tons displaces at least 100 tons of air. If it displaces more, it rises; if it displaces less, it descends. If it displaces exactly its weight, it hovers at constant altitude (Figure 7.16).

Figure 7.16 >

The weight of a floating object equals the weight of the water displaced by the submerged part.

Λ Figure 7.17

A ship loaded with cargo floats lower in the water. How does the weight of its load compare to the weight of additional water displaced?

CONCEPT **CHECK**

Fill in the blanks for these statements:

1. The volume of a submerged body is equal to the ___________ of the fluid displaced.
2. The weight of a floating body is equal to the ___________ of the fluid displaced.
3. Whether floating in salt or fresh water, the buoyant force is ___________.

CHECK YOUR ANSWERS

1. Volume.
2. Weight
3. The same. When you're floating, the weight of water you displace equals your weight. Salt water is denser, so you don't "sink" as far to displace your weight. You'd float even higher in mercury (density 13.6 g/cm^3), and you'd sink completely in alcohol (density 0.8 g/cm^3).

Floating Mountains

Since the buoyant force upon a body equals the weight of the fluid it displaces, denser fluids will exert more buoyant force upon a body than less-dense fluids of the same volume. A ship therefore floats higher in salt water than in fresh water because salt water is slightly denser than fresh water. In the same way, a solid chunk of iron will float in mercury even though it sinks in water.

Mountains float on Earth's semi-liquid mantle just as icebergs float in water. Both the mountains and icebergs are less dense than the material they float upon. Just as most of an iceberg is below the water surface (90%), most of a mountain (about 85%) extends into the dense semi-liquid mantle. If you could shave off the top of an iceberg, the iceberg would be lighter and be buoyed up to nearly its original height before its top was shaved. Similarly, when mountains erode they are lighter, and are pushed up from below to float to nearly their original heights. So, when a kilometer of mountain erodes away, some 85% of a kilometer of mountain returns. That's why it takes so long for mountains to weather away. Mountains, like icebergs, are bigger than they appear to be. The concept of floating mountains is *isostacy*—Archimedes' principle for rocks.

Notice in our discussion of liquids that Archimedes' principle and the law of flotation were stated in terms of *fluids*, not liquids. That's because although liquids and gases are different phases of matter, they are both fluids, with much the same mechanical principles. Next, we turn our attention to the mechanics of gases in particular.

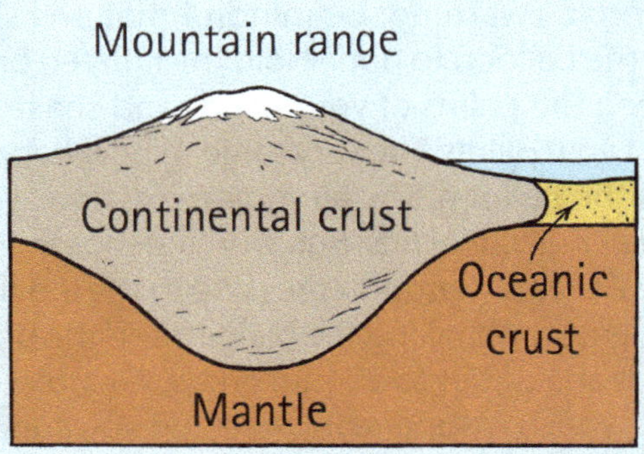

7.5 Gas Pressure

Liquids and gases are both fluids. A gas takes the shape of its container. A liquid does so only below its surface.

The main difference between a gas and a liquid is the distance between molecules. In a gas, the molecules are far apart and pretty much free from intermolecular forces. Molecular motions in a gas are less restricted. A gas expands, fills all space available to it, and exerts a pressure against its container. For only a very large quantity of gas, such as Earth's atmosphere or a star, do the gravitational forces limit the size or determine the shape of the mass of gas.

Boyle's Law

The air pressure inside the inflated tires of an automobile is much greater than the atmospheric pressure outside. The density of air inside is also more than the density of the air outside. To understand the relation between pressure and density, think of the molecules of air (primarily nitrogen and oxygen) inside the tire. The air molecules behave like tiny billiard balls, randomly moving and banging against the inner walls that produce a jittery force that appears to our coarse senses as a steady push. This pushing force, averaged over the wall area, provides the pressure of the enclosed air.

Suppose there are twice as many molecules in the same volume (**Figure 7.18**). Then the air density is doubled. If the molecules move at the same average speed—or, equivalently, if they have the same temperature—then the number of collisions will be doubled. This means that the pressure is doubled. So, pressure is proportional to density.

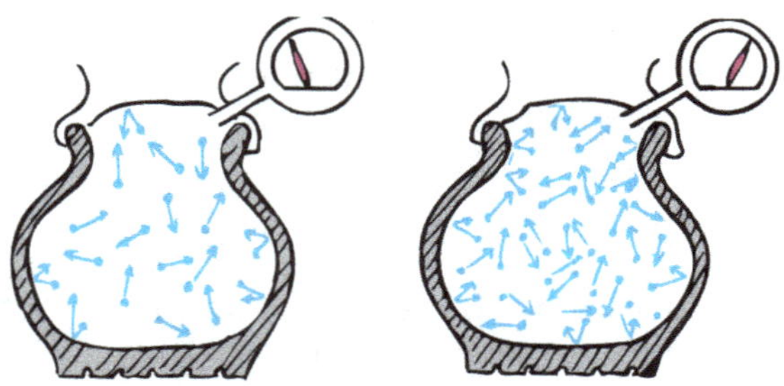

Figure 7.18 >

When the density of gas in the tire is increased, pressure is increased.

Quick Activity

How do Gas Molecules Exert Pressure?

Add a tablespoon of smooth small beans (or similar) to a balloon before inflating it. (Warning: Be mindful that any sharp edge to a bean might cause the balloon to pop.) Hold the inflated balloon with beans inside firm between the palms of your hands and shake. You'll feel the beans hitting against your palms. The very same thing is happening with the air molecules within the balloon. The air molecules, however, are so tiny that you don't feel their individual hits. But, working together, those many fast-moving air molecules apply enough pressure to the inner walls of the balloon to keep it inflated. Stop shaking the balloon and the beans will settle to the bottom, which is analogous to a liquid phase. Air molecules can also be slowed down to a liquid phase, but only at about -200°C, which is super cold. What do you suppose would happen to the inflated balloon if it were brought to that super low temperature? Look for an online video showing a balloon being submerged in liquid nitrogen.

< Figure 7.19

When the volume of gas is decreased, density and therefore pressure are increased.

We double the density of air in the tire by doubling the amount of air. We can also double the density of a fixed amount of air by compressing it to half its volume. Consider the cylinder with the movable piston in Figure 7.19. If the piston is pushed downward so that the volume is half the original volume, the density of molecules is doubled, and the pressure is correspondingly doubled. Decrease the volume to a third of its original value, and the pressure is increased by three, and so forth (provided the temperature remains the same).

Notice in these examples with the piston that the product of pressure and volume remains the same. For example, a doubled pressure multiplied by a halved volume gives the same value as a tripled pressure multiplied by a one-third volume. In general, we can state that the product of pressure and volume for a given mass of gas is a constant *as long as the temperature does not change*. "Pressure × volume" for a sample of gas at some initial time is equal to any "different pressure × different volume" of the same sample of gas at some later time. In shorthand notation,

$$P_1V_1 = P_2V_2$$

where P_1 and V_1 represent the original pressure and volume, respectively, and P_2 and V_2 the second pressure and volume. This relationship is called **Boyle's law**, after Robert Boyle, the seventeenth-century physicist who is credited with its discovery.

Boyle's law applies to ideal gases. An ideal gas is one in which the forces between molecules and the size of the individual molecules can be neglected. Air and other gases under normal pressures and temperatures approach ideal gas conditions.

READING CHECK

State Boyle's law in words.

CONCEPT **CHECK**

1. A piston in an airtight pump is withdrawn so that the volume of the air chamber is increased three times. What is the change in pressure?

2. A scuba diver breathes compressed air beneath the surface of water. If she holds her breath while returning to the surface, what happens to the volume of her lungs?

CHECK YOUR ANSWER

1. The pressure in the piston chamber is reduced to one-third. This is the principle that underlies a mechanical vacuum pump.

2. When she rises toward the surface, the surrounding water pressure on her body decreases, allowing the volume of air in her lungs to increase—ouch! A first lesson in scuba diving is to not hold your breath when ascending. To do so can be fatal.

7.6 The Atmosphere

∧ **Figure 7.20**

The famous "Magdeburg hemispheres" experiment of 1654, demonstrating atmospheric pressure. Two teams of horses couldn't pull the evacuated hemispheres apart. Were the hemispheres sucked together or pushed together? By what?

We live at the bottom of an ocean of air. The atmosphere, much like the water in a lake, exerts a pressure. One of the most celebrated experiments demonstrating the pressure of the atmosphere was conducted in 1654 by Otto von Guericke, burgermeister of Magdeburg and inventor of the vacuum pump. Von Guericke placed together two copper hemispheres about 1/2 meter in diameter to form a sphere, as shown in **Figure 7.20**. He set a gasket made of a ring of leather soaked in oil and wax between them to make an airtight joint. When he evacuated the sphere with his vacuum pump, two teams of eight horses each were unable to pull the hemispheres apart.

When the air pressure inside a similar cylinder like that shown in **Figure 7.21** is reduced, there is an upward force on the piston. This force is large enough to lift a heavy weight. If the inside diameter of the cylinder is 12 centimeters or greater, a person can be lifted by this force.

What do the experiments of Figures 7.20 and 7.21 demonstrate? Do they show that air exerts pressure or that there is a "force of suction"? If we say there is a force of suction, then we assume that a vacuum can exert a force. But what is a vacuum? It is an absence of matter; it is a condition of nothingness. How can nothing exert a force? The hemispheres are not sucked together, nor is the piston holding the weight sucked upward. The hemispheres and the piston are being pushed against by the pressure of the atmosphere.

READING CHECK

What is the cause of atmospheric pressure?

Just as water pressure is caused by the weight of water, atmospheric pressure is caused by the weight of air. We have adapted so completely to the invisible air that we sometimes forget it has weight. Perhaps a fish "forgets" about the weight of water in the same way. The reason we don't feel this weight crushing against our bodies is that the pressure inside our bodies equals that of the surrounding air. There is no net force for us to sense.

At sea level, 1 cubic meter of air at 20°C has a mass of about 1.2 kilograms. To estimate the mass of air in your room, estimate the number of cubic meters of the room, multiply by 1.2 kg/m^3, and you'll have the mass. Don't be surprised if it's heavier than your kid sister. If your kid sister doesn't believe air has weight, maybe it's because she's always surrounded by air. Hand her a plastic bag of water and she'll tell you it has weight. But hand her the same bag of water while she's submerged in a swimming pool, and she won't feel the weight. We don't notice that air has weight because we're submerged in air.

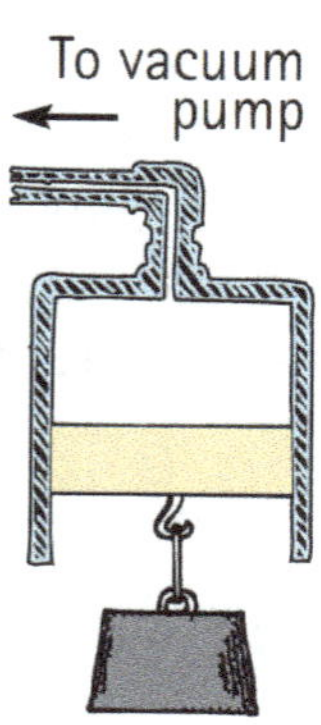

∧ **Figure 7.21**

Is the piston pulled up or pushed up?

The density of water is the same at any level (assuming constant temperature). Unlike water, air is easily compressed. The result is that air is more compact close to the ground but feathers out with increasing altitude. In other words, the density of air in the atmosphere becomes less with altitude. Although 1 cubic meter of air at sea level has a mass of about 1.2 kg, at 10 kilometers 1 cubic meter of air has a mass of about 0.4 kilograms. Air is thinner at higher altitudes. To compensate for this, airplanes are pressurized; the additional air needed to fully pressurize a jumbo jet is more than 1000 kilograms. Air is heavy, if you have enough of it.

Consider the mass of air in an upright 30-kilometer-tall hollow bamboo pole that has an inside cross-sectional area of 1 square centimeter. If the density of air inside the pole matches the density of air outside, the enclosed mass of air would be about one kilogram. The weight of this much air is about 10 newtons. So, air pressure at the bottom of the bamboo pole would be about 10 newtons per square centimeter (10 N/cm^2). Of course, the same is true without the bamboo pole. There are 10,000 square centimeters in 1 square meter, so a column of air 1-square meter in cross section that extends up through the atmosphere has a mass of about 10,000 kilograms. The weight of this air is about 100,000 newtons (10^5 N). This weight produces a pressure of 100,000 newtons per square meter.

Recall that 1 N/m^2 is called 1 pascal. So atmospheric pressure at Earth's surface is 100,000 pascals, or 100 kilopascals. To be more precise, the average atmospheric pressure at sea level is 101.3 kilopascals (101.3 kPa), which is often called 1 "atmosphere" (1 atm). In British units, this corresponds to 14.7 pounds per square inch (14.7 psi).

The pressure of the atmosphere is not uniform. Besides altitude variations, there are variations in atmospheric pressure at any one locality due to moving fronts and storms. Measurement of changing air pressure is important to meteorologists in predicting weather. For example, winds tend to blow from areas of relatively high atmospheric pressure to areas of relatively low atmospheric pressure. On weather maps you'll often see areas of high pressure indicated with a letter "H" and areas of low pressure with the letter "L". The lowest of low pressures occur within the eye of a hurricane. The high winds of a hurricane are formed as air rushes into its low-pressure center in a spinning fashion.

Λ Figure 7.22

You don't notice the weight of a bag of water while you're submerged in water. Similarly, you don't notice that the air around you has weight.

Λ Figure 7.23

The mass of air that would occupy a bamboo pole that extends to the "top" of the atmosphere is about 1 kg. This air has a weight of about 10 N.

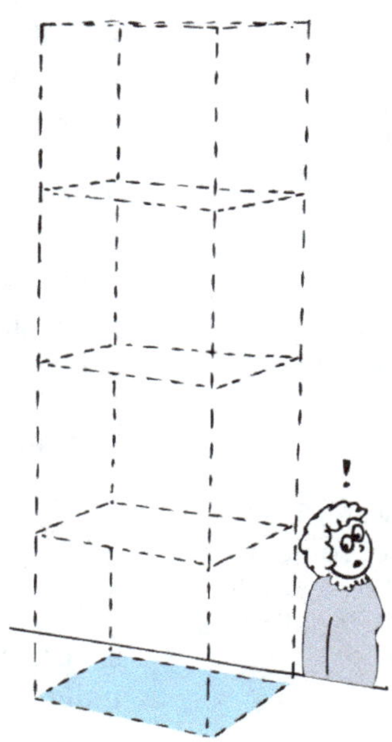

Λ Figure 7.24

The weight of air that presses down on a one-square-meter surface at sea level is about 100,000 newtons. So, atmospheric pressure is about 105 N/m^2, or about 100 kPa. In other units this can be expressed as 14.7 psi or 1 atm.

Quick Activity

Rock and Ruler

Drop a small rock onto the end of a ruler overhanging the edge of a table. As you might expect, the ruler flips off of the tabletop. Repeat this experiment with a sheet of paper flat over the portion of the ruler on the table. Of course, the paper weighs much less than the rock. Yet what happens? And, more importantly, why? Hint: the paper has weight, but so does the air above it.

CONCEPT **CHECK**

1. Estimate the mass of air in kilograms in a classroom that has a 200-m^2 floor area and a 4-m-high ceiling. (Assume a chilly 10°C temperature).
2. Why doesn't the pressure of the atmosphere break windows?

CHECK YOUR ANSWERS

1. The mass of air is 1000 kg. The volume of air is 200 m^2 × 4 m = 800 m^3; each cubic meter of air has a mass of about 1.25 kg, so 800 m^3 × 1.25 kg/m^3 = 1000 kg (about a ton—a hefty amount of air).
2. Atmospheric pressure is exerted on both sides of a window, so no net force is exerted on the window. If for some reason the pressure increases or decreases on one side only, as in a strong wind, then watch out!

7.7 Barometers

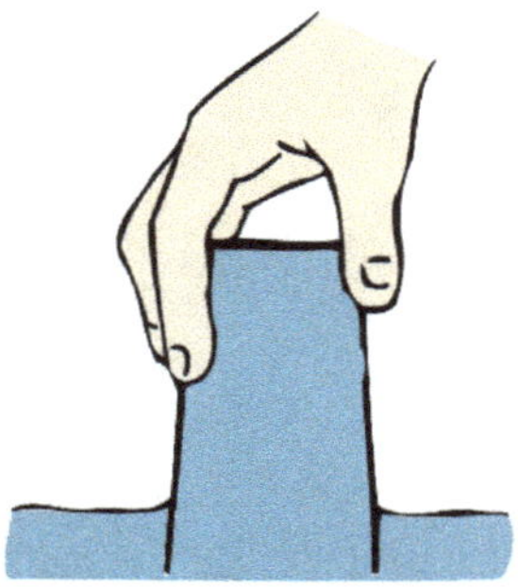

Λ Figure 7.25

Hold an inverted glass of water filled with water from the surface of some water. Why does this happen? Why wouldn't this occur if there were a hole drilled into the bottom of the glass?

READING CHECK

Why is the demonstration of Figure 7.25 not possible outside on the Moon?

Submerge a glass into a tub of water upright so that it fills with water. Now turn the glass upside down and slowly raise it above the water's surface as shown in **Figure 7.25**. Amazingly, the water stays within the glass—it doesn't flow out. But, why not?

First, think of the weight of the water being held up within the glass. This weight provides a downward pressure. Second, think of the weight of the atmosphere, which also provides a downward pressure upon the surface of the water. Which is greater? The water within the glass might provide a pressure of about one pound per square inch (psi). The atmosphere, by contrast, provides a pressure of about 14.7 psi, which is much greater. Think of it this way: The water within the glass would love to fall downward to seek its own level, but it can't because it's being pushed upward by the weight of the surrounding atmosphere. This demonstration would not work on the Moon where there is no atmosphere.

If the glass were super tall—over 10.3 meters tall—then the water within the glass would be heavy enough to push back against the atmosphere. But the water wouldn't drop to lower than 10.3 meters because if it did, the atmosphere would push it right back up to 10.3 meters. This is to say, a 10.3 meter column of water is able to apply the same pressure as the atmosphere. Yes, the atmosphere is heavy—heavy enough to match a column of water 10.3 meters tall!

A 10.3 meter column is quite tall (about 33 feet) and not very practical. Rather than using water, we can use mercury, which is a dense liquid metal. An inverted tube of mercury within a pool of mercury need only be 760 millimeters (0.76 m) tall in order to match atmospheric pressure. Such a device is called a **barometer**, which can be used to measure atmospheric pressure quite precisely. The standard atmospheric pressure at sea level is noted to equal 760 mm of mercury, which is 101 kPa or 1 atm. A high-pressure weather system might push the mercury to above 800 mm (107 kPa or 1.05 atm). Within the eye of a hurricane, this pressure can drop down to about 660 mm (87 kPa, or 0.86 atm). Meteorologists and other weather professionals trained in weather forecasting pay much attention to the relative highs and lows of atmospheric pressures across different regions.

What happens with a barometer is similar to what happens when you drink through a straw. By sucking, you reduce the air pressure in the straw when it is placed in a drink. Atmospheric pressure on the drink then pushes the liquid up into the reduced-pressure region in your mouth. Strictly speaking, the liquid is not sucked up; it is pushed up the straw by the pressure of the atmosphere. Imagine an opened bottle of water with a rubber stopper in the opening. Drill a hole in that stopper and insert a straw. Might you be able to suck up the water through that straw? The answer is no because the rubber stopper has sealed the water from the weight of the atmosphere. This makes for a nice party trick. An easier party trick is to try drinking water from a glass with two straws in your mouth—one straw in the beverage and the other out of the beverage, which makes it impossible for you lower the atmospheric pressure sufficiently within your mouth.

If you understand these ideas, you can understand why there is a 10.3-meter maximum height that water can be lifted with vacuum pumps. The old-fashioned farm-type pump, shown in **Figure 7.28**, operates by producing a partial vacuum in a pipe that extends down into the water below. Atmospheric pressure on the surface of the water simply pushes the water up into the region of reduced pressure inside the pipe. Can you see that, even with a perfect vacuum, the maximum height to which water can be lifted in this way is 10.3 meters?

Today, electronic barometers are commonly used. They work, however, by the same principle of detecting the pressure exerted by the weight of the atmosphere. At higher altitudes, atmospheric pressure is less. So, a barometer can be used to determine elevation. A barometer calibrated for altitude is called an *altimeter* (altitude meter), which can be sensitive enough to measure your change in elevation walking up a flight of stairs. Yes, the atmospheric pressure at the bottom of a flight of stairs is measurably greater than at the top of the stairs.

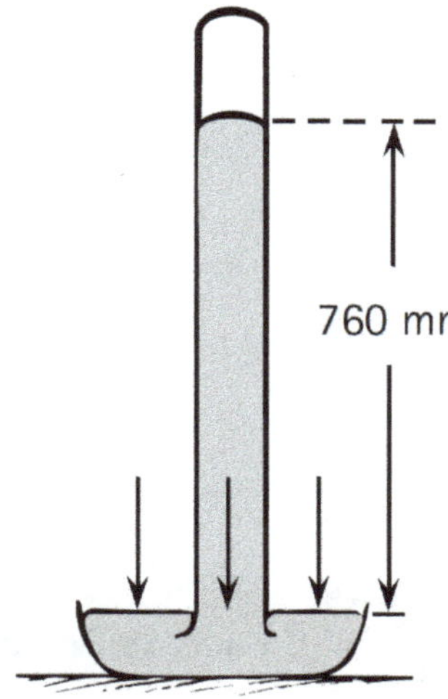

Λ Figure 7.26

A simple mercury barometer. Mercury is pushed up into the tube by atmospheric pressure.

Λ Figure 7.27

Strictly speaking, they do not suck the soda up the straws. They instead reduce pressure in the straws, which allows the weight of the atmosphere to press the liquid up into the straws. Could they drink a soda this way on the Moon?

< Figure 7.28

The atmosphere pushes water from below up into a pipe that is evacuated of air by the pumping action.

Quick Activity

Place a card over the open top of a glass filled to the brim with water, and then invert it. Why does the card stay in place? Try it sideways.

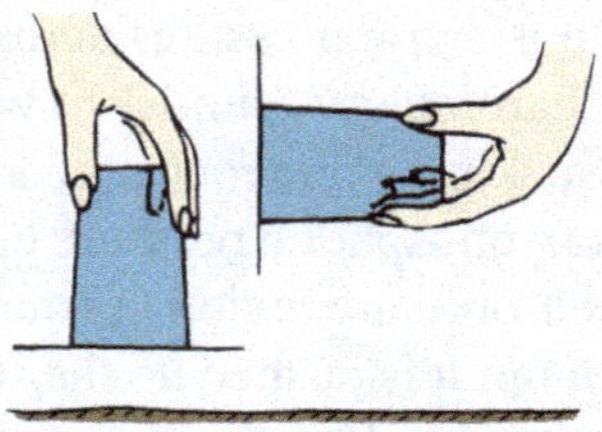

7.8 Pascal's Principle

The study of fluids whether at rest or in motion is *hydraulics*. An important fact about fluid pressure is that a change in pressure at one part of the fluid will be transmitted undiminished to other parts. For example, if the pressure of city water is increased at the pumping station by 10 units of pressure, the pressure everywhere in the pipes of the connected system will be increased by 10 units of pressure (providing the water is at rest). This rule of hydraulics is called Pascal's principle:

A change in pressure at any point in an enclosed fluid at rest is transmitted undiminished to all points in the fluid.

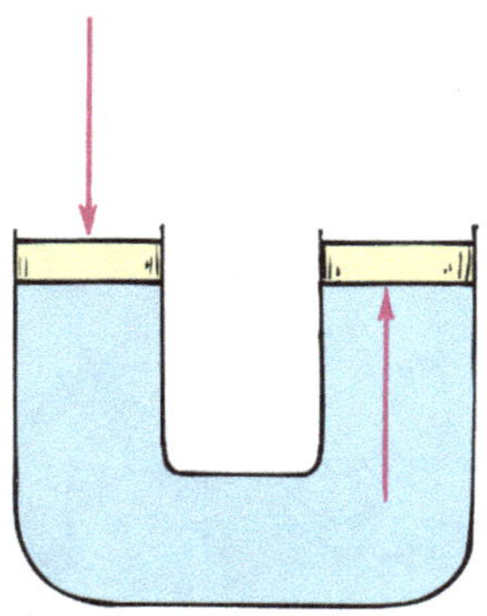

Figure 7.29

The force exerted on the left yellow piston increases the pressure in the liquid and is transmitted to the right yellow piston.

Pascal's principle was discovered in the seventeenth century by theologian and scientist Blaise Pascal, for whom the SI unit of pressure, the pascal (1 Pa = 1 N/m^2), is named.

Fill a U-tube with water and place pistons at each end, as shown in **Figure 7.29**. Pressure exerted against the left piston is transmitted throughout the liquid and against the bottom of the right piston. (The pistons are simply "plugs" that can slide freely but snugly inside the tube.) The pressure that the left piston exerts against the water will be exactly equal to the pressure the water exerts against the right piston. This is nothing to write to grandma about.

But suppose you make the tube on the right side wider and use a piston of larger area; then the result is impressive. In **Figure 7.30** the piston on the right has 50 times the area of the piston on the left (say the left has 100 square centimeters and the right 5000 square centimeters). Suppose a 10-kg load is placed on the left piston. Additional pressure due to the weight of the load is transmitted throughout the liquid and up against the larger piston. Here is where the difference between force and pressure comes in. The additional pressure is exerted against every square centimeter of the larger piston. Since there is 50 times as much area, 50 times as much force is exerted on the larger piston. Thus, the larger piston will support a 500-kg load—50 times the load on the smaller piston!

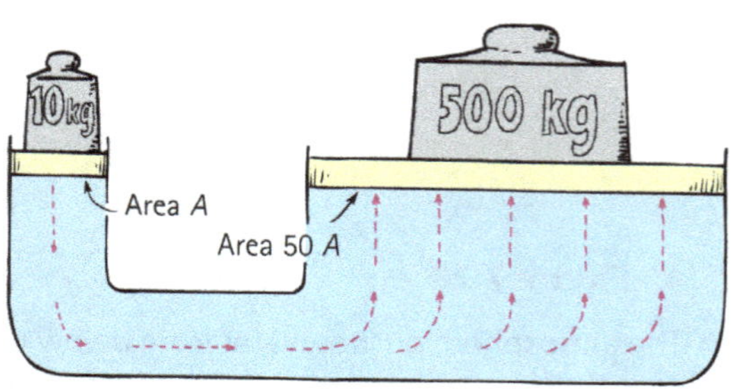

Figure 7.30

A 10-kg load on the left piston will support 500 kg on the right piston.

This *is* something to write to grandma about, for we can use such a device to multiply forces. With the right ratio, one newton input can produce 50 newtons output. By further increasing the area of the larger piston (or reducing the area of the smaller piston), we can multiply force, in principle, by any amount. Pascal's principle underlies the operation of the hydraulic press.

The hydraulic press does not violate energy conservation, because a decrease in the distance moved compensates for the increase in force. Expressed in equation form: $F\ d = F\ d$. Force is increased only at the expense of distance. When the small piston in **Figure 7.30** is moved downward 10 centimeters, the large piston will be raised only one-fiftieth of this, or 0.2 centimeter. The input force multiplied by the distance moved by the smaller piston is equal to the output force multiplied by the distance moved by the larger piston.

READING CHECK

Forces can be greatly multiplied with a hydraulic press. Why does this not violate the law of conservation of energy?

Pascal's principle applies to all fluids, whether gases or liquids. A typical application of Pascal's principle for gases and liquids is the hydraulic lift seen in many automotive service stations (**Figure 7.31**). Increased air pressure produced by an air compressor is transmitted through the air to the surface of hydraulic fluid in an underground reservoir. The fluid, in turn, transmits the pressure to the bottom of the large piston, which lifts the automobile. The air pressure required to lift the car is relatively small—about as much as is needed to fill the car tires.

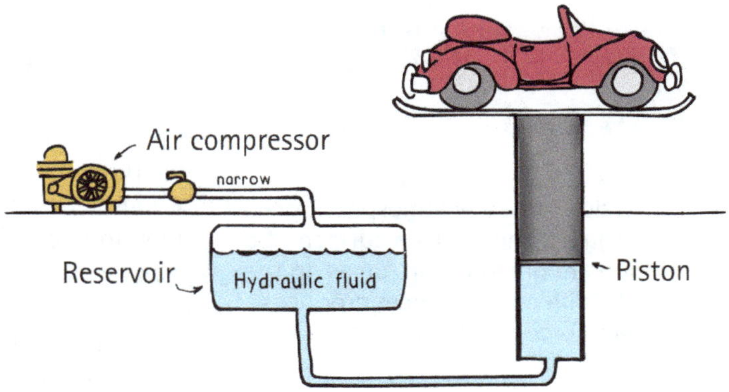

∧ Figure 7.31
Pascal's principle in a service station.

Hydraulics are employed in barber chairs, robotics, automobile brakes, power steering, and countless devices in industry and construction. Relatively small pressures through narrow piping can produce enormous forces. Look for the hydraulic pistons in almost all construction machines where heavy loads are involved (**Figure 7.32**). We can only wonder whether Pascal envisioned the extent to which his principle would allow huge loads to be so easily lifted.

Figure 7.32 >

Pascal's principle at work in the hydraulic devices on this common but incredible machine.

CONCEPT **CHECK**

1. If a pressure of 30 N/cm^2 (common in auto tires) is exerted against a piston with an area of 200 cm^2 (about the area of the palm of your hand), what force can be produced?
2. If a friend commented that a hydraulic device is a common way of multiplying energy, what would you say?

CHECK YOUR ANSWERS

1. $P = F \times A = 6000$ N (more than 1300 pounds). No wonder hydraulic devices are common in industry!
2. No, no, no! Although a hydraulic device, like a mechanical lever, can multiply force, it always does so at the expense of distance. Energy is the product of force and distance. Increase one, decrease the other. No device has ever been found that can multiply energy!

7.9 Buoyancy in a Gas

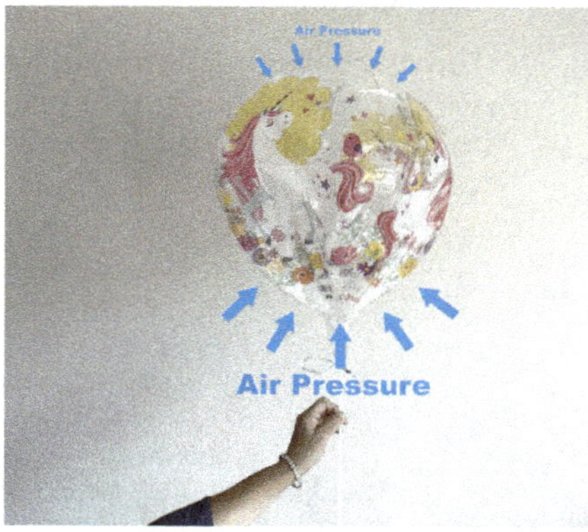

Λ Figure 7.33

Forces due to greater atmospheric pressure against the bottom surface of a balloon are greater than forces on the top. The difference in forces produces a buoyant force on the balloon.

A crab lives at the bottom of its ocean floor and looks upward at jellyfish and other lighter-than-water marine life drifting above it. Similarly, we live at the bottom of our ocean of air and look upward at balloons, blimps, and other lighter-than-air objects drifting above us. Balloons and jellyfish are suspended in fluids because they are buoyed upward by displaced weights of fluid equal to their weights. Objects in air are buoyed upward just as are objects in water. Archimedes' principle applies to air just as it does for water:

An object surrounded by air is buoyed up by a force equal to the weight of the air displaced.

The mass of 1 cubic meter of air at ordinary atmospheric pressure and room temperature is about 1.2 kilograms—a weight of about 12 newtons (about 2.7 pounds). Therefore, any 1-cubic-meter object in air is buoyed up with a force of 12 newtons. If the weight of the 1-cubic-meter object is *greater* than 12 newtons, it sinks to the ground when released. If an object of this size weighs *less* than 12 newtons, the buoyant force is greater than the weight and it rises in the air.

READING CHECK

How much buoyant force acts on a one-cubic-meter refrigerator in your school cafeteria?

Consider the balloon shown in **Figure 7.33**. For any balloon such as this, the atmospheric pressure is greater on the bottom side than on the top side. This difference in pressure results in a net force upward, which is the buoyant force. This particular balloon is filled with helium, a very light gas. The buoyant force on this balloon is greater than the combined weight of the helium and the balloon material itself. Therefore, the balloon rises. If you were to fill the balloon with a heavier gas, such as carbon dioxide, the total weight would be greater than the buoyant force and the balloon would sink to the ground. If you wanted to know the actual value of the buoyant force you would need to know the balloon's volume. As described in the previous paragraph, if the volume were 1 cubic meter, then the buoyant force would be 12 newtons. A balloon of 0.5 cubic meters would experience a buoyant force of 6 newtons. Again, for emphasis, the buoyant force equals the weight of the fluid displaced.

Λ Figure 7.34

All bodies are buoyed up by a force equal to the weight of air they displace. Why, then, don't all objects float like this balloon?

A low-density gas is used in a balloon for the same reason that foamed plastics are used in life preservers. The plastic foam possesses no strange tendency to be drawn toward the water's surface. Likewise, the low-density gas possesses no strange tendency to rise. Plastic foams and light gases are buoyed upward like anything else. They are simply light enough for the buoyancy to be greater than the weight.

Unlike water, there is no distinct surface at the "top" of the atmosphere that decreases to nothingness. Furthermore, unlike water, the atmosphere becomes less dense with altitude. Whereas a piece of wood will float to the surface of water, balloons do not rise to any atmospheric upper surface. How high will a balloon rise? A gas-filled balloon will rise until the weight of displaced air equals its weight. Then the sum of the forces on the balloon is zero. Rising rubber balloons expand when rising. They usually reach a height where expansion stretches the rubber until it ruptures.

CONCEPT CHECK

Is there a buoyant force acting on you? If there is, why are you not buoyed up by this force?

CHECK YOUR ANSWERS

There *is* a buoyant force acting on you, and you are buoyed upward by it. Without that buoyant force your weight would be about 0.7 newtons greater, which is about 2.6 ounces.

Concept Review

Chapter 7

Summary of Terms

Archimedes' principle An immersed body is buoyed up by a force equal to the weight of the fluid it displaces (applying to both liquids and gases).

Atmospheric pressure The pressure exerted against bodies immersed in the atmosphere resulting from the weight of air pressing down from above. At sea level, atmospheric pressure is about 101 kPa.

Barometer Any device that measures atmospheric pressure.

Boyle's law The product of pressure and volume is a constant for a given mass of confined gas regardless of changes in either pressure or volume individually, so long as temperature remains unchanged:

$$P_1V_1 = P_2V_2$$

Buoyant force The net upward force that a fluid exerts on an immersed object.

Density The amount of matter per unit volume.

Liquid pressure The pressure exerted by the weight of a fluid, which is equal to weight density × depth.

Pascal's principle A change in pressure at any point in an enclosed fluid at rest is transmitted undiminished to all points in the fluid.

Pressure The ratio of force to the area over which that force is distributed.

$$\text{Pressure} = \frac{\text{Force}}{\text{Area}}$$

Principle of flotation A floating object displaces a weight of fluid equal to its own weight.

Weight density is expressed as weight per unit volume.

Review Questions

1. Distinguish between mass density and weight density. What are the mass density and the weight density of water?
2. How does the pressure exerted by a liquid change with depth in the liquid? How does the pressure exerted by a liquid change as the density of the liquid changes?
3. How does water pressure one meter below the surface of a small pond compare to water pressure one meter below the surface of a huge lake?
4. If you punch a hole in the side of a container filled with water, in what direction does the water initially flow outward from the container?
5. Why does buoyant force act upward on an object submerged in water?
6. How does the volume of a completely submerged object compare with the volume of water displaced?
7. How does the buoyant force on a fully submerged object compare with the weight of water displaced?

8. What is the mass in kilograms of 1 L of water? What is its weight in newtons?

9. If a 1-L container is immersed halfway in water, what is the volume of water displaced? What is the buoyant force on the container?

10. Does the buoyant force on a floating object equal the weight of fluid displaced? Or on the weight of the object? When is buoyant force equal to both?

11. What weight of water is displaced by a 100-ton floating ship? What is the buoyant force that acts on this ship?

12. By how much does the density of air increase when it is compressed to half its volume? Its pressure?

13. How does the downward pressure of the 76-cm column of mercury in a barometer compare with the air pressure at the surface of the dish of mercury at that level?

14. How does the weight of mercury in a barometer tube compare with the weight of an equal cross-section of air from sea level to the top of the atmosphere?

15. Consider two tubes of equal cross-sectional area: One is a 76-cm tall tube of mercury and the other a 10.3-m tall tube of water. Which weighs more? Which exerts more pressure at the bottom of the tube?

16. What happens to the pressure in all parts of a confined fluid when the pressure in one part is increased?

17. Does Pascal's Principle provide a way to get more energy from a machine than is put into it? Defend your answer.

18. A balloon that weighs 1 N is suspended in air, drifting neither up nor down. How does the buoyant force acting on it compare with its weight? What happens if the buoyant force decreases? Increases?

19. A flame is inserted into a hot air balloon. This causes the air within the balloon to expand and the balloon inflates to a maximum size. Upon heating the air even further, the balloon itself has a hole on top that allows much of this heated air to escape. How does this affect the weight of the hot air balloon? How does this affect the buoyant force?

20. How does the weight of the air affect your weight as measured on a standard weighing scale?

Answers to Odd-Numbered Questions

1. Mass density is mass/volume; water's density is 1 kg/m^3. Weight density is weight/volume; water's weight density is 62.4 lb/ft^3.

3. Same pressure at same depths.

5. Greater pressure against the bottom than against the top. Correspondingly, greater force up on the bottom than down on the top.

7. Both are equal.

9. One half liter; 5 newtons.

11. Both weight of water displaced and buoyant force are 100 tons.

13. Both are the same.

15. Same weights and same pressures.

17. No, it can multiply forces or distances, but not energy. A conservation of energy no no!

19. As the hot air escapes, the weight of the balloon decreases, which allows for the hot air balloon to rise. The balloon itself stays about the same size, which means the buoyant force is also about the same.

Chapter 8

Heat

THE MAIN IDEA

Heat is the movement of thermal energy from high to low temperature.

Why can people walk barefoot without harm on red-hot coals? When white-hot sparks from a fireworks sparkler hit your skin, they don't burn you—why is this so? Why does a bite of the filling of a hot apple pie burn your tongue, while the crust does not, especially since both parts come from the same oven with the same temperature? Why do most materials expand as they heat up? Why does ice water do the opposite? And why does a tile floor feel cooler than a carpeted floor when both have the same temperature? Is it an incredible fact that everything continuously emits radiation? If so, then why doesn't everything get colder with time? Let's explore the awesome physics in this chapter!

8.1 Thermal Energy

Until a couple hundred years ago, people thought heat was a mysterious substance called *caloric* that passed from warmer objects to cooler ones. In 1798, scientist Count Rumford observed the drilling of a metal cylinder to make a cannon. He noticed that the metal got very hot as the drill turned. Rumford realized that the drill's motion must be producing the heat. He realized that motion produces heat and that heat and the energy of motion must be connected.

To understand how motion, heat, and energy are related, you need to know a bit about the particles of matter. Matter is made of tiny particles called *atoms*. Atoms join together to make clusters called *molecules*. Atoms and molecules constantly wiggle and jiggle. When jiggling slowly, they form solids. When they jiggle faster so they slide over one another, we have a liquid. When atoms and molecules move so fast that they disconnect and fly loose, we have a gas. So, whether a substance is a solid, liquid, or gas depends on the motion of its particles. All this is included in the *kinetic theory of matter*, which states:

> **Matter is composed of tiny particles called atoms and molecules. These particles are always moving. This random motion of atoms and molecules is present in different degrees in solids, liquids, and gases.**

Λ Figure 8.1

Count Rumford (1753–1814). Besides studying projectiles, setting up an early system of welfare, breeding horses, and acting as a spy, Count Rumford showed that heat is related to the motion of matter.

The warmer an object is, the faster its particles move. That means they have more kinetic energy. For example, when you strike a penny with a hammer, the penny becomes warmer. Why? Because the hammer's blow causes atoms in the coin to jiggle faster. When you put a flame to a liquid, the liquid becomes warmer as its particles move faster. When you rapidly compress air in a tire pump, the air gets warmer (**Figure 8.2**). In these cases, the molecules are made to move faster. They gain kinetic energy and the material becomes warmer.

The warmer an object is, the more *thermal energy* it contains. The **thermal energy** in a substance is the total energy of all of its atoms and molecules. Thermal energy consists of all the potential and kinetic energy of the particles in a substance as they travel about, twist and turn, and vibrate back and forth (**Figure 8.3**).

< Figure 8.2

Push down on the piston and you do work on the air inside. Air molecules gain kinetic energy, move faster, and the air is warmed.

READING CHECK

How does thermal energy relate to the kinetic and potential energies of its particles?

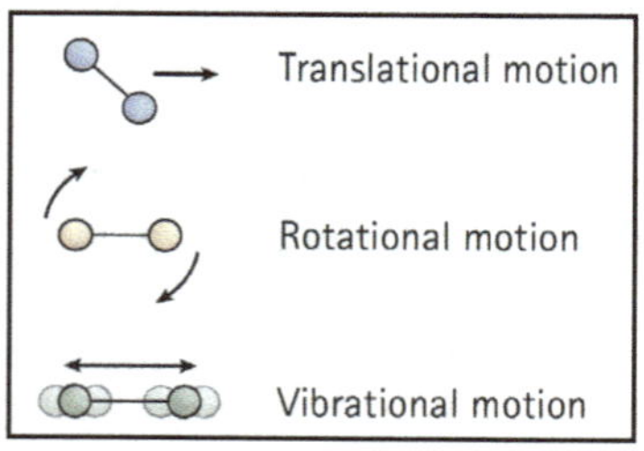

< Figure 8.3

Types of motion in matter. Increases in translational motion increase kinetic energy, and therefore temperature. Increases in rotational and vibrational motion increase potential energy.

8.2 Temperature

To tell how warm or cold an object is we measure its **temperature**. A common thermometer measures temperature by expansion or contraction of a liquid, usually colored alcohol.

The most common thermometer in the world is the Celsius thermometer, named after the Swedish astronomer Anders Celsius (1701–1744). Celsius was the first person to suggest the scale of 100 degrees between the freezing point and boiling point of water. The number 0 represents the temperature at which water freezes and the number 100 represents the temperature at which water boils (at standard atmospheric pressure). In between are 100 equal parts called degrees.

An earlier scale was developed by the German physicist Gabriel Fahrenheit (1686-1736). Fahrenheit assigned the number 0 to freezing sea water and the number 100 to the average internal temperature of a human. On this scale, fresh water happens to freeze at 32°F and boil at 212°F. Today we know a normal human temperature ranges from 97°F to 99°F. Nonetheless, the Fahrenheit scale remains popular in the United States.

Arithmetic formulas are used for converting from one temperature scale to the other and are common in classroom exams. Because such arithmetic exercises are not really physics, we won't be concerned with these conversions. (This may be important in a math class, but not here.) Besides, the conversion between Celsius and Fahrenheit temperatures is closely approximated in the side-by-side scales of **Figure 8.4**. With such an illustration no math is needed—just look!

As **Figure 8.5** indicates, temperature is proportional to the average translational kinetic energy per particle that makes up a substance. By *translational* we mean to-and-fro linear motion. A substance with a high temperature has atoms or molecules with high average translational kinetic energies. (To be brief, from now on in this chapter, we'll simply say molecules to mean atoms and molecules.) So, what's the difference between a hot cup of coffee and a cold cup of coffee? Answer: Molecules in the hot cup of coffee are moving faster.

READING CHECK

How does temperature relate to the average kinetic energy of molecular motion?

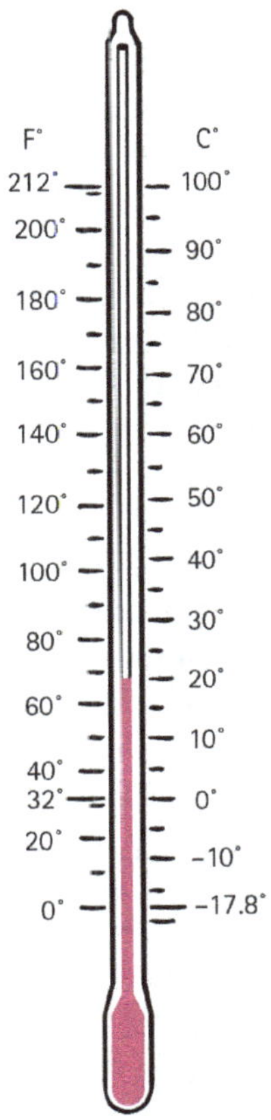

< Figure 8.4

Fahrenheit and Celsius scales on a thermometer.

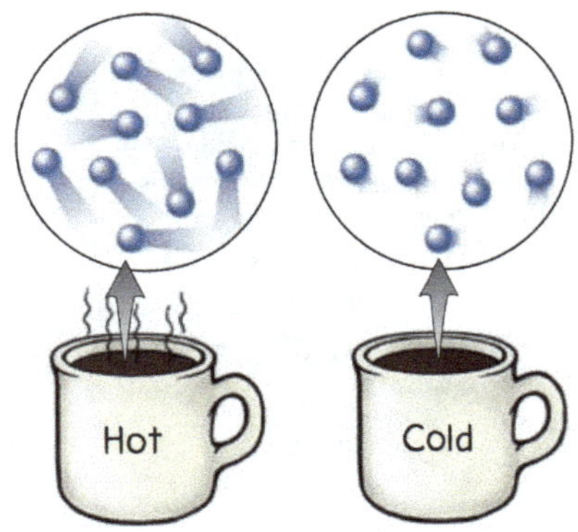

∧ Figure 8.5

The difference between hot coffee and cold coffee is the average speed of the molecules. In the hot coffee, the molecules are moving faster on average than they are in the cold coffee. (The "motion trails" on the molecules of hot coffee indicate their higher speed.)

Quick Activity

Can You Trust Your Senses?

Put some hot water, some warm water, and some cold water in three open containers. Place a finger in the hot water and a finger of the other hand in the cold water. How do they feel? After a few seconds, place both fingers in the warm water. How do they feel now? Will both fingers feel the same temperature when they are put in the warm water? Our senses are a good guide at times—but not at all times. To measure the hotness or coldness of things, use a thermometer.

CONCEPT CHECK

True or false: Temperature is a measure of the total kinetic energy in a substance.

CHECK YOUR ANSWER

False. Temperature is a measure of the *average* (not the total!) kinetic energy of the molecules in a substance. For example, there is twice as much total molecular kinetic energy in 2 liters of boiling water as in 1 liter—but the temperatures of the two volumes of water are the same because the average kinetic energy per molecule in each is the same.

8.3 Absolute Zero

Plasma is found in stars, where the temperature is many millions of degrees. Cooler plasmas are in fluorescent lamps and in the pixels of plasma TV screens. Plasma is also known as the fourth state of matter.

In principle, there is no upper limit of temperature. As thermal motion keeps increasing, a solid object melts to a liquid and then evaporates to a gas. Further heating of a gas breaks molecules up into atoms that lose some or all of their electrons—a cloud of electrically charged particles called a *plasma*, as found within the Sun or even a fluorescent light bulb. Temperature has no upper limit.

In contrast, there is a definite lower limit to temperature. Here's how experimenters in the 19th century found that limit. Gases expand when heated and contract when cooled. All gases were found to shrink by 1/273 of their volume for each Celsius degree lowering in temperature. This occurs when the gas pressure is held constant. So, if a gas were cooled down to 273°C, it would be reduced to zero volume.

In practice, every gas turns to a liquid before it gets this cold. Nevertheless, these decreases by 1/273 increments suggested the idea of a lowest temperature: –273°C. We call this lower limit of temperature **absolute zero**. At this temperature molecules have lost all available kinetic energy. No more energy can be taken from a substance at absolute zero. It can't get any colder. More than this, absolute zero itself is a theoretical lower limit of temperature that cannot be attained, only approached.

READING CHECK

How do the sizes of division compare on the Kelvin and Celsius scales?

The absolute temperature scale is called the Kelvin scale, named after the famous British physicist, Lord Kelvin. Absolute zero is 0 K (short for "0 kelvin,"; note that the word "degrees" is not used with Kelvin temperatures). There are no negative numbers on the Kelvin scale. Degrees on the Kelvin scale are the same size as divisions on the Celsius scale. Thus, the melting point of ice is 273 K, and the boiling point of water is 373 K. Using crafty laboratory techniques, the record low temperature ever achieved has been about 0.0000000003 K. These super cold temperatures allow physicists to explore the fundamental nature of matter and energy.

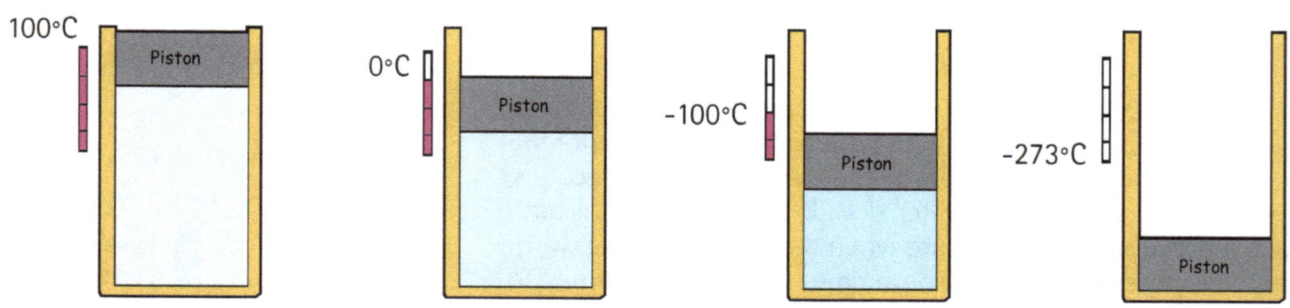

Figure 8.6

As the temperature of a gas decreases so does its volume. The temperature at which the volume reaches zero is -273°C. This is known as absolute zero, which is the theoretical lower limit of temperature.

A waiter brings you a slice of apple pie at 0°C. Not wanting such a cold slice of apple pie you kindly ask the waiter to warm it to twice as hot. To what temperature should the waiter heat the apple pie? To answer this question we need to recognize that even at 0°C the molecules within the pie are jostling to-and-fro quite fast. To double the temperature we need to double how fast, on average, they are moving. For this, we need to use the absolute temperature scale. At 0°C the slice of apple pie is at 273K. Double the temperature you'll have 546 K, which translates to 273°C (523°F). A knowledgeable waiter would suggest a not so hot temperature.

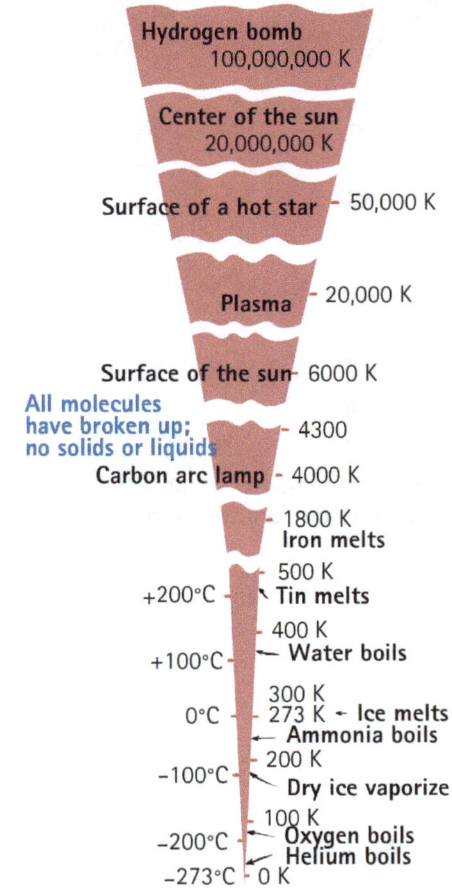

Λ Figure 8.7

Some temperatures in the kelvin scale.

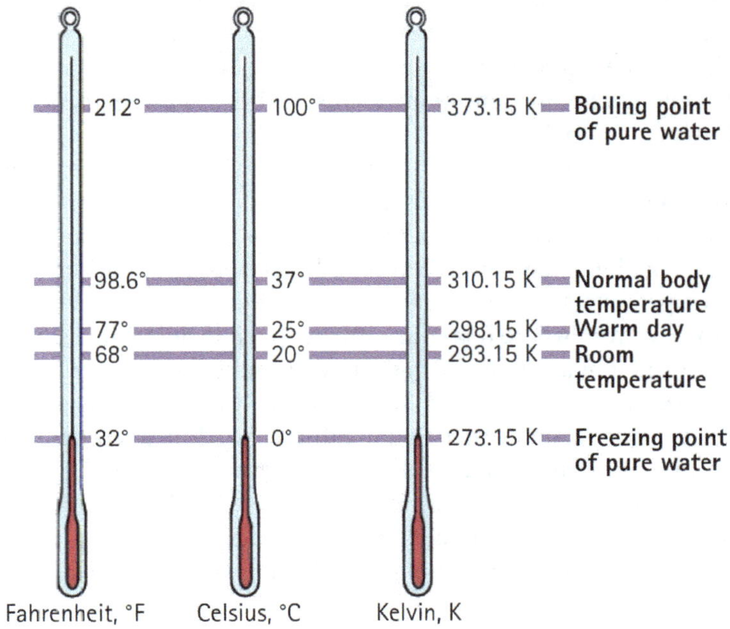

Λ Figure 8.8

Some familiar temperatures measured on the Fahrenheit, Celsius, and Kelvin scales.

Figure 8.9 >

The temperature of the sparks is very high, about 2000°C. That's a lot of thermal energy per molecule of spark. Because there are only a few molecules per spark, however, the total amount of thermal energy in the sparks is safely small. Temperature is one thing; transfer of thermal energy is another.

CONCEPT **CHECK**

1. Which is larger, a Celsius degree or a kelvin?
2. A sample of hydrogen gas has a temperature of -5°C. If the gas is heated until its molecules have doubled their kinetic energy, what is its temperature?

CHECK YOUR ANSWERS

1. Neither. They are equal.
2. The -5°C gas has an absolute temperature of 268 K. Twice as much kinetic energy means that it has twice the absolute temperature, or two times 268 K. This would be 536 K, or 263°C.

8.4 Heat

If you touch a hot stove, thermal energy enters your hand because the stove is warmer than your hand. When you touch a piece of ice, however, thermal energy passes out of your hand and into the ice. The direction of thermal energy flow is normally from a warmer substance to a cooler one. A scientist defines **heat** as the thermal energy transferred from one substance to another due to a temperature difference between the two substances.

READING CHECK

What becomes of heat after it is transferred to another substance?

According to this definition, matter does not *contain* heat. Matter contains *thermal energy.* Heat is thermal energy in transit. After heat has been transferred to an object or substance, it ceases to be heat—and becomes thermal energy. So, by definition, heat is the flow of thermal energy. If it's not flowing, it's not heat.

For substances in contact, thermal energy flows from the higher-temperature substance into the lower-temperature one until they reach the same temperature, which is *thermal equilibrium.* There is a fundamental reason for this. At a higher temperature, thermal energy is more concentrated. At a lower temperature, the thermal energy is more dilute. Thermal energy always tends to flow from where it is concentrated to where it is dilute. By analogy, the sweet smell of an apple pie in the oven is quite concentrated within that oven. Open the oven door and that aroma will naturally disperse around the house—from concentrated to more dilute. So, drop a red-hot metal tack into a bowl of warm water. Heat will flow from where it is concentrated within the red-hot tack to where it is more dilute within the warm water.

It's important to distinguish between the temperature of a substance and the total amount of thermal energy within that substance. A red-hot tack may have a high temperature, but its total amount of thermal energy is relatively small. A bucket of warm water may have a lower temperature, but its total amount of thermal energy is quite large compared to the hot tack.

So, which has a higher temperature: a cup of boiling water or a heated swimming pool? Answer: The cup of boiling water has a higher temperature. But which has more thermal energy? Compare your utility bills after boiling a cup of water to heating an Olympic-sized swimming pool to a comfortable temperature. The swimming pool has much more thermal energy, very much so. This thermal energy, however, is diluted across a large mass of water as evidenced by its lower temperature.

Figure 8.10 >

The water molecules within a hot cup of coffee are moving faster than the water molecules within a cooler swimming pool. The total energy, however, depends on the quantity of water. Consider your electric bill after heating a swimming pool full of water by even a few degrees.

Heat is directional. When it comes toward us, we experience a warming effect. When it leaves us, we experience a cooling effect. Consider a freezing winter day when you're waiting at the bus stop, you feel cold not because something called cold gets to you. Rather, you feel cold because you are losing heat. As a result, your total thermal energy is decreasing. Brrr! A nicely insulated winter jacket helps to slow the rate you lose heat, which helps to keep you feeling warm.

READING CHECK

How does something that is cold relate to thermal energy?

CONCEPT CHECK

When you touch a cold surface, does cold travel from the surface to your hand or does energy travel from your hand to the cold surface?

CHECK YOUR ANSWER

The direction of energy travel is from hot to cold—from your hand to the cold surface. There is no "cold" that travels in the other direction.

Heat is a form of energy and is measured in *joules*. It takes 4.18 joules of heat to change 1 gram of water by 1 Celsius degree. A unit of heat still common in the United States is the calorie. A calorie is defined as the amount of heat needed to change the temperature of 1 gram of water by 1 Celsius degree. (The relationship between calories and joules is that 1 calorie = 4.18 joules.)

The energy ratings of foods and fuels are measured by the energy released when they are burned. (Metabolism is really "burning within the body" at a slow rate.) The heat unit for labeling foods is the kilocalorie, which is 1000 calories (the heat needed to change the temperature of 1 kilogram of water by 1°C). The food unit of energy is usually called a Calorie with a capital C. So, 1 Calorie = 1 kilocalorie = 1000 calories.

Figure 8.11

To the weight watcher, the peanut contains 10 Calories; to the physicist, it releases 10,000 calories (41,800 joules) of energy when burned or digested.

CONCEPT CHECK

Which will raise the temperature of water more, adding 4.18 joules or 1 calorie?

CHECK YOUR ANSWER

Both the same. This is like asking which is longer, a 1-mile long track or a 1.6-kilometer long track. They're the same in different units.

8.5 Entropy

Energy tends to disperse. It flows from where it is concentrated to where it is spread out. The energy of a hot pan, for example, does not stay concentrated in the pan once the pan is taken off the stove. Instead, the energy spreads away from the pan into the cooler surroundings.

READING CHECK

What is one of the central reasons for the occurrence of just about anything?

Scientists consider this tendency of energy to disperse to be one of the central reasons for why just about anything ever happens. In other words, changes that result in energy spreading outward tend to occur on their own—nature favors them.

The opposite holds true, too. Changes in the natural world that result in the concentration of energy do *not* tend to occur—they are not favored. Heat from the room, for example, will never spontaneously move into a pan to heat it up. The natural flow of energy is always a one-way trip from where it is concentrated to where it is less concentrated, or "spread out." That energy tends to disperse is spelled out by the **second law of thermodynamics**, which can be paraphrased as follows:

> **Any natural process that happens with no outside assistance results in a net dispersal of energy.**

Entropy is a measure of this natural spreading of energy. Wherever there is a spreading of energy, there is a corresponding *increase* in entropy. Interestingly, there is a close relationship between entropy and our psychological sense of time. That energy tends to spread out is part of our human experience. We *expect* a hot pan to cool, which is entropy in action, just as we *expect* hot gases to expel through an exhaust pipe, which is also entropy in action.

Figure 8.12

The Sun is truly a "hothouse"—dispersing enormous amounts of energy via thermonuclear reactions. A tiny fraction of the Sun's energy is used to drive photosynthesis, which is vital to both plants and plant-eating creatures like us.

But what if we saw the reverse? For example, if we saw smoke moving *into* a smokestack? Or what if we saw the warmth of the water concentrate itself to push a diver flying back upward to the platform? If we were watching these energy-concentrating events on video, we would quickly conclude that the video was running backward. However, if in real life we actually saw such things—if we could survive the shock—we would immediately sense that time itself was running backward. Thus, when we watch smoke spewing out of a smokestack and a diver diving into water, as shown in **Figure 8.13**, we have the sense that time is moving forward. The second law of thermodynamics, therefore, gives us our psychological sense of time; it is the "arrow of time."

CONCEPT **CHECK**

The energy of a swimmer jumping into water is dispersed in the water, which is warmed a bit. Can the reverse occur—energy in the water propelling the swimmer to her starting position?

CHECK YOUR ANSWER

No. Nature doesn't behave that way. If the swimmer is to return to her starting position it will be via energy from the food she eats that comes from photosynthesis, which is a biochemical process where plants collect and temporarily store energy dispersed from the Sun.

< Figure 8.13

As she jumps, her potential energy is converted into kinetic energy. When splashing into the water, this kinetic energy is spread to make the water molecules move faster and to heat up just a little. Will this dissipated energy reconcentrate itself to push her back through the air up to the cliff? Our common experience with the second law of thermodynamics says no.

8.6 Specific Heat

When you're eating, have you noticed that some foods remain hotter much longer than others? For example, just after an apple pie has been taken out of an oven, the filling burns your tongue while the crust doesn't. Or you can take a bite of a piece of hot toast a few seconds after coming from the hot toaster, but you have to wait several minutes before eating soup from a pot on a stove as hot as the toaster.

Different substances have different capacities for storing thermal energy. When you heat a pot of water on a stove you find that it takes about 15 minutes to bring it to a boil. If you put an equal mass of iron on the same stove, you'd find it rising through the same temperature range in only about 2 minutes. For silver, the time would be less than a minute. Different materials require different amounts of thermal energy to raise temperature. This is because different materials absorb energy in different ways. The added energy may increase the translational motion of molecules, which raises the temperature. Or added energy may increase the amount of internal vibration or rotation within the molecules and therefore become potential energy. It's mainly the translation motion of its atoms and molecules that raises the temperature of a substance. Each substance has its own characteristic specific heat capacity.

The specific heat capacity of any substance is defined as the quantity of heat required to change the temperature of a unit mass of the substance by 1 degree.

∧ Figure 8.14

The filling of hot apple pie may be too hot to eat, even though the crust is not.

READING CHECK

What kind of molecular motion affects the temperature of substances?

fyi FOR YOUR INFORMATION

If we know the specific heat capacity c of a substance, the formula for the quantity of heat Q involved when a mass m of the substance undergoes a change in temperature ΔT is $Q = cm\Delta T$. In words, heat transferred equals specific heat capacity times mass times the temperature change.

Specific heat capacity is a measure of *thermal inertia*. Recall in our study of Newton's laws that inertia is the property of matter to resist changes in motion. Specific heat capacity is a similar property of matter to resist a change in temperature.

Water has a much higher capacity for storing energy than most all other substances. A lot of heat energy is needed to change the temperature of water. This explains why water is very useful in the cooling system of automobiles and other engines. It absorbs a great quantity of heat for small rises in temperature. Likewise, water also takes longer to cool.

Water's high specific heat capacity affects the world's climate. Very much so. Look at a world globe and notice the high latitude of Europe. Water's high specific heat keeps climate there milder than regions of the same latitude in northeastern regions of Canada. Both Europe and Canada receive about the same amount of sunlight per square kilometer. What happens is that the Atlantic Ocean current known as the Gulf Stream carries warm water northeast from the Caribbean. It holds much of its thermal energy long enough to reach the North Atlantic Ocean off the coast of Europe. Then it cools, releasing 4.18 joules of energy for each gram of water that cools 1°C. The released energy is carried by westerly winds over the European continent.

A similar effect occurs in the United States. The winds in North America are mostly westerly. On the West Coast, air moves from the Pacific Ocean to the land. In winter months, the ocean water is warmer than the air. Air blows over the warm water and then moves over the coastal regions. This warms the climate. In summer, the opposite occurs. The water cools the air and the coastal regions are cooled. Temperature changes are moderate rather than extreme. The East Coast does not benefit from the moderating effects of water because the direction of air is from the land to the Atlantic Ocean. Land, with a lower specific heat capacity, gets hot in the summer but cools rapidly in the winter.

Islands and peninsulas do not have the extremes of temperatures that are common in interior regions of a continent. The high summer and low winter temperatures common in Manitoba and the Dakotas, for example, are largely due to the absence of large bodies of water. Europeans, islanders, and people living near ocean air currents find that their climates are greatly moderated by water's high specific heat capacity.

On a global scale the planet is warming due mostly to greater amounts of greenhouse gases being produced by human activities. This includes carbon dioxide, CO_2, as well as methane, CH_4, which you will learn more about in a chemistry program. For now, you should know that about 90 percent of the added greenhouse heat to this planet has been absorbed by the oceans. So, water's high specific heat is key in protecting us from global warming.

CONCEPT **CHECK**

Which has a higher specific heat capacity, water or sand? In other words, which takes longer to warm in sunlight (or longer to cool at night)?

CHECK YOUR ANSWER

Water has a higher specific heat capacity. In the same sunlight, the temperature of water increases more slowly than the temperature of sand. And water cools more slowly at night. Sand and soil's low specific heat capacity, as evidenced by how quickly they warm in the morning Sun and how quickly they cool at night, affects local climates.

Unfortunately, as the oceans absorb this heat they also expand, which leads to a slow but sure rise in sea level. Furthermore, the oceans will naturally hold onto this heat for many years to come, which means those sea level rises are long-term. On top of this, warmer ocean temperatures also mean potential changes to ocean currents, including the Gulf Stream that helps to keep Europe relatively warm. And more so, warmer ocean temperatures are already causing changes in marine ecosystems, such as the bleaching of tropical coral, which, in turn, can have profound effects in the food chain. Much attention is given to potential changes in our terrestrial climates. For the full picture, we need also focus our attention on what global warming is doing to our oceans. While our oceans protect us, we need to do what we can to protect them.

FOR YOUR INFORMATION

Water is very useful in the cooling systems of automobiles and other engines because it absorbs a great quantity of heat for small increases in temperature. Water also takes longer to cool.

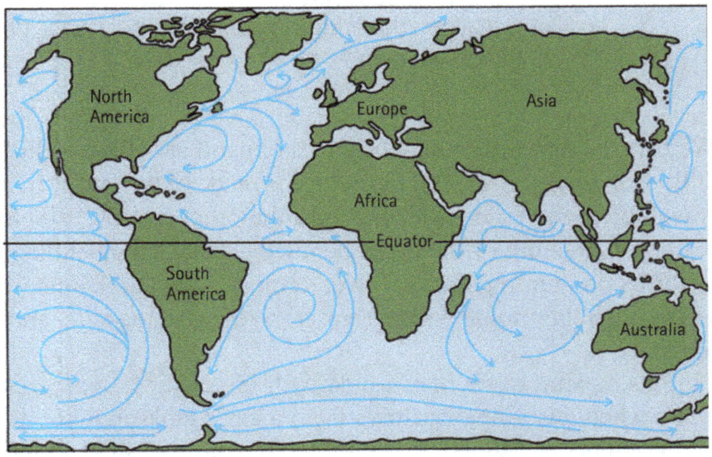

< Figure 8.15

Many ocean currents, shown in blue, distribute heat from the warmer equatorial regions to the colder polar regions.

CONCEPT CHECK

Bermuda is close to North Carolina, but unlike North Carolina, it has a tropical climate year round. Why?

CHECK YOUR ANSWER

Bermuda is an island. The surrounding water warms it when it's too cold, and cools it when it's too warm.

8.7 Thermal Expansion

As the molecules in a hot substance jiggle faster they move farther apart. The result is **thermal expansion**. Most substances expand when heated and contract when cooled. Power lines are longer and sag more on a hot summer day than in winter. Figure 8.16 shows an example of railroad tracks having expanded too much on a hot summer day.

Thermal expansion must be taken into account in structures and devices of all kinds. A dentist uses filling material with the same rate of expansion as teeth. A civil engineer uses reinforcing steel with the same expansion rate as concrete. A long steel bridge usually has one end fixed while the other rests on rockers (Figure 8.17). Notice also tongue-and-groove gaps called expansion joints on bridges (Figure 8.18). We see the effects of expansion all around us.

∧ Figure 8.16

Thermal expansion. Extreme heat on a summer day caused the buckling of these railroad tracks.

Λ Figure 8.17

One end of the bridge is fixed, but the end shown rides on rockers to allow for thermal expansion.

Λ Figure 8.18

This gap in the roadway of a bridge is called an expansion joint; it allows the bridge to expand and contract. (Was this picture taken on a warm or a cold day?)

We can see that different substances expand at different rates with a bimetallic strip (**Figure 8.19**). This device is made of two strips of different metals welded together, one of brass and the other of iron. When heated, the greater lengthening of the brass bends the strip. This bending may be used to turn a pointer, regulate a valve, or close a switch. Bimetallic strips are used in oven thermometers, refrigerators, electric toasters, and various other devices.

READING CHECK

Which expands more with increases in temperature, liquids or solids?

Liquids expand more than solids with increases in temperature. We notice this when gasoline overflows from a car's tank on a hot day. If the tank and contents expanded at the same rate, no overflow would occur. This is why a gas tank being filled shouldn't be "topped off" on a hot day.

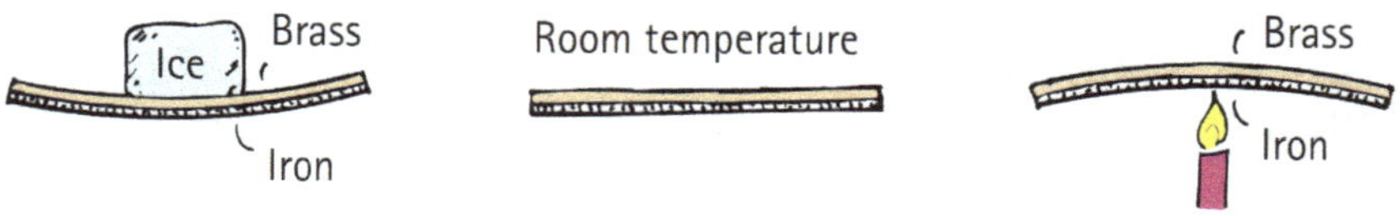

Λ Figure 8.19

A bimetallic strip. Brass expands more when heated than iron does and contracts more when cooled. Because of this behavior, the strip bends as shown.

CONCEPT **CHECK**

1. When you can't loosen a metal lid on a glass jar, how can you use the concept of thermal expansion to rescue the situation?
2. Supersonic aircraft can be 20 cm longer when in flight than when parked on the ground. Offer an explanation.

CHECK YOUR ANSWERS

1. Hold the lid of the jar under hot water for a few seconds. The metal should expand more than the glass, making it easier to loosen.
2. At cruising speed (faster than the speed of sound), air friction against the aircraft raises its temperature dramatically, resulting in this significant thermal expansion.

Expansion of Water

Water, like most other substances, expands when heated. But interestingly, it doesn't expand in the temperature range between 0°C and 4°C. Something quite fascinating happens in this range. Ice has a crystalline structure, with six-sided open-structured crystals. Water molecules in this open structure occupy a greater volume than they do in the liquid phase (**Figure 8.20**). This means that ice is less compact (less dense) than water.

When ice melts, not all the six-sided crystals collapse. Some remain in the ice-water mixture, making up a microscopic slush that slightly "bloats" the water—increases its volume slightly. This results in ice water being less dense than slightly warmer water. As the temperature of water near 0°C is increased, more of the remaining ice crystals collapse. This further decreases the volume of the water. This contraction continues only up to 4°C. That's because two things happen at the same time; volume decreases due to ice crystal collapse, and, volume increases due to greater molecular motion. The collapsing effect

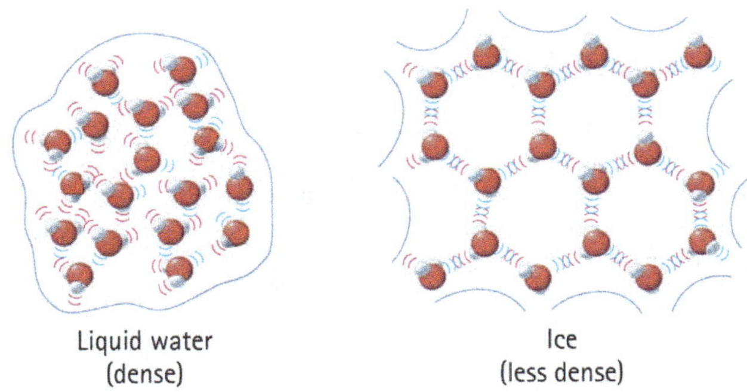

Figure 8.20

Water molecules in a liquid are denser than water molecules frozen in ice, where they have an open crystalline structure.

Figure 8.21

The six-sided structure of snow crystals is a result of the six-sided ice crystals that make it up.

READING CHECK

What two processes occur for water as its temperature increases from 0°C?

CONCEPT CHECK

What's inside the open spaces of the water crystals shown in Figures 8.20 and 8.22 and the cartoon shown here? Is it air, water vapor, or nothing?

CHECK YOUR ANSWER

There's nothing at all in the open spaces. It's empty space—a void. If there were air in the spaces, the illustration would have to show air molecules. If there were water vapor in the spaces, H_2O molecules would have to be shown.

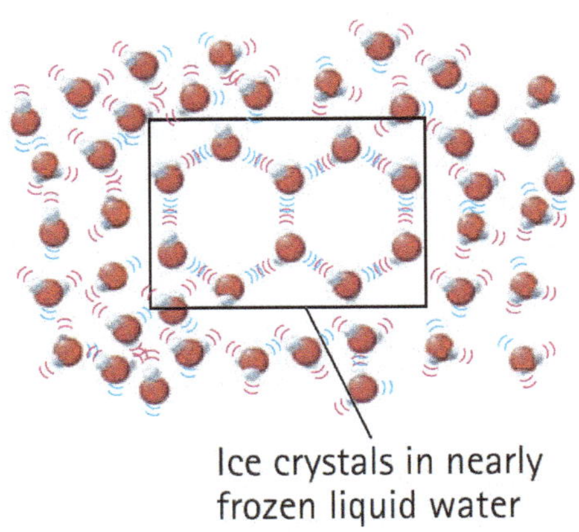

Figure 8.22

Close to 0°C, liquid water contains crystals of ice. The open structure of these crystals increases the volume of water slightly.

Figure 8.23 >

Between 0°C and 4°C, the volume of liquid water decreases as temperature increases. Above 4°C, water behaves the way other substances do. Its volume increases as its temperature increases. The volumes shown here are for a 1-gram sample.

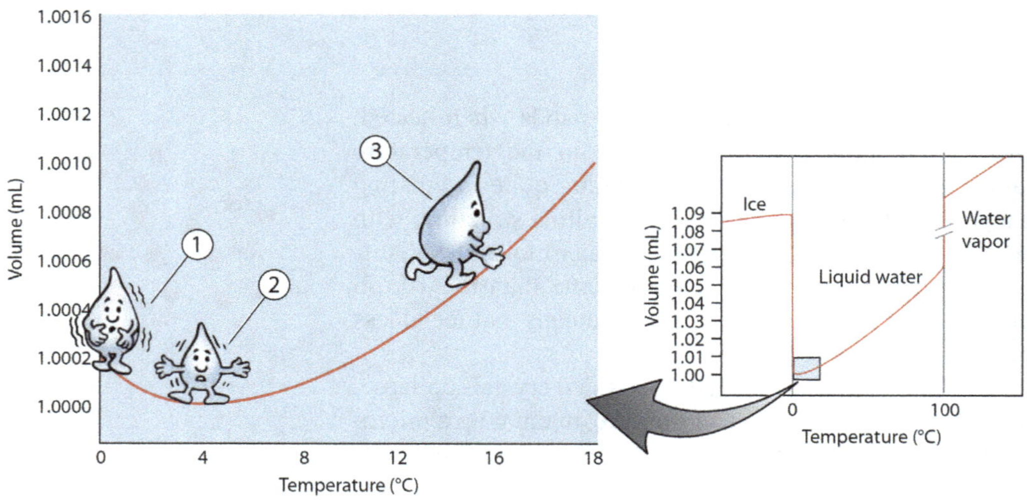

(1) Liquid water below 4°C is bloated with ice crystals.

(2) Upon warming, the crystals collapse, resulting in a smaller volume for the liquid water.

(3) Above 4°C, liquid water expands as it is heated because of greater molecular motion.

dominates until the temperature reaches 4°C. After that, expansion overrides contraction because most of the ice crystals have melted. (**Figure 8.23**).

When ice water freezes to become solid ice, its volume increases tremendously—and its density is much lower. That's why ice floats on water. Like most other substances, solid ice contracts with further cooling, but never enough to become more dense than liquid water, which is why ice always floats in liquid water—even boiling liquid water.

This behavior of water is very important in nature. If water were most dense at 0°C it would settle to the bottom of a pond or lake. Liquid water at 0°C is less dense and "floats" at the surface. That's why ice forms at the surface.

So, a pond freezes from the surface downward. In a cold winter, the ice will be thicker than in a milder winter. Water at the bottom of an ice-covered pond remains at 4°C, relatively warm for organisms that live there.

Interestingly, very deep bodies of water are not ice-covered even in the coldest of winters. This is because all the water must be cooled to 4°C before lower temperatures can be reached. For deep water, the winter is not long enough to reduce an entire pond to 4°C. Any 4°C water lies at the bottom. Because of water's high specific heat and poor ability to conduct heat, the bottom of deep bodies of water in cold regions remains at a constant 4°C year-round. Fish should be glad that this is so.

Figure 8.24 >

As water cools, it sinks until the entire pond is at 4°C. Then, as water at the surface is cooled further, it floats on top and can freeze. Once ice is formed, temperatures lower than 4°C can extend down into the pond.

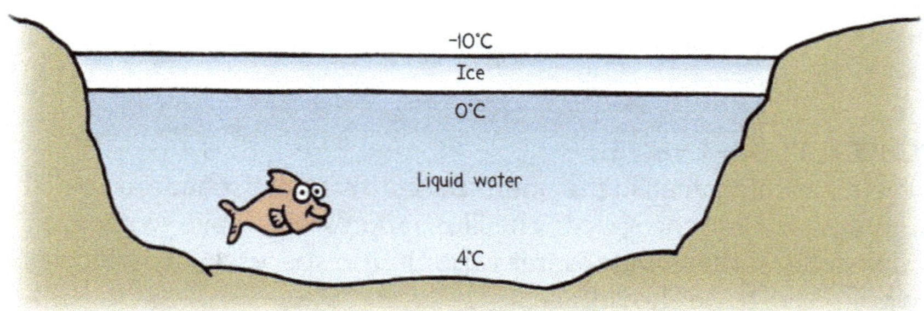

8.8 Conduction

There are three main ways that heat moves from one place to another. These are conduction, convection, and radiation. We investigate each, starting with conduction.

If you hold one end of an iron nail in a flame, the nail quickly becomes too hot to hold. If you hold one end of a short glass rod in a flame, the rod takes much longer before it becomes too hot to hold. In both cases heat at the hot end travels along the entire length. This method of heat transfer is called **conduction**. Thermal conduction occurs by collisions between particles and their immediate neighbors. Because the heat travels quickly through the nail we say that it is a good *conductor* of heat. Materials that are poor conductors are called *insulators*.

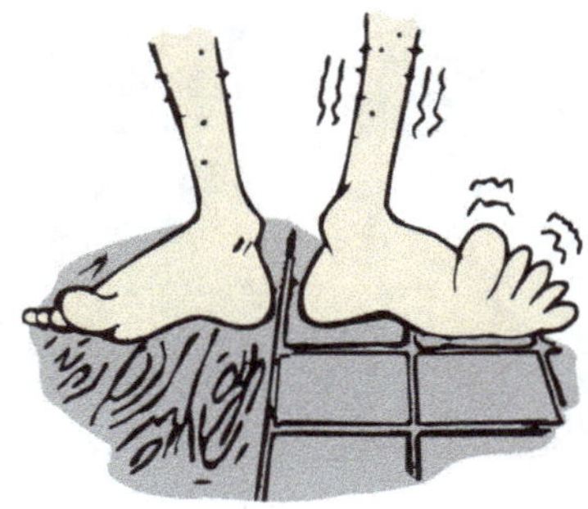

Λ Figure 8.25

The tile floor feels colder than the wooden floor, even though both are at the same temperature. Tile is a better heat conductor than wood, and it more quickly conducts internal energy from your feet.

Solids (such as metals) whose atoms or molecules have loosely held electrons are good conductors of heat. These mobile electrons move quickly and transfer energy to other electrons, which migrate quickly throughout the solid. Poor heat conductors (such as glass, wool, wood, paper, cork, and plastic foam) are made up of molecules that hold tightly to their electrons. In these materials, energy is transferred much more slowly.

Materials that conduct poorly are insulators. Wood is a good insulator and is used for cookware handles. Even when a pot is hot, you can quickly grasp the wooden handle with your bare hand without harm. An iron handle of the same temperature would surely burn your hand. Wood is a good insulator even when it's red hot. This explains how people can walk barefoot on red-hot wooden coals without burning their feet. (CAUTION: Even experienced firewalkers sometimes receive bad burns when conditions aren't just right.) The main factor here is the poor conductivity of wood—even red-hot wood. Although its temperature is high, very little heat is conducted to the feet. A firewalker must be careful that no iron nails or other good conductors are among the hot coals. Ouch!

Air is a *very* poor conductor. That's why you can briefly put your hand in a hot pizza oven without harm. The hot air doesn't conduct heat well. But don't touch the metal in the hot oven. Ouch again! The good insulating properties of such things as wool, fur, and feathers are largely due to the air spaces they contain. Porous substances are also good insulators because of their many small air spaces. Be glad that air is a poor conductor; if it weren't, you'd feel quite chilly on a 20°C (68°F) day!

READING CHECK

What accounts for the good insulating properties of wool, fur, feathers, and snow?

Snow is a poor conductor of heat. Snowflakes trap air and provide insulation. That's why a blanket of snow keeps the ground warm in winter. Animals in the forest find shelter from the cold in snow banks and in holes in the snow. The snow doesn't provide them with thermal energy—it simply slows down the loss of body heat generated by the animals. Then there are the igloos in Arctic dwellings that are shielded from the cold by their snow covering.

Homes are insulated with rock wool or fiberglass. Interestingly, insulation doesn't prevent the flow of heat. Insulation simply slows down the rate at which heat flows. Even a well-insulated warm home gradually cools. Insulation merely delays the rate at which heat conducts from a warmer region to a cooler one. In winter, we wish to slow conduction from inside to outside. But on hot summer days we wish to slow down conduction in the other direction, from outside to inside. Insulation slows conduction in either direction.

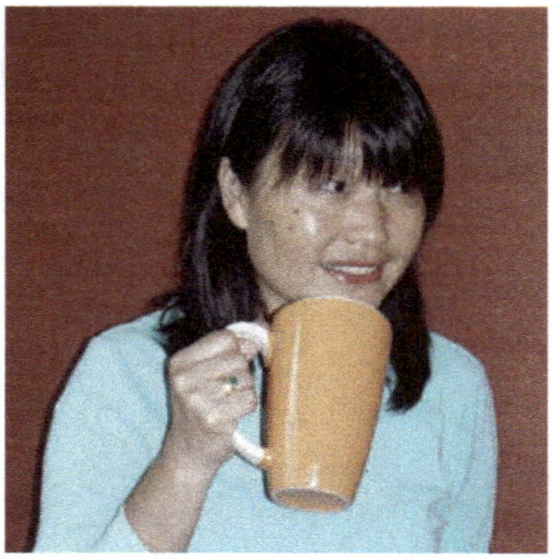

Λ Figure 8.26

Lil can comfortably hold the handle of a cup of hot chocolate because the cup is a good insulator.

Λ Figure 8.27

Snow patterns on the roof of a house show areas of conduction and insulation. Bare parts show where heat from inside has conducted through the roof and melted the snow.

CONCEPT **CHECK**

1. In desert regions that are hot in the day and cold at night, the walls of houses are often made of mud. Why is it important that the mud walls be thick?

2. Wood is a better insulator than glass. Yet fiberglass is commonly used to insulate buildings. Why?

CHECK YOUR ANSWERS

1. A wall of correct thickness keeps the house warm at night by slowing heat conduction from warmer inside to cooler outside. In the daytime conduction is slowed from warmer outside to cooler inside. Such a wall has "thermal inertia."

2. Fiberglass is a good insulator, many times better than glass, because of the air that is trapped among its fibers.

8.9 Convection

READING CHECK

Why does a warm fluid, such as heated air, rise?

FOR YOUR INFORMATION

Convection ovens are simply ovens with a fan inside. Cooking is speeded up by the circulation of heated air.

Liquids and gases transfer heat mainly by **convection**, which is transfer by motion of a fluid—by currents. Convection occurs in all fluids. Whether we heat water in a pot, or warm air in a room, the process is the same (**Figure 8.28**). As the fluid is heated from below, the molecules at the bottom begin moving faster. They spread apart and become less dense. Then they are buoyed upward. Denser, cooler fluid migrates to the bottom. In this way, convection currents keep the fluid stirred up. Warmer fluid moves away from the heat source and cooler fluid moves toward the heat source and is warmed.

As warm air expands, it becomes less dense, and then rises in the cooler surrounding air—like a balloon buoyed upward. When the rising air reaches an altitude where air density is the same, it no longer rises. We see this occurring when smoke from a fire rises and then settles off as it cools and its density matches that of the surrounding air.

Interestingly, as air rises, it expands. And as rising air expands, it also cools. The cooling occurs because in expanding, the molecules are receding away from each other, which has the effect of slowing them down. By analogy, what happens to the speed of a tennis ball when hit by a racquet not moving toward the ball but away from the ball? Answer: The tennis ball slows down. Likewise, when air molecules can only collide with other air molecules moving away from them, the overall effect is a slowing down of the molecules, as indicated in **Figure 8.29**. For this and other reasons, the expansion of a gas is a cooling process. Thus, any moisture within a rising parcel of air also cools. This causes the moisture to condense into tiny water droplets forming a cloud, as shown in **Figure 8.30a**.

Cooling by expansion is the opposite of what occurs when air is compressed, which is a warming process. A sinking parcel of air gets squeezed by greater atmospheric pressure as it lowers in elevation and thus also heats up, as can occur at the foot of a tall mountain range, as shown in **Figure 8.30b**.

Some parts of the Earth's surface absorb heat from the Sun more readily than others. This results in uneven heating of the air near the ground. We see this at the seashore, as **Figure 8.31** shows. In daytime, the ground warms up more than the water, and air above the ground that is warmed then rises. It is replaced by cooler air that moves in from above the water. The result is a sea breeze. At night, the process reverses because the shore cools off more quickly than the water, and then the warmer air is over the sea. Build a fire on the beach and you'll notice that the smoke sweeps inward during the day and seaward at night.

∧ Figure 8.28

Convection currents in a gas (air) and a liquid.

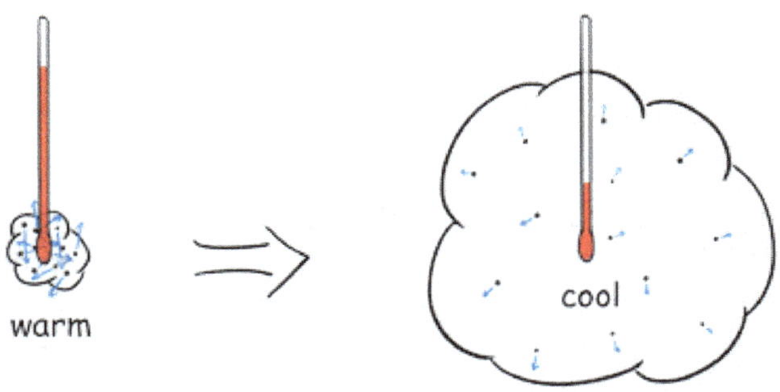

< Figure 8.29

Molecules in a region of expanding air collide more often with receding molecules than with approaching ones. Their average rebound speeds are therefore less, which results cooler expanding air.

(a) (b)

∧ Figure 8.30

(a) A parcel of rising air expands and cools. Any moisture within the air will condense to form a cloud. (b) A parcel of sinking air is compressed and warms. Beneath a mountain range this can lead to warm and fast-moving winds, also known as Chinook winds.

Figure 8.31 >

Convection currents produced by unequal heating of land and water. During the day, warm air above the land rises, and cooler air over the water moves in to replace it. At night the direction of air flow is reversed because now the water is warmer than the land.

Quick Activity

Breath-On Exploration

Do the following experiment right now. With your mouth open, blow on your hand. Your breath is warm. Now repeat, but this time pucker your lips to make a small hole so your breath expands as it leaves your mouth. Note that your breath is appreciably cooler! Expanding air cools.

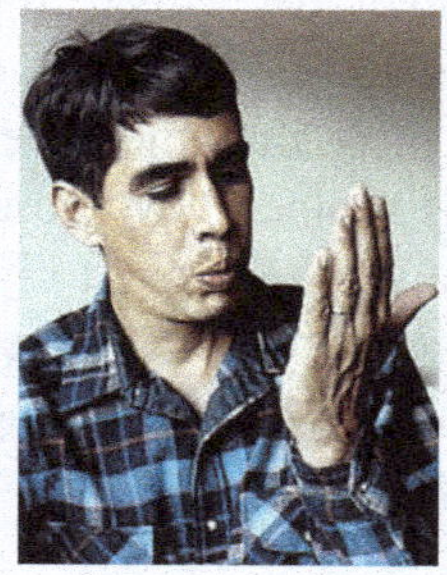

8.10 Radiation

READING CHECK

In what form does radiant energy travel?

Energy from the Sun passes through space and then through Earth's atmosphere before it warms Earth's surface. This heat transfer passes through the vacuum of space—no solid, liquid, or gas is involved. Therefore, this form of energy movement is not conduction, nor convection. Instead, it represents a third mode of energy transfer, which is **radiation**. Any energy transferred in this way is called *radiant energy*.

Radiant energy is in the form of electromagnetic waves. It includes a wide span of waves that begin with radio waves and infrared waves, continue through to visible-light waves, and to gamma rays. The lengths of waves differ. The waves of infrared (below-the-red) waves, for instance, are longer than those of visible-light waves. The longest visible wavelengths are for red light, and the shortest are for violet light. Shorter waves can't be seen by the eye. We'll treat waves in general and electromagnetic waves specifically in later chapters.

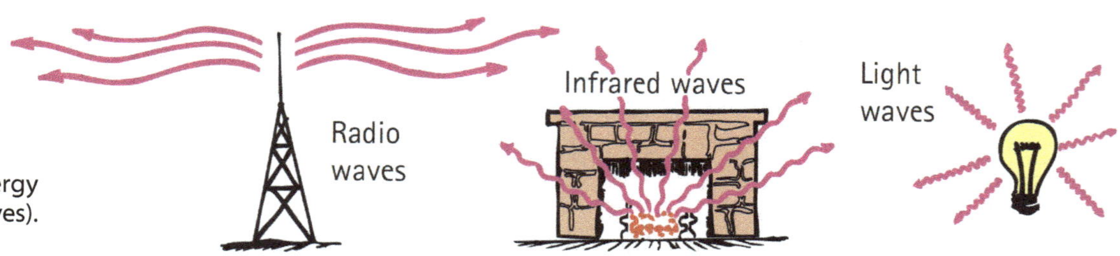

Figure 8.32 >

Types of radiant energy (electromagnetic waves).

The wavelength of radiation is related to the frequency of radiation. Frequency is the rate of vibration of a wave source. The girl in **Figure 8.33** shakes a rope at a low frequency (top), and a higher frequency (bottom). Note that shaking at a low frequency produces a long lazy wave, and the higher-frequency shake produces shorter waves. Likewise, with electromagnetic waves. As we'll discuss later, electromagnetic waves arise from vibrating electrons. High-frequency vibrations produce short waves and low frequency vibrations produce longer waves.

Figure 8.33

A wave of long wavelength is produced when the rope is shaken gently (at a low frequency). When shaken more vigorously (high frequency), a wave of shorter wavelength is produced.

Emission of Radiant Energy

All substances at any temperature above absolute zero emit radiant energy. The average frequency f of the radiant energy is directly proportional to the absolute temperature T of the emitter:

$$f \sim T$$

The Sun has a very high temperature and therefore emits radiant energy at a high frequency—high enough to stimulate our sense of sight. All those frequencies coming from the Sun are called *solar radiation*. The Earth, in comparison, is relatively cool. So, the radiant energy it emits has a frequency lower than that of visible light. The radiation emitted by Earth is in the form of infrared waves. This radiant energy emitted by Earth is called **terrestrial radiation.**

Earth has a naturally warm interior, which is a source of terrestrial radiation. Earth's surface, however, is warmed primarily as it absorbs solar radiation. Warmed by the Sun, the surface then radiates infrared waves. So, most terrestrial radiation has its origins with solar radiation.

We find that Earth's atmosphere permits the passage of incoming high-frequency solar radiation. Yet, at the same time, it blocks the passage of lower-frequency terrestrial radiation. In other words, the atmosphere

READING CHECK

What is terrestrial radiation?

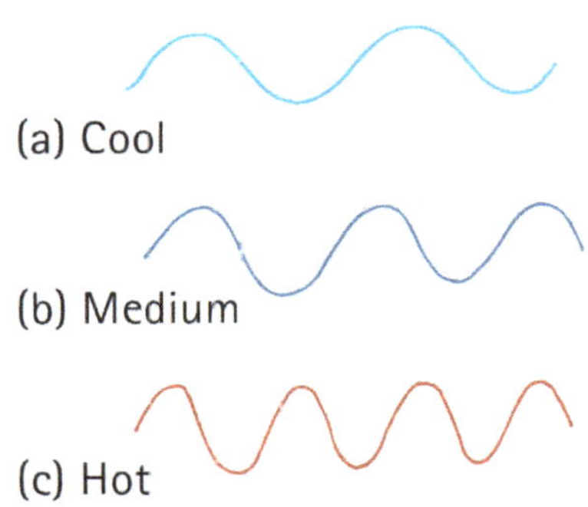

Figure 8.34

(a) A low-temperature (cool) source emits primarily low-frequency, long-wavelength waves. (b) A medium-temperature source emits primarily medium-frequency, long-wavelength waves. (c) A high-temperature source (hot) emits primarily high-frequency, short-wavelength waves.

Figure 8.35

Both the Sun and the Earth emit radiant energy. The Sun's glow is at a high enough frequency to be visible to the eye; Earth's glow consists of longer waves and so is not visible to the eye.

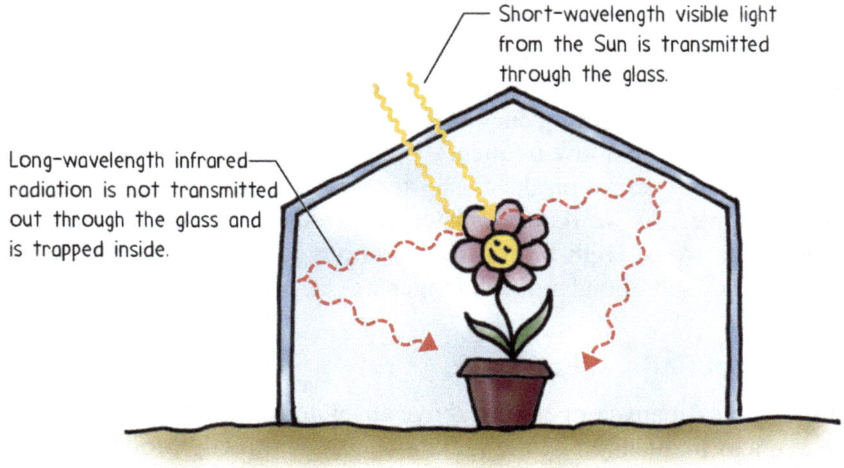

Figure 8.36 >

Glass acts as a one-way valve, letting visible light in and preventing infrared energy from exiting. This is known as the greenhouse effect.

behaves like a one-way valve: The radiation can enter, but it has a hard time going back out. In this way, the atmosphere acts to keep the surface of planet Earth warm. This is called the "greenhouse effect" because a gardener's glass-enclosed greenhouse works by the same principle (**Figure 8.36**). If it weren't for the greenhouse effect, Earth's surface would be a chilly -18°C, which would freeze all the oceans. Hooray for the greenhouse effect!

Human activities, however, are resulting in increasing amounts of carbon dioxide, CO_2, methane, CH_4, and other gases that enhance the greenhouse effect. This, in turn, is leading to higher global temperatures - what climate change is all about, indicated in **Figure 8.37**. At the heart of worldwide increasing temperatures is that greater amounts of greenhouse gases make it all the more difficult for terrestrial radiation to escape into space. As an example of things gone wrong, we need only look to planet Venus where an atmosphere of mostly carbon dioxide (a greenhouse gas) results in surface temperatures in excess of 475°C (900°F). Any oceans that Venus may have once had have long since boiled away.

All objects—you, your teacher, and everything in your surroundings—continually emit radiant energy in a mixture of frequencies (because temperature corresponds to a mixture of molecular kinetic energies). Objects of everyday temperatures mostly emit low frequency infrared waves. When the higher-frequency infrared waves are absorbed by your skin, you feel the sensation of heat. So, it is common to refer to infrared radiation as heat radiation.

Heat radiation underlies infrared thermometers. You simply point the thermometer at something whose temperature you want, press a button, and a digital temperature reading appears. The radiation emitted by the object in question provides the reading. Typical classroom infrared thermometers operate in the range of about -30°C to 200°C.

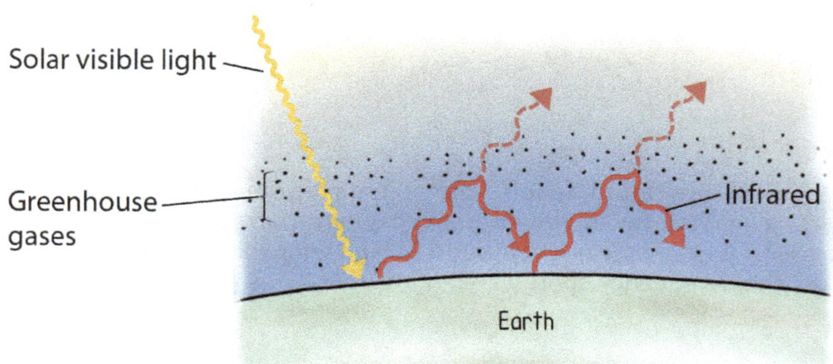

Figure 8.37 >

The greenhouse effect in Earth's atmosphere. Visible light from the Sun is absorbed by the ground, which then emits infrared radiation. Carbon dioxide, water vapor, and other greenhouse gases in the atmosphere reflect most of that heat back downward.

Common hotter sources that give the sensation of heat are the burning embers in a fireplace, a lamp filament, and the Sun. All of these emit both infrared radiation and visible light. When this radiant energy falls on other objects, it is partly reflected and partly absorbed. The part that is absorbed increases the thermal energy of the objects.

Figure 8.38

An infrared thermometer measures the infrared radiant energy emitted by a body and converts it to temperature.

CONCEPT **CHECK**

Which of the following do not give off radiant energy? (a) The Sun; (b) Lava from a volcano; (c) Red-hot coals; (d) A book.

CHECK YOUR ANSWER

Did you answer (d), the book? Sorry, wrong answer. The book, like the other things listed, has temperature—though not as high as the others listed. By the rule $f \sim T$, it therefore emits radiation. Since its temperature is low, the frequency of radiation is also low. Everything with any temperature above absolute zero emits electromagnetic radiation. That's right—everything!

Absorption of Radiant Energy

If everything is emitting energy, why doesn't everything finally run out of it? The answer is, all things also *absorb* energy. Good emitters of radiant energy are also good absorbers; and poor emitters are poor absorbers. For example, a radio antenna constructed to be a strong emitter of radio waves is also, by its very design, a strong receiver (absorber) of them. The same holds true with the antennas for cell phones. Any poorly designed transmitting antenna is also a weak receiver. An object that absorbs radiant energy looks dark. If it absorbs all the radiant energy on it, it looks perfectly black.

Every surface, hot or cold, both absorbs and emits radiant energy. If the surface is hotter than its surroundings the surface will be a net emitter and will cool. If it's colder than its surroundings, it will be a net absorber and will become warmer.

READING CHECK

Does the surface of a substance absorb, or does it emit radiant energy?

Reflection of Radiant Energy

Radiant energy can be reflected. There are no perfect reflectors, so some absorption always occurs. When you look at the open ends of pipes in a stack, the holes appear black. Look at open doorways or windows of distant houses in the daytime, and they, too, look black. Openings appear black because the light that enters them is reflected back and forth on the inside walls many times and is partly absorbed at each reflection. As a result, very little or none of the light returns back out the opening (**Figure 8.39**).

The pupil of your eye is another example. The pupil is a hole in your iris that allows light to enter with no reflection. That's why it appears black. (An exception occurs in flash photography when pupils appear red or pink, which occurs when very bright light is reflected off the eye's inner pink surface and back through the pupil.)

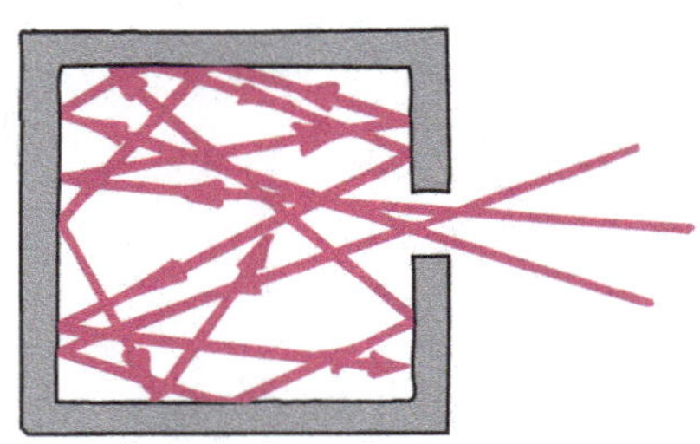

Figure 8.39

Radiation that enters the opening has little chance of leaving because most of it is absorbed. For this reason, the opening to any cavity looks black to us.

Figure 8.40 >

The hole looks perfectly black and indicates a black interior, when in fact the interior has been painted a bright white.

CONCEPT **CHECK**

Which would be more effective in heating the air in a room, a heating radiator painted black, or silver?

CHECK YOUR ANSWER

Interestingly, the color of paint is a small factor, so either color can be used. That's because radiators do very little heating by radiation. Their hot surfaces warm surrounding air by conduction, the warmed air rises, and warmed convection currents heat the room. (A better name for this type of heater would be a convector.) Now if you're interested in optimum efficiency, a silver-painted radiator will radiate less, become and remain hotter, and do a better job of heating the air.

Concept Review

Chapter 8

Summary of Terms

Absolute zero The theoretical temperature at which the atoms and molecules that make up a substance possess no kinetic energy.

Conduction The transfer of internal energy by molecular and electronic collisions within a substance (especially a solid).

Convection The transfer of internal energy in a gas or liquid by means of currents in the heated fluid. The fluid flows, carrying energy with it.

Entropy A measure of the natural spreading of energy.

Heat The thermal energy that flows from a substance of higher temperature to a substance of lower temperature, commonly measured in calories or joules.

Insulator A material that is a poor conductor of heat.

Radiation The transfer of energy by means of electromagnetic waves.

Second law of thermodynamics Any natural process that happens with no outside assistance results in a net dispersal of energy.

Specific heat capacity The quantity of heat required to raise the temperature per unit mass of a substance by 1 degree Celsius.

Temperature A measure of the hotness or coldness of substances, related to the average kinetic energy per molecule in a substance, measured in degrees Celsius, or in degrees Fahrenheit, or in kelvins.

Terrestrial radiation The radiant energy emitted by Earth.

Thermal energy The total energy (kinetic plus potential) of the submicroscopic particles that make up a substance.

Thermal expansion The expansion of a substance due to increased molecular motion in that substance.

Review Questions

1. Why does a penny become warmer when struck by a hammer?
2. What are the temperatures for freezing water on the Celsius and Fahrenheit scales? For boiling water at sea level?
3. What is meant by the statement, "a thermometer measures its own temperature"?
4. What is the temperature of melting ice on the Kelvin scale? Of boiling water at atmospheric pressure?
5. When you touch a cold surface, does "coldness" travel from the surface to your hand or does thermal energy travel from your hand to the cold surface? Explain.
6. Distinguish between temperature and heat.
7. Is cold the opposite of thermal energy or the lack of it?
8. Distinguish between a joule and a calorie.
9. Does a substance that heats up quickly have a high or a low specific heat capacity?
10. Why is the West Coast of the United States warmer in winter than the East Coast?
11. Why does a bimetallic strip bend with changes in temperature?
12. When the temperature of ice-cold water is increased slightly, does it undergo a net expansion or net contraction?
13. At what temperature do the combined effects of contraction and expansion produce the smallest volume for water?
14. What is the explanation for a barefoot firewalker being able to walk safely on red-hot wooden coals?

15. Does a good insulator prevent heat from getting through it, or does it simply slow its passage?

16. What happens to the temperature of air when it expands?

17. Why does the direction of coastal winds change from day to night?

18. How does the frequency of radiant energy relate to the absolute temperature of the radiating source?

19. What is terrestrial radiation? How does it differ from solar radiation?

20. Is energy added, or is it released, when boiling water is changed to steam?

Answers to Odd-Numbered Questions

1. Atoms are set into motion.

3. Both the thermometer and whatever it measures reach a common temperature—this is thermal equilibrium. In this sense, a thermometer measures its own temperature.

5. Thermal energy, not "coldness," can travel. So, when your hand touches a cold surface, thermal energy travels from hand to the cold surface.

7. Cold is the lack of thermal energy (like black is the absence of light).

9. Low.

11. One side changes length more than the other, bending the strip, which can trip a switch and activate a thermal device.

13. 4°C.

15. Insulation slows heat transfer, not stop it.

17. Reverses at night from daytime direction.

19. Terrestrial radiation comes from Earth; solar from the Sun. Both are electromagnetic waves. Earth's waves are lower frequency than those of the Sun.

Chapter 9

Electricity

THE MAIN IDEA

The electrical nature of atoms allows electrical energy to flow through them.

Why do we sometimes get a shock when we scuff our shoes on the rug? Why does a balloon rubbed on our hair stick to a wall? What causes shock—electric current or electric voltage? And what's the difference between current and voltage? What's the difference between direct current and alternating current? We'll discuss the answers to these questions and more in this chapter.

Electricity is a part of just about everything around us. It's in the lightning from the sky, what holds atoms together, and what charges your smartphone and lights up the pixels on its viewing screen. Harnessing electricity has enormously changed the world. It's worth serious study. Let's begin with the concept of electric charge.

9.1 Electric Charge

When you rub an inflated rubber balloon on your hair, you charge the balloon. In charging it, you scrape *electrons* off the atoms of your hair onto the balloon. The rubber has more "grab" on electrons than your hair does. If you rub your hair with two balloons, you'll find that when you bring these two balloons close to each other they repel.

The first rule of electricity is: *Like charges repel one another*. For historical reasons we say that the charge on an electron is negative. Since there are extra electrons on the balloons rubbed with hair, the balloons are

Quick Activity

Lighting a Light Bulb?

Try to light a small light bulb with just a battery and two pieces of wire. Now try to get the same result with different arrangements of the wires, battery, and light bulb. What conditions are necessary in order for the bulb to light?

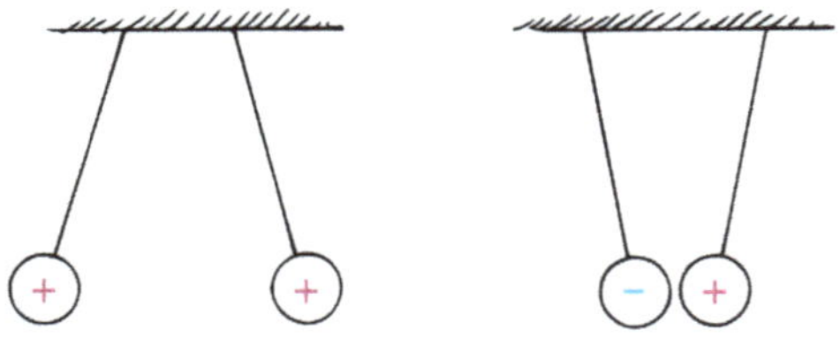

^ Figure 9.1

Like charges repel. Unlike charges attract.

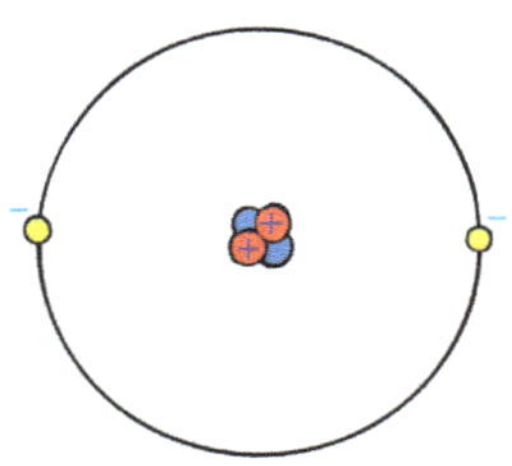

^ Figure 9.2

Model of a helium atom. The atomic nucleus is made up of two protons and two neutrons. The positively charged protons attract two negative electrons. What is the net charge of this atom?

Figure 9.3 >

Electrons are transferred from the fur to the rod. The rod is then negatively charged. Is the fur charged? How much compared to the rod? Positively or negatively?

negatively charged. Experience shows us that a pair of negatively-charged balloons repel each other. Likewise, any two objects repel each other when they are electrically charged in the same way. This can be summarized as follows:

Rule 1: Like charges repel one another.

Consider that kitchen plastic wrap is even better at grabbing electrons than is a rubber balloon. If you charge a balloon by rubbing it with a piece of kitchen plastic wrap, the plastic grabs electrons from the balloon. The balloon then has an opposite charge to the one rubbed on your hair. When you bring the two oppositely charged balloons together, they attract—they'll stick to one another. Here's the second rule of electricity:

Rule 2: Unlike charges attract one another.

In **Figure 9.1**, we see a pair of like charges repelling, and a pair of unlike charges attracting. In **Figure 9.2**, we see a simple model of an atom. Particles called *protons* in the nucleus carry the positive charge. The protons attract the whirling electrons and hold them in orbit.

Electric charge is a quality of matter. To understand electric charge, we look at a preview of atomic parts, which we'll be exploring further in a later chapter. Here are some basic facts about atoms:

1. Every atom has a positively charged *nucleus* surrounded by negatively charged electrons.
2. The electrons of all atoms are identical. Each electron has the same quantity of negative charge and the same mass.
3. Protons and neutrons make up the nucleus. (The common form of hydrogen has no neutron and is the only exception.) Protons are about 1800 times more massive than electrons, but carry an amount of positive charge equal to the negative charge of electrons. Neutrons have slightly more mass than protons and have no charge.
4. While electrons can be readily lost or gained by an atom, protons and neutrons within the nucleus remain tightly bound.

An atom in its neutral state has the same number of electrons as protons. When an atom loses one or more electrons, it has a positive net charge. When it gains one or more electrons, it has a negative net charge. A charged atom is called an ion. A positive ion has a net positive charge. A negative ion, with one or more extra electrons, has a net negative charge.

It is important to note that when we charge something, no electrons are created or destroyed. They are simply transferred from one material to another. Charge is conserved. In every event, whether on a large-scale or small-scale, the principle of *conservation of charge* has always proved true. No case of the creation or destruction of net electric charge has ever been found. Conservation of charge ranks with conservation of energy and momentum as a significant fundamental principle in physics.

READING CHECK

What does it mean to say that charge is conserved?

CONCEPT CHECK

If you walk across a rug and scuff electrons from your feet, are you negatively or positively charged?

CHECK YOUR ANSWER

You have fewer electrons after you scuff your feet, and so you are positively charged (and the rug is negatively charged).

A build-up of electric charge can be dangerous. Two hundred years ago, young boys called powder monkeys ran below the decks of warships to bring sacks of black gunpowder to the cannons above. It was ship law that this task be done barefoot. Why? Because it was important that no build up of static charge accumulate on the powder on their bodies as they ran to and fro. Bare feet scuffed the decks much less than shoes and assured no charge buildup that might produce an igniting spark and an explosion.

Static charge is a danger in many industries today. Not because of explosions but because delicate electronic circuits may be destroyed by static charge. Some sensitive circuit components can be "fried" by static electric sparks. Electronics technicians often wear clothing of special fabrics with ground wires between their sleeves and their socks. Some wear special wrist bands that are connected to a grounded surface to prevent static charge build-up—when moving a chair, for example. The smaller the electronic circuit, the more hazardous are sparks that may short-circuit their elements.

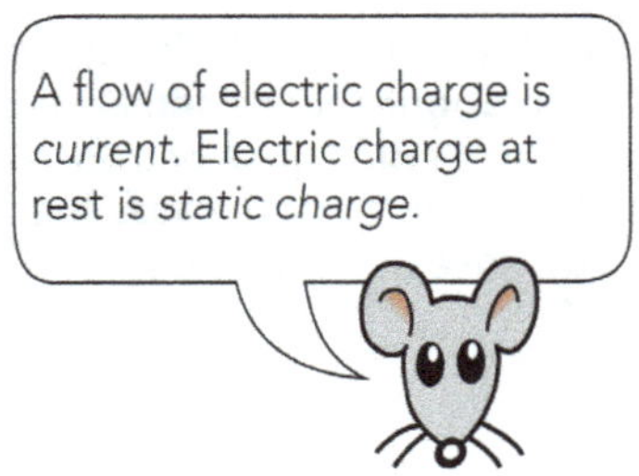

9.2 Coulomb's Law

The electrical force has a pattern much like gravitational force. It depends on the quantity of charge and is inversely proportional to the square of the distance between charged particles. This relationship was discovered by Charles Coulomb in the 18th century and so is called **Coulomb's law**. It states that the force between two charged particles varies directly as the product of their charges and inversely as the square of the separation distance. The force acts along a straight line between the particles. Coulomb's law can be expressed as

$$F = k\frac{q_1 q_2}{r^2}$$

READING CHECK

State Coulomb's law in words.

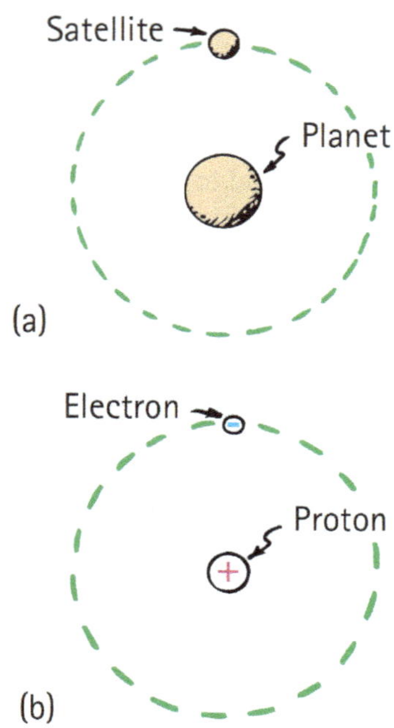

Λ Figure 9.4

(a) A gravitational force holds the satellite in orbit about the planet, and (b) an electrical force holds the electron in orbit about the proton. In both cases the force follows the inverse-square law.

where k is the proportionality constant, q_1 represents the quantity of charge of one particle, q_2 represents the quantity of charge of the other particle, and r is the distance between the charged particles.

The unit of charge is the coulomb, abbreviated C. It turns out that a charge of 1 C is the charge on 6.25 billion billion electrons. This might seem like a great number of electrons, but it represents only the amount of charge that passes through a common household LED in little more than a second.

The proportionality constant k in Coulomb's law is similar to G in Newton's law of gravity. Instead of being a very small number like G, k is a very large number, approximately

$$k = 9{,}000{,}000{,}000\ \mathrm{N{\cdot}m^2/C^2}$$

In scientific notation, $k = 9 \times 10^9\ \mathrm{N{\cdot}m^2/C^2}$. The unit $\mathrm{N{\cdot}m^2/C^2}$ is not important to learn here. It simply converts the right-hand side of the equation to the unit of force, the newton (N). What is important is the large magnitude of k. For example, if a pair of like charged particles each carrying a charge of 1 coulomb were 1 meter apart, the force of repulsion between them would be 9 billion newtons. That would be about ten times the weight of a battleship!

Obviously, such amounts of net charge do not usually exist in our everyday environment. If such charges were common, we would see things attracting and repelling quite often. Like charges repel and separate from one another before much charge can build up. The electrical force between even slightly charged objects greatly overwhelms the gravitational force between them.

So, Newton's law of gravitation for masses is similar to Coulomb's law for electrically-charged objects. The most important difference between gravitational and electrical forces is that electrical forces may be either *attractive* or *repulsive*, whereas gravitational forces are only attractive. Coulomb's law underlies the bonding forces within molecules, which is a focus of the science of chemistry.

Charge Polarization

We began this chapter by discussing an inflated balloon rubbed on your hair. If you place the balloon against a wall you'll see that it sticks. This is because the negative charge on the balloon pulls the positive part of atoms in the wall closer to it. The balloon has the effect of "inducing" an opposite charge in the wall. Although the atoms in the wall don't move, their "centers of charge" are moved. The positive part of each atom is attracted toward the balloon and the negative part is repelled. We say the distorted atoms (**Figure 9.5**) are **electrically polarized**.

Figure 9.5 >

(a) The center of the negative "cloud" of electrons coincides with the center of the positive nucleus in an atom. (b) When an external negative charge is brought nearby to the right, as on a charged balloon, the electron cloud is distorted so the centers of negative and positive charge no longer coincide. The atom is electrically polarized.

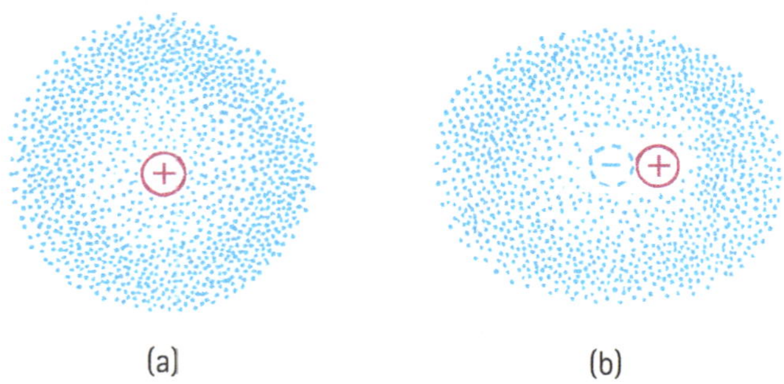

A polarized object has no net charge. Only the distribution of charge in the material is altered. Charge polarization is an important concept in chemistry as it explains how molecules come to be "sticky." Through polarization, water molecules are particularly apt to stick to other water molecules. The stickiness of water molecules can be used to explain many of water's unusual properties, such as its high specific heat and high boiling point.

READING CHECK

Does an electrically polarized object have a net charge, or is its distribution of charge altered?

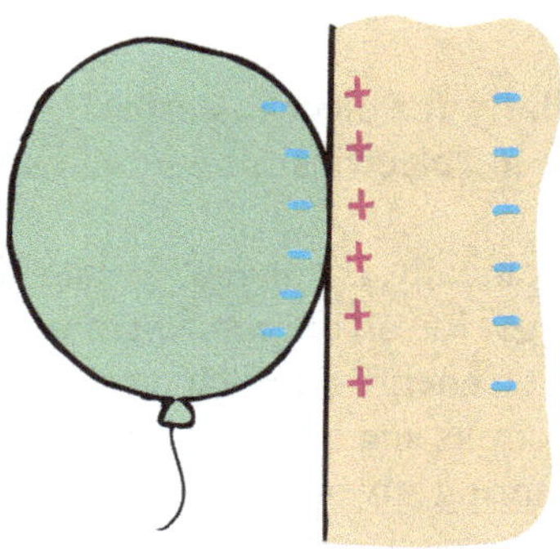

Figure 9.6

Because the negatively charged balloon polarizes atoms in the wall and creates a positively charged surface, the balloon sticks to the wall.

9.3 Electric Current from Voltage

Recall from Chapter 8 that loose electrons in metals are responsible for the good heat conduction in metals. The same is true for electrical conduction. Loosely-held outer electrons in the atoms of a metal are called *conduction electrons*. Protons don't move about in a metal because they are bound inside the atomic nucleus. Conduction electrons, however, can freely migrate through a metal.

READING CHECK

Why do electrons in a wire, rather than protons, make up the flowing charge that gives us electric current?

Electrons flow in a way similar to water flow. Just as water current is the flow of H_2O molecules, **electric current** is the flow of electrons. But there are differences between water flow and electron flow. If you buy a water pipe at a hardware store, the clerk doesn't sell you the water to go with it. You provide that yourself. By contrast, when you buy "an electron pipe," an electric wire, you get the electrons too. Every bit of matter, wires included, contains enormous numbers of electrons that swarm about in random directions. When they are set in motion in one direction, a net direction, we have an electric current.

The rate of electrical flow is measured in *amperes* (abbreviation A). An **ampere** is the rate of flow of 1 coulomb of charge per second. (That's a flow of 6.25 billion billion electrons per second.) In a wire that carries 5 amperes, 5 coulombs of charge pass any cross section in the wire each second. In a wire that carries 10 amperes, twice as many coulombs pass any cross section each second.

When water flows in a pipe, there is more pressure at one end than the other. There must be a pressure difference to keep the water flowing. Also recall from our study of heat in Chapter 8 that heat flow depends on a temperature difference. Heat flows from zones of high temperature to

Figure 9.7 >

(a) Water flows from the reservoir of higher pressure to the reservoir of lower pressure. The flow ceases when the difference in pressure ceases. (b) Water continues to flow because a difference in pressure is maintained with the pump.

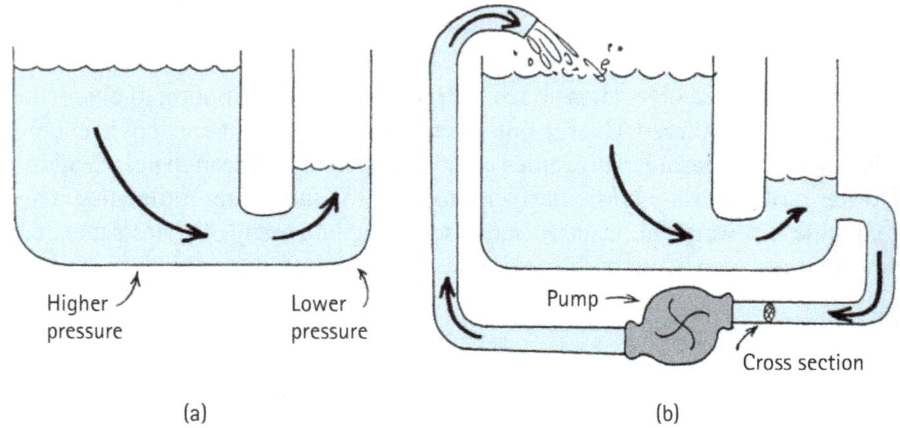

low temperature. Similarly for electric current. Electrons flow in a wire only when a difference in "electrical pressure" exists. The name for this pressure is *voltage*.

So, what, more specifically, is voltage? Voltage is directly proportional to *electric potential energy*. Recall the concept of potential energy discussed in Chapter 5. Things have energy due to their positions (**Figure 9.9**).

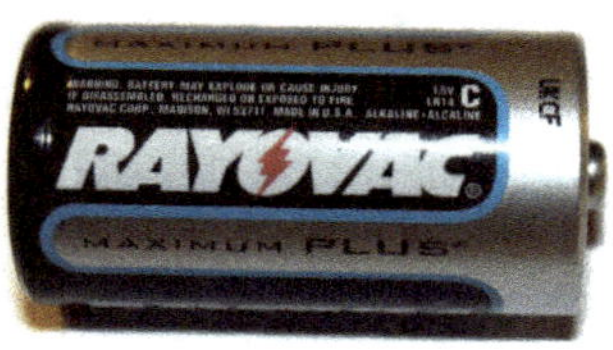

Λ Figure 9.8

Each coulomb of charge made to flow with this 1.5 V battery is energized with 1.5 J.

Likewise, with electrons and other charged particles. In **Figure 9.10a**, we see that pushing a spring gives it potential energy, because the compressed spring can do work on something when released (apply a force over a distance). We also see that pushing a charged particle toward another charge can increase its electric potential energy (**Figure 9.10b**).

How much electric potential energy depends upon how much charge. If you pushed 2 coulombs of charge to a particular position then you would have twice the energy compared to pushing only 1 coulomb of charge. This is analogous to lifting weights. If you lift twice the mass to the same height, then you would have twice the gravitational potential energy. This is intuitive.

Figure 9.9 >

(a) Gravitational potential energy converts to kinetic energy. (b) Similarly, electric potential energy converts to kinetic energy of electrical charge.

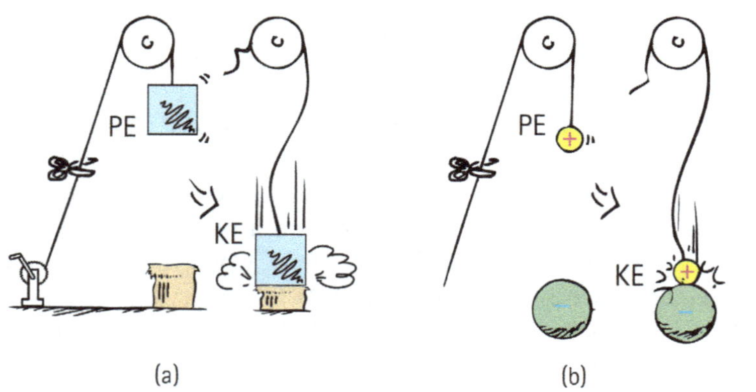

Figure 9.10 >

(a) The spring has more mechanical PE when compressed. (b) The charged particle similarly has more electrical PE when pushed closer to the charged sphere. In both cases the increased PE is the result of work input.

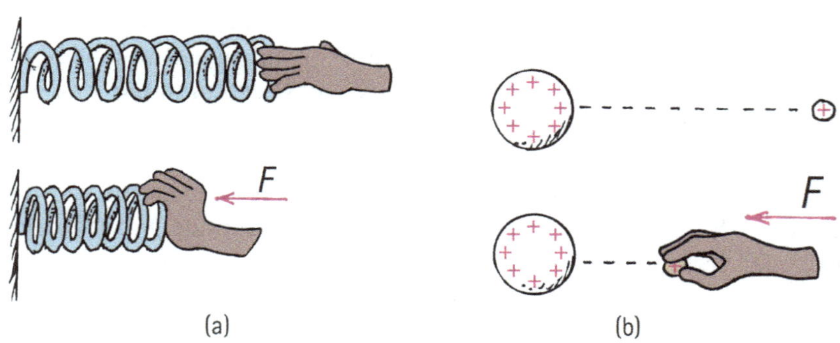

Less intuitive is the idea of electric potential. Take the phrase "electric potential energy" and remove the word "energy." What you end up with is a very related but distinct concept called *electric potential*. We define **electric potential** as the ratio of energy per charge. While electric potential energy depends upon the amount of charge, *electric potential* depends only upon the *position* of that charge. For example, you might have 2 coulombs of charge at a certain position and this gives you 10 joules of energy. Or you might have 20 coulombs in that same position, which will give you 100 joules of energy. In either case, the ratio of energy to charge (electric potential) is the same because the position is the same, as shown in **Figure 9.11**.

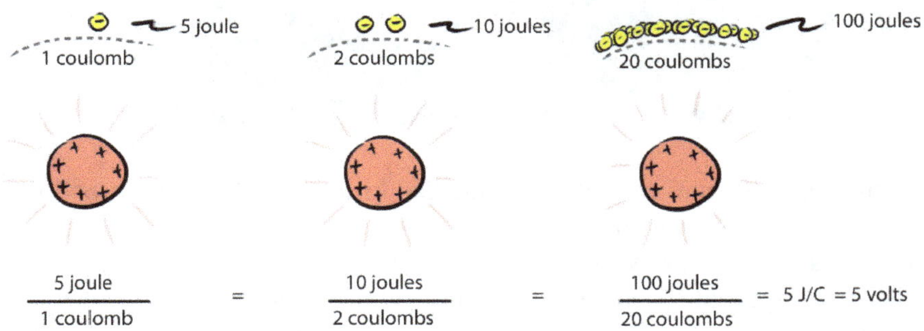

< Figure 9.11

At a certain *position* within an electric field one coulomb might hold 5 joules, as shown on the far left. Double the coulombs at this position and the number of joules will also double. Add 20 times the number of couloumbs and the joules of energy will also be 20 times. In all cases the electric potential is 5 volts.

We measure electric potential in units of joules per coulomb, which is the amount of energy per charge. By definition the joule per coulomb is also known as the **volt**, in honor of the Italian physicist Alessandro Volta (1745-1827) who invented one of the first electric batteries.

Electric current flows in a wire when and only when there is a difference in voltage across the ends of the wire. To achieve a steady current of charge (current) you'll need a pumping device to provide a sustained difference in voltage. This is the role of chemical batteries or electric generators, which work as electron pumps—electrons that have fallen to the lower position (lower voltage) are pumped back up to the higher position (higher voltage).

READING CHECK

When does an electric current flow?

A common automobile battery provides a voltage difference of 12 volts. When each of its terminals are attached to ends of a wire, there is a voltage difference of 12 volts across the wire. That means 12 joules of energy are supplied to each coulomb of charge flowing in the wire. The wire is usually a part of an electric circuit.

There is often some confusion about charge flowing *through* a circuit and voltage *across* a circuit. We can see the difference by thinking of a long pipe filled with water. Water flows *through* the pipe if there is a difference in pressure across, or between, its ends; it flows from the high-pressure to the low-pressure end. Only the water flows, not the pressure. Similarly, electrons flow because of a difference in electrical pressure (voltage difference). Electrons flow *through* a circuit because of an applied voltage across the circuit. Voltage doesn't flow through a circuit—it doesn't go anywhere. They are the electrons that flow. Voltage produces current (if there is a complete circuit).

How might a voltage be created? A battery produces a voltage by chemical reactions between electrodes inside the battery. For a lead battery, one electrode is lead oxide, and the other what's called sponge lead. In a

Figure 9.12 >

An unusual source of voltage. The electric potential between the head and tail of the electric eel (Electrophorus electricus) can be up to 600 V.

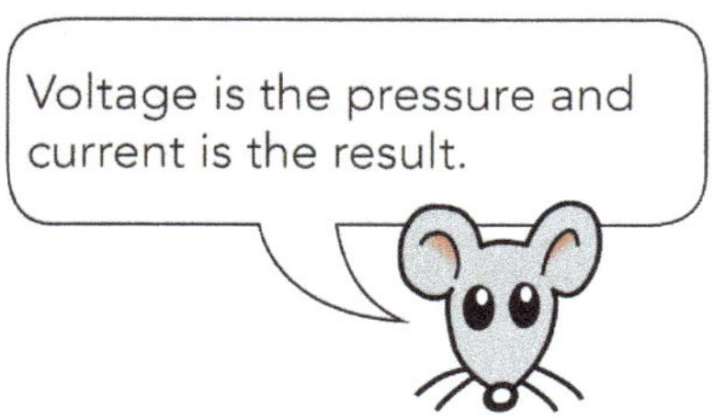

lithium battery, one terminal is connected to intercalated lithium and the other typically to graphite. For both, a chemical reaction between these materials sustains a flow of electrons. A generator produces a voltage difference by spinning loops of wires within a magnetic field (or vice versa). The energy source is whatever gets the loops spinning, such as water rolling downhill, steam rising upward, or winds blowing sideways. Photovoltaic cells produce a voltage difference directly from the energy of sunlight. In the world of biology, our nervous system operates from voltage differences created from the energy of the food we eat. The electric eel takes voltage creation to a biological extreme as shown in **Figure 9.12**.

9.4 Electrical Resistance

A voltage source, such as a battery or generator, can move electrons in a circuit. How much current there is depends on two things—the voltage, and the **electrical resistance** of the circuit. Just as narrow pipes resist water flow more than wide pipes, narrow wires resist electrical current more than wider wires. And length contributes to resistance also. Just as long pipes have more resistance than short ones, long wires offer more electrical resistance. And most important is the kind of material. Copper has a low electrical resistance, while a strip of rubber has an enormous resistance. So we see that the electrical resistance of a wire depends on the kind of material used in the wire and the thickness and length of the wire.

READING CHECK

Cite three factors that determine the electrical resistance of a wire.

Temperature also affects electrical resistance. The greater the jostling of atoms within a conductor (in other words, the higher the temperature), the greater resistance a conductor has. The resistance of some materials reaches zero at very low temperatures. These are *superconductors*.

Electrical resistance is measured in units called *ohms*. The Greek letter *omega*, Ω, is commonly used as the symbol for the ohm. This unit is named after Georg Simon Ohm, a German physicist who in 1826 discovered a simple and very important relationship among voltage, current, and resistance.

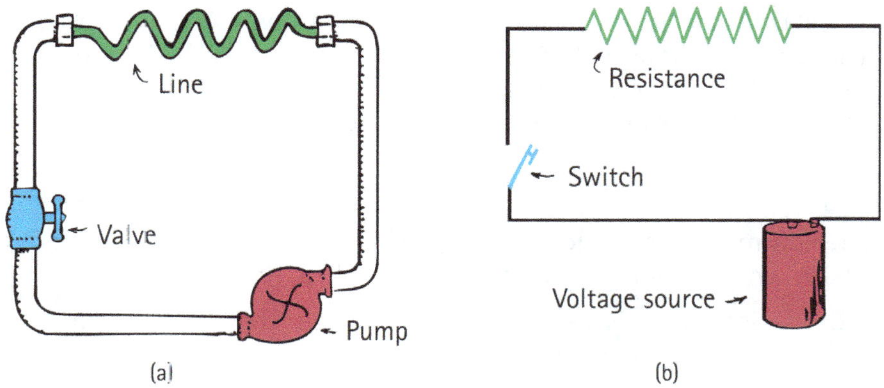

Figure 9.13

(a) In the hydraulic circuit the narrow pipe (green) offers resistance to water flow. (b) In the electric circuit a light bulb or other device (shown by the zigzag symbol for resistance) offers resistance to electron flow.

READING CHECK

What three concepts are tied together in Ohm's law?

Ohm's Law

The relationship between current, voltage, and resistance is **Ohm's law**. Ohm discovered that the amount of current in a circuit is directly proportional to the voltage across the circuit and inversely proportional to the resistance of the circuit:

$$\textbf{Current} = \frac{\textbf{voltage}}{\textbf{resistance}}$$

Or, in units form:

$$\textbf{Amperes} = \frac{\textbf{volts}}{\textbf{ohms}}$$

Ohm's law tells us that 1 volt across a circuit with a resistance of 1 ohm produces a current of 1 ampere. If there are 12 volts across the same circuit, the current is 12 amperes. So, for a given circuit of constant resistance, current and voltage are proportional to each other. This means we get twice the current for twice the voltage. The greater the voltage, the greater the current. But if the resistance is doubled for a circuit, the current is reduced to half. The higher the resistance, the lower the current. Ohm's law makes good sense.

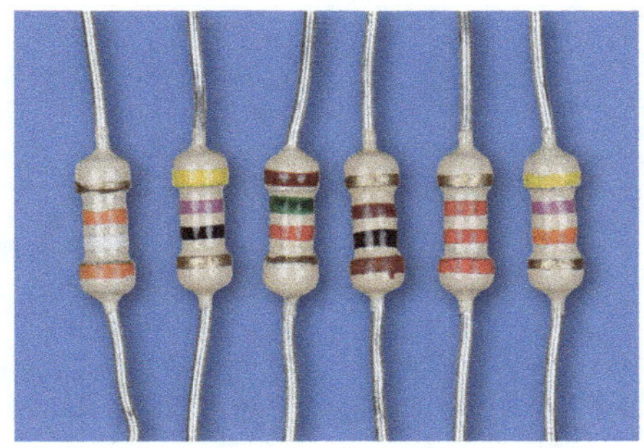

Figure 9.14

Resistors. The symbol of resistance in an electric circuit is Ω

Figure 9.15

More water flows through a thick hose than through a thin one connected to a city's water system (same water pressure). Likewise for electric current in thick and thin wires connected across the same voltage.

The resistance of a typical lamp cord is much less than 1 ohm, and a typical light bulb has a resistance of more than 100 ohms. An iron or electric toaster has a resistance of 15 to 20 ohms. The current inside these and all other electrical devices is regulated by circuit elements called *resistors*, whose resistance may be a few ohms or millions of ohms.

FOR YOUR INFORMATION

If we use the letter *V* for voltage, *I* for current, and *R* for resistance, Ohm's law is expressed $I = V/R$. It can also be written, $V = IR$, or $R = V/I$, so if any two variables are known, the third can be found. Units are abbreviated V for volts, A for amperes, and Ω for ohms.

CONCEPT CHECK

1. How much current flows through a light bulb with a resistance of 60 W when the voltage across the light bulb is 12 V?

2. What is the resistance of an electric frying pan that draws a current of 12 A when connected to a 120-V circuit?

CHECK YOUR ANSWER

1. This is calculated from Ohm's law:

$$\text{Current} = \frac{\text{voltage}}{\text{resistance}} \quad \text{so} \quad \frac{12\ \text{V}}{60\ \Omega} = 0.2\ \text{A}$$

2. Rearrange Ohm's law to read:

$$\text{Resistance} = \frac{\text{voltage}}{\text{current}} = \frac{120\ \text{V}}{12\ \text{A}} = 10\ \Omega$$

In ordinary conductors, such as household wiring, the moving electrons that make up current often collide with atomic nuclei in the wire. The colliding electrons transfer their kinetic energy to the wire and the wire heats up. Energy is wasted. However, experiments show that certain metals lose all their electrical resistance when placed in a bath of liquid helium at a temperature of 4 K. All of it! The electrons in these extremely cold conductors travel pathways that avoid collisions. Hence the electrons can flow indefinitely. The materials that work this way are called *superconductors*. **Superconductors** have zero electrical resistance to the flow of charge. In superconductivity, no current is lost and no heat is generated.

Various ceramic oxides are superconducting at temperatures above 100 K. Steady currents have been observed to persist for years in some superconductors without any apparent loss in energy. There is presently enormous interest in the technology of superconductors. Imagine the energy-saving potential of superconducting devices! Explanations of superconductivity involve the wave nature of matter (quantum mechanics) and are being vigorously researched.

9.5 Electrical Shock

Which causes electric shock in the human body—current or voltage? The damaging effects of shock result from current through the body. From Ohm's law, we see current in a body depends on the voltage applied and also on the body's electrical resistance. A person's resistance ranges from about 100 ohms if the body is soaked with salt water to about 500,000 ohms if the skin is very dry. If you touch the two electrodes of a battery with dry fingers, you make up a circuit with a resistance of about 100,000 ohms. You usually cannot feel 12 volts. Even 24 volts just barely tingles. If your skin is moist, 24 volts can be quite uncomfortable. Table 9.1 describes the effects of different amounts of current on the human body.

READING CHECK

Upon what two factors does electric current depend when getting a shock?

Table 9.1 Effect of Electric Current on the Body

Current (A)	Effect
0.001	Can be felt
0.005	Is painful
0.010	Causes involuntary muscle contractions (spasms)
0.015	Causes loss of muscle control
0.070	Goes through the heart; serious damage, probably fatal for if current lasts for more than 1 s

CONCEPT CHECK

1. At 100,000 ohms, how much current will flow through your body if you touch the terminals of a 12-V battery?
2. If your skin is very moist, so that your resistance is only 1000 W, and you touch the terminals of a 12-V battery, how much current do you receive?

CHECK YOUR ANSWERS

1. $\frac{12\ \text{V}}{100{,}000\ \Omega} = 0.00012\ \text{A}$

2. $\frac{12\ \text{V}}{1000\ \Omega} = 0.012\ \text{A}$ Ouch!

To receive a shock, there must be a difference in voltage between one part of your body and another part. Electron flow will pass along the path of least electrical resistance connecting these two points. Suppose you fell from a bridge and grabbed onto a high-voltage power line, halting your fall. If you touch nothing else of different voltage, you receive no shock. Even if the wire is a few thousand volts and even if you hang by two hands, no significant electron flow will occur between your hands. This is because there is no voltage difference between your hands. If, however, you reach over with one hand and grab onto a wire of different voltage...zap! We have all seen birds perched on high-voltage wires. Every part of their bodies is at the same high voltage as the wire, and so they have no problem. Current only flows when there is a *difference* in voltage.

∧ Figure 9.16

The bird can stand harmlessly on one wire of high voltage, but it had better not reach over and grab a neighboring wire! Why not?

∧ Figure 9.17

The round prong connects the body of the appliance directly to ground (the Earth). Any charge that builds up on an appliance is therefore conducted to the ground—preventing accidental shock.

Most electric plugs and sockets are wired with three connections. The two flat prongs on a plug are for the current-carrying double wire inside the socket. One part is "live" (energized) and the other is neutral. The larger round prong connects to a wire in the electrical system that is grounded—connected directly to the ground (**Figure 9.17**). If the live wire of the plugged-in appliance accidentally comes in contact with the metal surface of the appliance, and you touch the appliance, you could receive a dangerous shock. This won't occur when the appliance casing is grounded via the ground wire. Then the voltage of the appliance is the same voltage as the ground—relatively speaking, zero. You can't get shocked unless there is a voltage *difference*.

CONCEPT CHECK

1. Which most often causes electric shock—current or voltage?
2. What is the source of electrons that produce a shock in your body?
3. What is the source of a simple battery-powered circuit?

CHECK YOUR ANSWERS

1. Electric shock occurs when current is produced in the body, which is caused by an applied voltage. Voltage is the cause; current is the effect.
2. The source of electrons are those already in your body. Like any conductor, the electrons are already there. A voltage across your body will set them in motion.
3. The circuit elements themselves. Just as a water pump moves water in pipes, a battery supplies energy to move electrons already in a circuit.

Many people are killed each year by current from common 120-volt electric circuits. If you touch a faulty 120-volt light fixture with your hand while your feet are on the ground, there may be a 120-volt "electrical pressure" between your hand and the ground. Resistance to current is usually greatest between your feet and the ground, and so the current is usually not enough to do serious harm. But if your feet and the ground are wet, there is a low-resistance electrical path between you and the ground. The 120 volts across this lowered resistance may produce a current greater than your body can withstand.

Pure water is not a good conductor. But the ions normally found in water make it a fair conductor. More dissolved materials, especially small amounts of salt, lower the resistance even more. There is usually a layer of salt left from perspiration on your skin, which when wet lowers your skin resistance to a few hundred ohms or less. Handling electrical devices while taking a bath is a definite no-no.

Injury by electric shock comes in three forms: (1) Burning of tissues by heating, (2) muscle contraction, and (3) disruption of cardiac rhythm. These conditions are caused by excess electric power delivered in critical body regions for too long a time.

Electric shock can upset the nerve center that controls breathing. In rescuing shock victims, the first thing to do is clear them from the electric supply. Use a dry wooden stick or some other nonconductor so that you don't get electrocuted yourself. Then apply artificial respiration, which can help to trigger self-breathing within the victim.

9.6 AC/DC

Electric current may be *dc* or *ac*. By *dc*, we mean **direct current**. Direct current is current made up of electrons that flow in one direction. A battery produces direct current in a circuit because the terminals of the battery always have the same opposite signs. Electrons move from the repelling negative terminal toward the attracting positive terminal, always moving through the circuit in the same direction.

It is interesting to note that the speed of electrons as they drift through a wire is surprisingly slow. This is because electrons continually bump into atoms in the wire. The drift speed of electrons in a typical circuit is much less than one centimeter per second! The electric signal, however, travels at nearly the speed of light. That's the speed at which the electric field in the wire is established. (An electric field exists around charged particles like a gravitational field exists around massive bodies—both fields travel from one point to another at the speed of light.)

READING CHECK

How does electron drift speed differ from the speed of an electric signal?

Alternating current (*ac*) acts as the name implies. Electrons in the circuit flow initially in one direction and then in the opposite direction. This is done by switching the sign at the terminals of the power-station generator. The net drift speed of electrons in an *ac* circuit is zero. They vibrate back and forth about fixed positions and travel nowhere!

How can an electron traveling nowhere transmit energy from one place to another? As we'll be exploring later, sound results from the vibration of molecules. A vibrating tuning fork, for example, sets all the surrounding air molecules into the same vibration, including those adjacent to your ear drum. There are no air molecules that travel from the tuning fork to your ear. What travels from the tuning fork to your ear is acoustical energy transmitted from one vibrating air molecule to the next. Likewise, what travels down a wire is electrical energy transmitted by one vibrating electron to the next. In both cases, it's the energy that moves from point A to point B. The carriers of that energy (vibrating electrons or air molecules) actually remain in place.

Nearly all commercial *ac* circuits involve currents that alternate back and forth at a frequency of 60 cycles per second. This is 60-hertz current (a cycle per second is called a hertz).

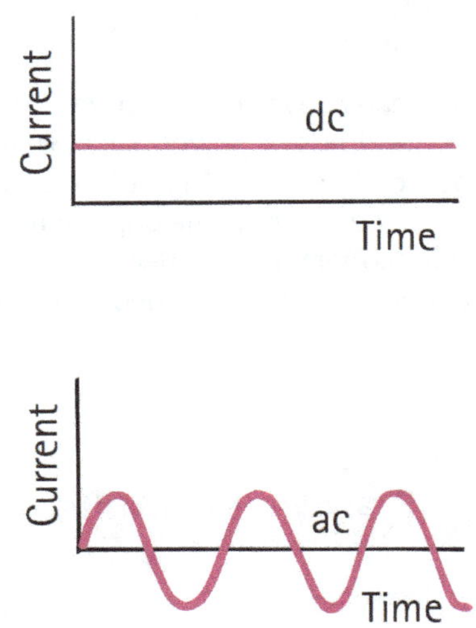

Figure 9.18

Time graphs of *dc* and *ac*.

Electric Power—The Rate of Doing Work

Moving charges in an electric current can do work. They can heat a circuit or turn a motor. Recall from Chapter 5 how we defined *power* as the rate of using energy. Power is the energy transformed divided by the elapsed time. Electrical energy may be transformed to mechanical energy (as in a motor), to light (as in a lamp), to thermal energy (as in a heater), or to other forms. In electrical terms, **power** is equal to current multiplied by voltage.

$$\textbf{Power} = \textbf{current} \times \textbf{voltage}$$

When current is in amperes and voltage is in volts, then power is expressed in watts. So, in units form:

$$\textbf{Watts} = \textbf{amperes} \times \textbf{volts.}$$

READING CHECK

Mechanical power is the rate at which work is done, energy per charge. How is electrical power expressed?

ʌ **Figure 9.19**

The power and voltage on this classic light bulb read "100 W 120 V." How many amperes will flow through the bulb?

If a lamp rated at 120 watts operates on a 120-volt line, it draws a current of 1 ampere (120 W = 1 A × 120 V). A 60-watt light bulb draws 1/2 ampere on a 120-volt line. **Figure 9.19** shows the simpler innards of an incandescent bulb that was common before LEDs.

CONCEPT **CHECK**

What power is needed to operate an electrical radio if it draws a current of 0.05 amperes from your household circuit?

CHECK YOUR ANSWER

Power = current × voltage = 0.05 A × 120 V = 6 W.

FOR YOUR INFORMATION

In the early days of electricity, high voltages burned out electric light filaments, and so low voltages were more practical. The hundreds of power plants built in the U.S. prior to 1900 adopted 110 volts (or 115 or 120 volts) as their standard. Tradition has it that 110 volts was agreed upon because it made bulbs of the day glow as brightly as a gas lamp. By the time electricity became popular in Europe, engineers had figured out how to make light bulbs that would not burn out so fast at higher voltages. Because power transmission is more efficient at higher voltages, Europe adopted 220 volts as their standard. The U.S. remained with 110 volts (today officially 120 volts) because of the installed base of 110-volt equipment.

9.7 Series Circuits

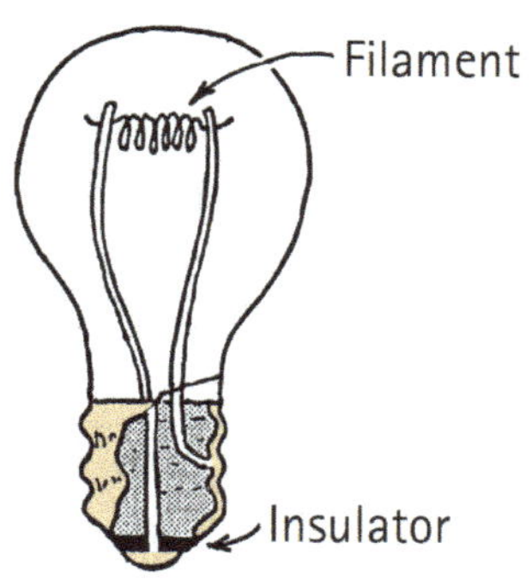

ʌ **Figure 9.20**

The conduction electrons that surge back and forth in the filament of a classic incandescent light bulb do not come from the voltage source. They are in the filament to begin with. The voltage source simply provides them with surges of energy. Similarly, for modern LED bulbs.

Any path along which electrons can flow is a *circuit*. For a steady current there must be a complete circuit with no gaps. A gap is usually provided by an electric switch that can be opened or closed. With that switch there is control to either stop or allow energy flow.

Most circuits contain more than one device that receives electric energy. It may be several lamps, for example. These devices are commonly connected in one of two ways, *series* or *parallel*. When connected in series, the devices and the wires connecting them form a single pathway for electron flow between the terminals of the battery, generator, or wall socket. When connected in parallel, the devices and wires connecting them form branches, each providing separate paths for electron flow. Series and parallel connections each have their own distinctive characteristics.

A simple **series circuit** is shown in Figure 9.21, where three light bulbs are connected in series with a battery. When the switch is closed, the same current exists almost immediately in all three light bulbs. The current does not "pile up" in any light bulb but flows through each light bulb. Remember that the electrons in a circuit are inside the conductors themselves. They flow in a direction from the negative terminal of the battery, through each light bulb, then toward the positive terminal of the battery. Inside the battery the electrons move in a direction toward the negative terminal (the energy for

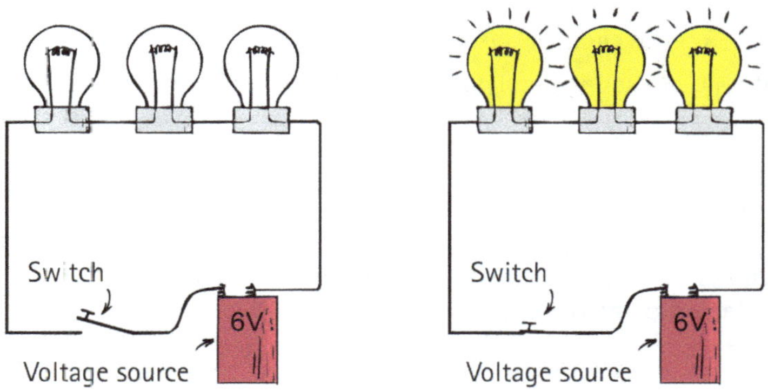

< Figure 9.21

A simple series circuit. The 6-V battery provides 2 V across each light bulb.

this to happen is supplied by chemical reactions). This means the amount of current passing through the battery is the same that passes through the rest of the circuit. This is the only path of the electrons through the circuit. If a break occurs anywhere in the path, the flow of electrons stops. Burning out of one of the light bulb filaments or simply opening the switch causes such a break.

The circuit shown in **Figure 9.21** illustrates the following important characteristics of series connections:

1. Electric current has but one pathway. This means that the current is the same in every part of the circuit.
2. This current is resisted by the resistance of the first device, the resistance of the second, and that of the third also. So, the total resistance to current in the circuit is the sum of the individual resistances along the circuit path (assuming the resistance of the connecting wires is negligible).
3. The current in the circuit is numerically equal to the voltage supplied by the source divided by the total resistance of the circuit. This is in accord with Ohm's law.
4. The total voltage established across a series circuit divides among the electrical devices in the circuit so that the sum of the "voltage drops" across each device is equal to the total voltage supplied by the source. (This follows from the fact that the amount of energy given to the total current is equal to the sum of energies given to each device.)
5. The voltage drop across each device is proportional to its resistance. (This follows from the fact that more energy is wasted as heat when a current passes through a high-resistance device than through a low-resistance device.)

It is easy to see the main disadvantage of a series circuit. If one device fails, current in the entire circuit ceases. Some inexpensive Christmas tree lights are connected in series. When one bulb burns out, it's fun and games (or frustration) trying to locate which bulb to replace.

Most circuits are wired so that it is possible to operate several electrical devices, each independently of the others. In your home, for example, a light switch can be turned on or off without affecting other electrical appliances on the same circuit. This is because these devices are connected not in series, but in parallel, which we explore next.

READING CHECK

How does the amount of current in a battery compare with the current in the rest of the circuit?

CONCEPT CHECK

1. What happens to the current in other light bulbs in a series circuit if one bulb burns out?
2. What happens to the light intensity of each light bulb in a series circuit when more bulbs are added to the circuit?

CHECK YOUR ANSWERS

1. If one bulb burns out, the circuit path is broken and current ceases. All bulbs go out.
2. The addition of more bulbs in a series circuit results in more circuit resistance. This lowers the current in the circuit and therefore in each bulb. That's why the bulbs dim. The same amount of energy is divided among more light bulbs, which means less energy per bulb. In other words, less voltage drop across each bulb.

9.8 Parallel Circuits

A simple parallel circuit is shown in **Figure 9.22**. Three light bulbs are connected to the same two points A and B. Electrical devices connected to the same two points of an electrical circuit are connected in parallel. Electrons leaving the negative battery terminal need to travel through only one light bulb filament before returning to the positive terminal of the battery. In this case, current branches into three separate pathways from A to B. How much current is in each branch depends of the resistance in that branch. A break in any one path does not interrupt the flow of charge in the other paths. The current in one branch isn't affected by the current in any other branch. Each device operates independently of the others.

READING CHECK

In a three-bulb series circuit, current in one light bulb is the same in the other light bulb. How does current differ in the branches of a three-bulb parallel circuit?

The circuit shown in **Figure 9.22** illustrates the following major characteristics of parallel connections:

1. Each device connects the same two points A and B of the circuit. The voltage is therefore the same across each device.
2. The total current in the circuit divides among the parallel branches. Since the voltage across each branch is the same, the amount of current in each branch is inversely proportional to the resistance of the branch.
3. The total current in the circuit equals the sum of the currents in its parallel branches.
4. As the number of parallel branches is increased, the overall resistance of the circuit is lowered (just as more check-out cashiers at a supermarket lowers people-flow resistance). With each added parallel path, the overall circuit resistance is lowered. This means the overall resistance of the circuit is less than the resistance of any one of the branches.

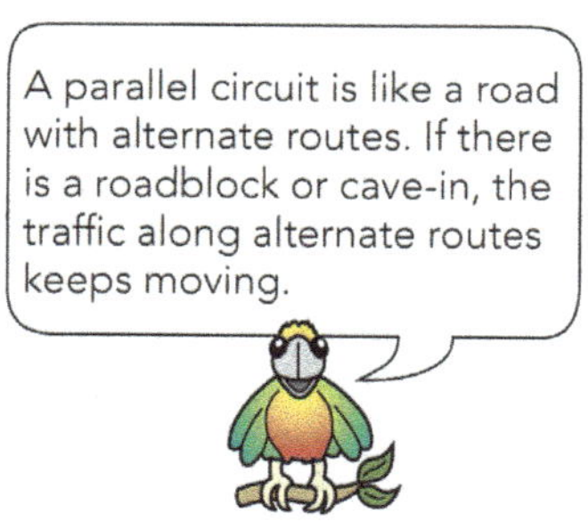

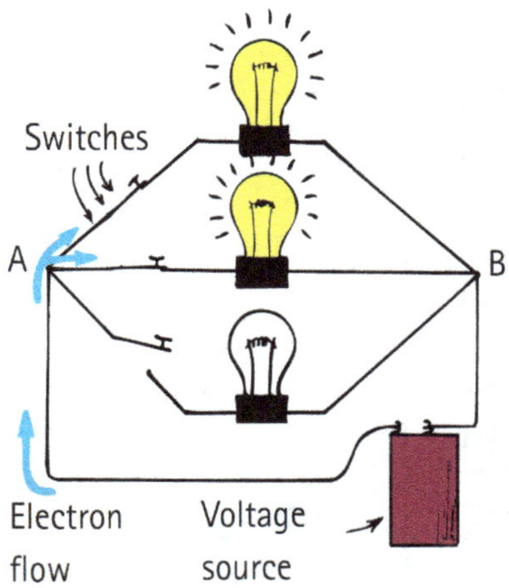

< **Figure 9.22**

A simple parallel circuit. A 6-V battery provides 6 V across each light bulb.

Overloading

Electricity is usually fed into a home by way of two wires called lines. These lines are very low in resistance and are connected to wall outlets in each room—sometimes through two or more separate circuits. About 110–120 volts are applied across these lines by a transformer in the neighborhood. (A transformer is a device that steps down the higher voltage supplied by the power utility.) As more devices are connected to a circuit, more pathways for current result. More pathways lower the combined resistance of the parallel circuit. Therefore, more current exists in the circuit, which is sometimes a problem. Circuits that carry more than a safe amount of current are said to be overloaded.

READING CHECK

What is the effect of more pathways on the amount of current in a household circuit?

We can see how overloading occurs in **Figure 9.23**. The supply line is connected to an electric toaster that draws 8 amperes, an electric heater that draws 10 amperes, and an electric lamp that draws 2 amperes. When only

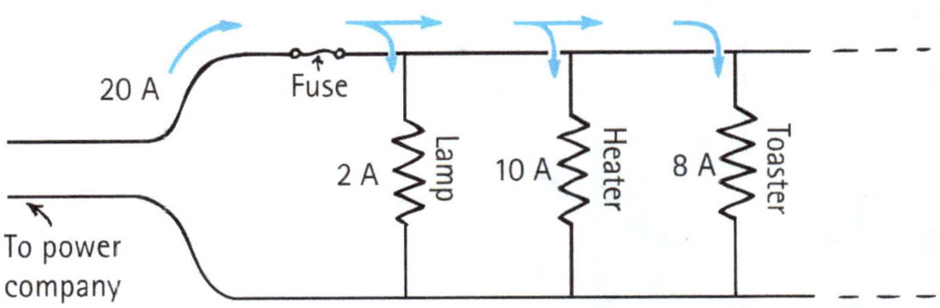

< **Figure 9.23**

Circuit diagram for appliances connected to a household circuit.

CONCEPT **CHECK**

1. What happens to the current in other light bulbs in a parallel circuit when one bulb burns out?

2. What happens to the light intensity of each light bulb in a parallel circuit when more bulbs are added?

CHECK YOUR ANSWERS

1. When one bulb burns out, the others are unaffected. The current in each branch, according to Ohm's law, is equal to voltage/resistance. Since neither voltage nor resistance is affected in the other branches, the current in those branches is unaffected. The total current in the overall circuit (the current through the battery), however, is lowered. It is reduced by an amount equal to the current drawn by the bulb in question before it burned out. But the current in any other single branch is unchanged.

2. The light intensity of each bulb is unchanged as other bulbs are introduced (or removed). Although changes of resistance and current occur for the circuit as a whole, no changes occur in any individual branch in the circuit.

PARALLEL CIRCUIT
I'M GOING TO EXTEND THE TERMINALS OF THIS 12-VOLT BATTERY WITH THIS PAIR OF METAL RODS.
A LAMP WILL GLOW JUST AS BRIGHT WHEN CONNECTED HERE . . .
AS HERE!
OF COURSE... IT'S STILL ACROSS 12 VOLTS
NOW I FASTEN ANOTHER IDENTICAL LAMP --WHAT HAPPENS TO THE BRIGHTNESS OF THE FIRST?
SAME SAME
SAME BRIGHTNESS BECAUSE 12 VOLTS IS ACROSS EACH!
THAT'S WHAT I SAID!
AND A THIRD LAMP ALSO --- SAME BRIGHTNESS!
WHY NOT? THERE'S STILL 12 VOLTS ACROSS EACH LAMP.
HOW DOES THE BATTERY "KNOW" TO SUPPLY 3 TIMES AS MUCH CURRENT TO 3 LAMPS??
PHYSICS OHM'S LAW PHYSICS PHYSICS
I KNOW! WITH 3 SEPARATE PATHS, THE BATTERY HAS 1/3 THE RESISTANCE BETWEEN ITS TERMINALS--THAT'S WHY IT WILL DELIVER 3 TIMES AS MUCH CURRENT FOR ITS 12 VOLTS!
VERY GOOD! BUT DOESN'T THIS MEAN THE BATTERY IS PUTTING OUT 3 TIMES AS MUCH ENERGY AS WHEN LIGHTING ONLY ONE LAMP?
YES, AND YOU CAN EXPECT THE BATTERY TO WEAR DOWN 3 TIMES AS FAST!
SUCH EXCELLENCE
SIGH

the toaster is operating and drawing 8 amperes, the total line current is 8 amperes. When the heater is also operating, the total line current increases to 18 amperes (8 amperes to the toaster plus 10 amperes to the heater). If you turn on the lamp, the line current increases to 20 amperes. Connecting any more devices increases the current still more. Connecting too many devices into the same circuit results in overheating the wires, which can cause a fire.

To prevent overloading in circuits, fuses are connected in series along the supply line. In this way the entire line current must pass through the fuse. The fuse shown in **Figure 9.24** is constructed with a wire ribbon that heats up and melts at a given current. If the fuse is rated at 20 amperes, it will pass 20 amperes but no more. A current above 20 amperes melts the fuse ribbon, which "blows out" and breaks the circuit. Before a blown fuse is replaced, the cause of overloading should be determined and remedied. Often, insulation that separates the wires in a circuit wears away and allows the wires to touch each other. This is called a short circuit.

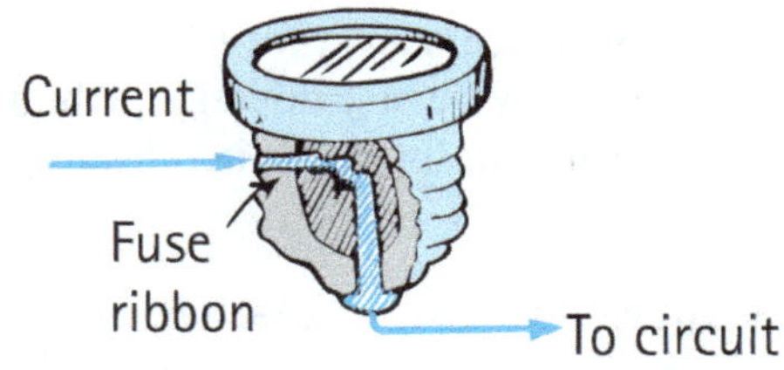

Figure 9.24

A safety fuse.

In modern buildings, fuses have been largely replaced by circuit breakers, which use magnets or bimetallic strips to open a switch when the current is too much. Utility companies use circuit breakers to protect their lines all the way back to the generators.

Electrical Energy and Technology

Try to imagine life before electrical energy was something that humans could control. Imagine home life without electric lights, refrigerators, heating and cooling systems, your smart phone and other electronic devices. We may romanticize a better life without these, but only if we overlook the hours of daily toil doing laundry, cooking, and heating homes. We'd also have to overlook how difficult it was getting a doctor in times of emergency before the advent of the telephone—when all the doctor had in his bag were laxatives, aspirins, and sugar pills—and when infant death rates were staggering.

We have become so accustomed to the benefits of technology that we are only faintly aware of our dependency on dams, power plants, mass transportation, electrification, modern medicine, and modern agricultural science for our very existence. When we dig into a good meal, we give little thought to the technology that went into growing, harvesting, and delivering the food on our table. When we turn on a light we give little thought to the power grid that links the widely separated power stations by long-distance transmission lines. These lines serve as the productive life force of industry, transportation, and the electrification of civilization. Anyone who thinks of science and technology as "inhuman" fails to grasp the ways in which they make our lives more human.

Chapter 9

Summary of Terms

Alternating current (*ac*) Electrically charged particles that repeatedly reverse direction, vibrating about relatively fixed positions. In the United States the vibrational rate is 60 Hz.

Ampere The unit of electric current, equivalent to 1 coulomb per second (the flow of 6.25×10^{18} electrons per second). In symbols, 1 A= 1 C/s.

Coulomb The unit of electrical charge. It is equal in magnitude to the total charge of 6.25×10^{18} electrons.

Coulomb's law The electrical force between two charged bodies is directly proportional to the product of the charges and inversely proportional to the square of the distance between them:

$$F = k\frac{q_1 q_2}{r^2}$$

Direct current (*dc*) Electrically charged particles flowing in one direction only.

Electric current The flow of electric charge that transports energy from one place to another. Measured in amperes.

Electric potential The electric potential energy per unit charge, measured in volts; often called voltage.

Electric power The rate of energy transfer, or rate of doing work. Measured by:

$$\textbf{Power} = \textbf{current} \times \textbf{voltage}$$

Measured in watts (or kilowatts), where 1 A × 1 V = 1 W.

Electrical resistance The property of a material that resists the flow of charged particles through it. Measured in ohms (Ω).

Electrically polarized Term applied to an atom or molecule in which the charges are aligned so that one side has a slight excess of positive charge and the other side a slight excess of negative charge.

Ohm's law The statement that the current in a circuit varies in direct proportion to the voltage across the circuit and inversely with the circuit's resistance:

$$\text{Current} = \frac{\textbf{voltage}}{\textbf{resistance}}$$

Parallel circuit An electric circuit in which electrical devices are connected so that the same voltage acts across each one and any single one completes the circuit independently of all the others.

Series circuit An electric circuit in which electrical devices are connected so that the same electric current exists in all of them.

Superconductor A material that can acquire zero electric resistance to the flow of charge.

Voltage A form of "electrical pressure":

$$\textbf{Voltage} = \frac{\textbf{potential energy}}{\textbf{charge}}$$

Review Questions

1. Which part of an atom is positively charged and which part is negatively charged?
2. How does the charge of one electron compare with that of another electron?
3. How does the number of protons in the atomic nucleus normally compare with the number of electrons that orbit the nucleus?
4. What is meant by saying charge is conserved?
5. How is Coulomb's law similar to Newton's law of gravitation? How is it different?
6. How does an electrically polarized object differ from an electrically charged object?
7. Why do electrons rather than protons make up the flow of charge in a metal wire?

8. How much energy is given to each coulomb of charge passing through a 6-V battery?

9. Does electric charge flow across a circuit or through a circuit? Does voltage flow across a circuit or is it impressed across a circuit? Explain.

10. Which has the greater resistance, a thick wire or a thin wire of the same length?

11. When the voltage across the ends of a piece of wire is doubled, what effect does this have on the current in the wire?

12. When the resistance of a circuit is doubled, and no other changes occur, what effect does this have on the current in the circuit?

13. What is the function of the third prong on the plug of an electric appliance?

14. Distinguish between *dc* and *ac*.

15. Does a battery produce *dc* or *ac*? Does the generator at a power station produce *dc* or *ac*?

16. Which draws more current, a 40-W bulb or a 100-W bulb?

17. In a circuit consisting of two lamps connected in series, if the current through one lamp is 1 A, what is the current through the other lamp?

18. In a circuit consisting of two lamps connected in parallel, if there is 6 V across one lamp, what is the voltage across the other lamp?

19. How does the total current through the branches of a parallel circuit compare with the current through the voltage source?

20. Are household circuits normally wired in series or in parallel?

Answers to Odd-Numbered Questions

1. Nucleus (with protons) is positive; electrons are negative.

3. Same. (When the numbers don't match, we have an ion.)

5. Similar in form, both an inverse-square law. Different in that there are attractive and repelling forces in Coulomb's law, but only attractive in Newton's law of gravitation.

7. Protons are anchored in the nucleus. Electrons, at least the outermost ones in metals, are not strongly tied to atoms and can easily flow.

9. Charge flows through a circuit. Voltage is impressed across a circuit. Just as water flows through a pipe when a difference in pressure exists across its ends, charge flows in a circuit when a voltage difference exists across it.

11. The current doubles.

13. To ground the appliance by conducting away unwanted current.

15. Battery, *dc*. Generators, *ac*.

17. Same, 1 A.

19. Adds up to equal the current in the source.

After failing more than 6000 times before perfecting the first electric lightbulb, Thomas Edison stated that his trials were not failures, because he successfully discovered 6000 ways that don't work.

Chapter 10

Magnetism

THE MAIN IDEA

A moving electric charge is surrounded by a magnetic field

In the previous chapter we discussed a charged balloon that sticks to a wall. Is the physics similar for metal washers sticking to a magnet? And what causes magnetism—is it electrical? What are electromagnets? Are electromagnets essential for the operation of electric generators and electric motors? Does the fact that a compass needle points north tell us that Earth is a giant magnet? Does magnetism about Planet Earth's shield us from harmful cosmic rays? Is Earth's magnetism responsible for the spectacular colors of the aurora borealis? These questions and more will be answered in this chapter.

10.1 Magnetic Poles

The term magnetism comes from the region of Magnesia, a province of Greece. Certain magnetic stones were found by the Greeks who lived there more than 2000 years ago. These stones, called lodestones, could attract pieces of iron. When lodestones were rubbed on pieces of iron, magnets were made. Magnets were first made into compasses and used for navigation by the Chinese in the twelfth century. Today much of modern industry—from motors to computers—relies on magnetism.

If you've played around with magnets, you know they exert forces on one another. Magnetic forces are similar to electrical forces, for both kinds of forces cause objects to attract and repel. Also, both electrical and magnetic forces act between objects that are not touching. And similar to electrical force, the strength of a magnetic interaction depends on the distance between the two magnets.

Hang a bar magnet at its center by a piece of string and you've got a compass. One end, called the north-seeking pole, points northward. The opposite end, called the south-seeking pole, points southward. More simply, these are called the north and south magnetic poles. All magnets have both a north and a south pole (some have more than one of each). Refrigerator magnets have narrow strips of alternating north and south poles. In a simple bar magnet the magnetic poles are located at the two ends. A common horseshoe magnet, as shown in **Figure 10.1**, is a bar magnet bent into a U shape—the poles are at its two ends facing each other.

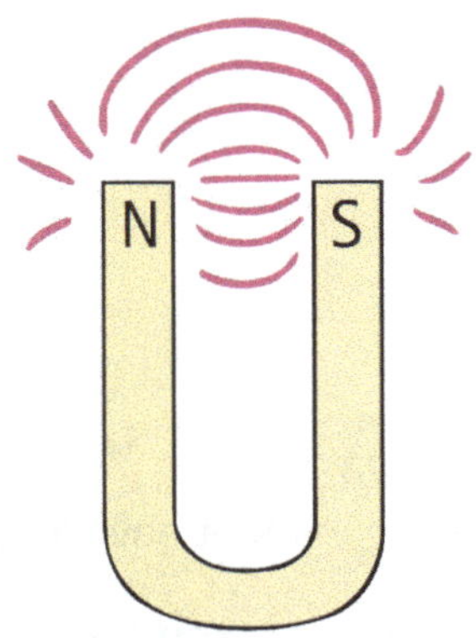

Figure 10.1

A horseshoe magnet.

When the north pole of one magnet is brought near the north pole of another magnet, they repel each other. The same is true of a south pole near a south pole. If opposite poles are brought together, however, attraction occurs. The rule is:

Like poles repel each other; opposite poles attract.

This rule is similar to the rule for electric charges, where like charges repel one another and unlike charges attract. But there is a very important difference between magnetic poles and electric charges. Electric charges can be isolated, but magnetic poles cannot. Electrons and protons are entities by themselves. A cluster of electrons doesn't need the company of a cluster of protons, and vice versa. But a north magnetic pole never exists without the company of a south pole, and vice versa. The north and south poles of a magnet are like the head and tail of the same coin.

READING CHECK

In electricity, a negative charge doesn't need a positive companion. In magnetism, can a north pole exist without a south pole?

If you break a bar magnet in half, as shown in **Figure 10.2**, each half still behaves as a complete magnet. Break the pieces in half again, and you have four complete magnets. You can continue breaking the pieces in half and never isolate a single pole. Even when your piece is one atom thick, there are two poles. This suggests that atoms themselves are magnets.

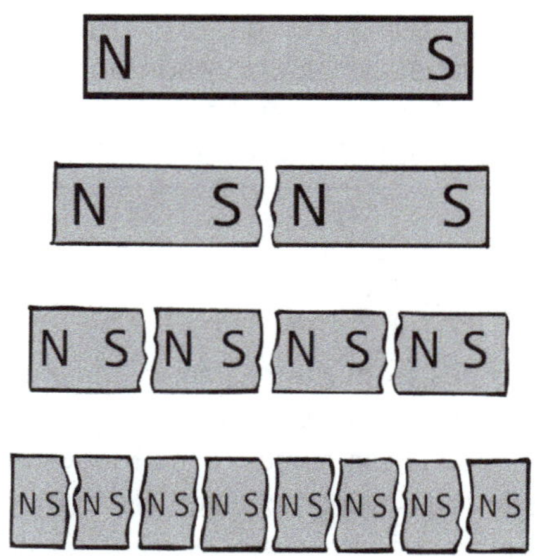

∧ Figure 10.2

Break a magnet in half, and you have two magnets. Break these in half, and you have four magnets, each with a north and south pole. Continue breaking the pieces further and further, and you find the same results. Magnetic poles always exist in pairs.

CONCEPT **CHECK**

Must every magnet have a north and south pole?

CHECK YOUR ANSWER

Yes, just as every coin has two sides, a head and a tail. Some trick magnets may have more than one pair of poles, but nevertheless poles occur in pairs.

10.2 Magnetic Fields

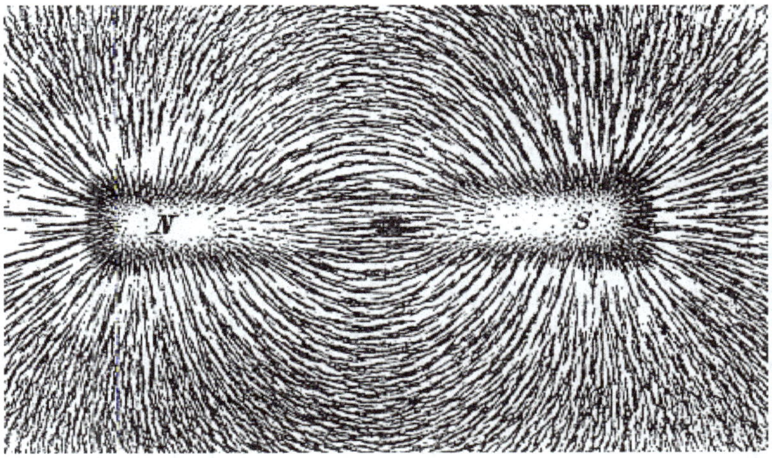

< Figure 10.3

Top view of iron filings sprinkled around a magnet. The filings trace out a pattern of magnetic field lines in the space surrounding the magnet. Interestingly, the magnetic field lines continue inside the magnet (not revealed by the filings) and form closed loops.

When you sprinkle some iron filings on a sheet of paper placed on a magnet, you'll see that the filings trace out an orderly pattern of lines that surround the magnet. Just as a gravitational field surrounds a massive body, a **magnetic field** occupies the space around the magnet. The shape of the field is shown by the filings. Note that the filings line up with the magnetic field lines that spread out from one pole and return to the other.

The direction of the field outside a magnet is from the north to the south pole. Where the lines are closer together, the field is stronger. The concentration of iron filings at the poles of the magnet in **Figure 10.3** shows the magnetic field strength is greater there. If we place another magnet or a small compass anywhere in the field, its poles line up with the magnetic field.

A magnetic field is produced by the motion of electric charge. Where, then, is this motion in a common bar magnet? The answer is, in the electrons of the atoms that make up the magnet. These electrons. like the electrons in all matter, are in constant motion. Investigation reveals that electrons spin about their own axes like tops. The spinning of an electron's charge creates north and south poles about the electron. Thus, every spinning electron behaves as a tiny magnet.

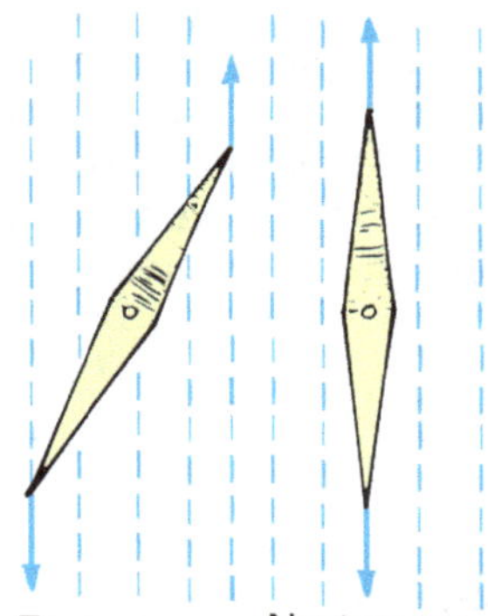

∧ Figure 10.4

(Left) When the compass needle is not aligned with the magnetic field, the oppositely directed forces on the needle produce a pair of torques (twisting forces) that (right) twist the needle into alignment.

READING CHECK
What, exactly, produces a magnetic field?

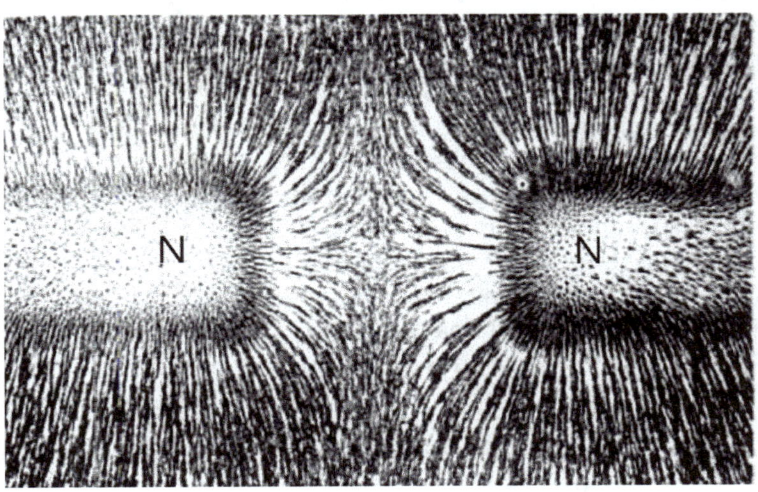

< Figure 10.5

The magnetic field patterns for a pair of magnets with the same poles are nearest each other.

Λ Figure 10.6

Because every electron is charged, the spinning motion of every electron produces a magnetic field. Note: the electron is represented here as a yellow sphere, which is inaccurate because 1) electrons are not yellow, and 2) electrons are points in space, not spheres in space. That said, it has been found that all electrons have this property called spin that generates a tiny magnetic field.

Typically, the electrons within a material spin in all directions, which means their magnetic poles work against each other. With atoms such as iron, nickel, and cobalt, however, the fields do not cancel each other entirely. Most common magnets are therefore made from alloys containing iron, nickel, and cobalt in various proportions. Rather strong magnetic alloys can be made by including the rare earth metal neodymium, Nd. These neodymium magnets have many useful applications. Wind turbines fitted with neodymium magnets, for example, are able to produce electricity much more efficiently.

Magnetic Domains

Large clusters of magnetic atoms can line up with one another. These clusters of aligned atoms are **magnetic domains**. Each domain is made up of numerous aligned atoms. The domains are microscopic (**Figure 10.7**), and there are many of them in a crystal of a magnetic metal, such as iron. Domains themselves tend to align with one another.

In a strong magnet, most all domains are aligned. The reason most pieces of iron aren't magnets is because the domains in ordinary iron are not lined up. In a common iron nail, for example, the domains are randomly oriented. But when you bring a magnet nearby, they can be induced into alignment. The domains align themselves much as electrical charges in a piece of paper align themselves (become polarized) in the presence of a charged rod. When you remove the nail from the magnet, ordinary thermal motion causes most or all of the domains in the nail to return to a random arrangement. Hence, the magnetism induced within the nail is temporary.

Relatively permanent magnets can be made by placing pieces of iron or other magnetic materials in strong magnetic fields. Iron alloys differ in their ability to become magnetized; soft iron is easier to magnetize than steel. It helps to tap the material to nudge any stubborn domains into alignment. Another way is to stroke the material with a magnet. The stroking motion lines up the domains. If a permanent magnet is dropped or heated, some of the domains are jostled out of alignment and the magnet becomes weaker.

READING CHECK

Why are some pieces of iron magnets, while most pieces of iron are not magnetic?

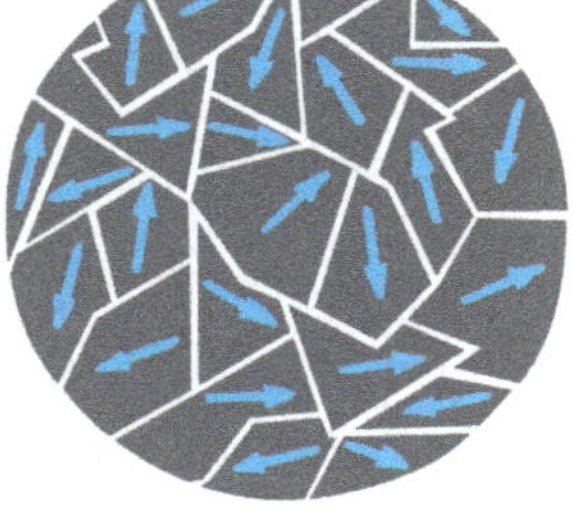

Λ Figure 10.7

A microscopic view of magnetic domains in a crystal of iron. Each domain consists of numerous aligned iron atoms. In this view the domains are not aligned.

Quick Activity

What Does a Magnetic Field Look Like?

Place a bar magnet on a level surface. Cover the magnet with a thin sheet of paper, glass, clear plastic, or cardboard. Sprinkle iron filings onto the sheet above the magnet. Gently tap the sheet and observe the pattern formed by the filings. Try doing the same with multiple magnets in various orientations. In some places the filings might bunch up and in others they might be more spread out. What do you think this indicates about the magnetic field strengths?

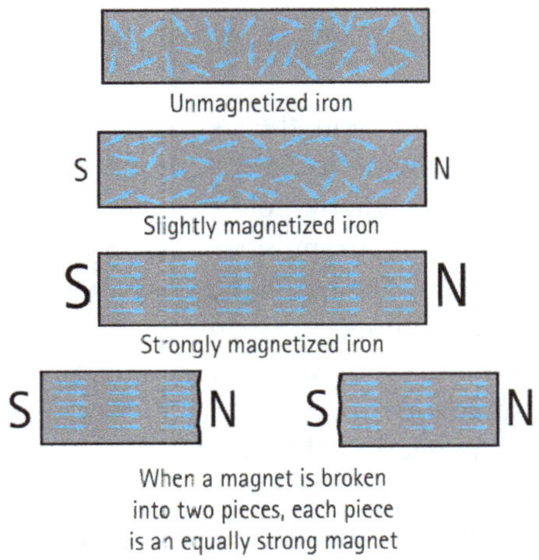

Λ Figure 10.8

Pieces of iron in successive stages of becoming magnetized. The arrows represent domains; each arrowhead represents a north pole and each tail a south pole. Poles of neighboring domains neutralize each other's effects, except at the two ends of the piece of iron.

Λ Figure 10.9

The iron nails become induced magnets.

FOR YOUR INFORMATION

Earth's magnetic field is known to flip orientation. Over the past 83 million years there have been about 183 such reversals—the last complete flip occurring about 700,000 years ago with a more temporary flip occurring about 42,000 years ago. This likely results from changes in the direction of electric currents deep within Earth's dynamic interior.

Quick Activity

Can Compass

Most iron objects around you are magnetized to some degree. A filing cabinet, a refrigerator, or even cans of food on your pantry shelf have north and south poles induced by the Earth's magnetic field. Pass a compass from their bottoms to their tops and their poles are easily identified. Turn cans upside down and see how many days it takes for the poles to reverse themselves!

CONCEPT CHECK

1. How can a magnet attract a piece of iron that is not magnetized?
2. Why will a magnet not pick up a penny or a piece of wood?

CHECK YOUR ANSWERS

1. Like the compass needle in Figure 10.4, domains in the unmagnetized piece of iron are induced into alignment by the magnetic field of the magnet. One domain pole is attracted to the magnet and the other domain pole is repelled. Does this mean the net force is zero? No, because the force is slightly greater on the domain pole closest to the magnet than on the farther pole. That's why there is a net attraction. In this way a magnet attracts non-magnetized pieces of iron (Figure 10.9).
2. A penny and a piece of wood have no magnetic domains that can be induced into alignment.

10.3 Electromagnets

If a single moving charge produces a magnetic field, then a flow of charges also produces a magnetic field. Thus, a magnetic field is found around any wire carrying an electric current. The magnetic field that surrounds a current-carrying wire can be demonstrated by arranging an assortment of compasses around the wire (**Figure 10.10**). The magnetic field about the current-carrying wire makes up a pattern of concentric circles. When the current reverses direction, the compass needles turn around, showing that the direction of the magnetic field also changes.

If the wire is bent into a loop, the magnetic field lines bunch together inside the loop (**Figure 10.11**). If the wire is bent into another loop, overlapping the first, the bunching of field lines in the loops is doubled. More loops mean more magnetic field intensity. The magnetic field intensity is strong for a current-carrying coil of many loops.

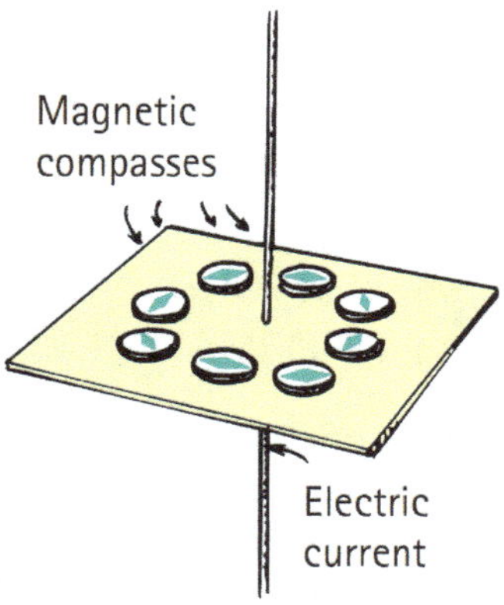

∧ Figure 10.10

The compasses show the circular shape of the magnetic field surrounding the current-carrying wire.

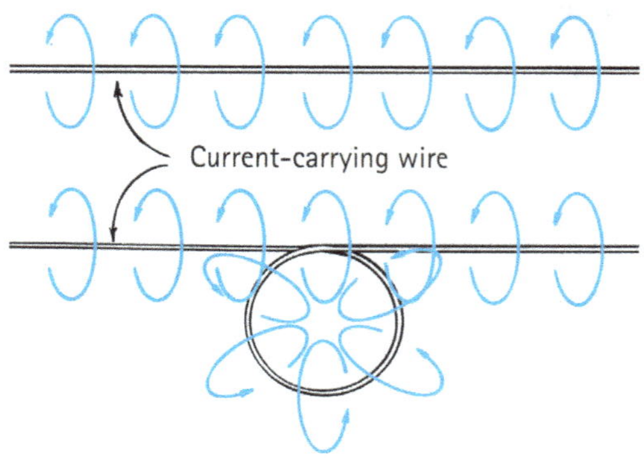

< Figure 10.11

Magnetic field lines about a current-carrying wire crowd up when the wire is bent into a loop.

READING CHECK

What pattern occurs in the magnetic field about a current-carrying wire?

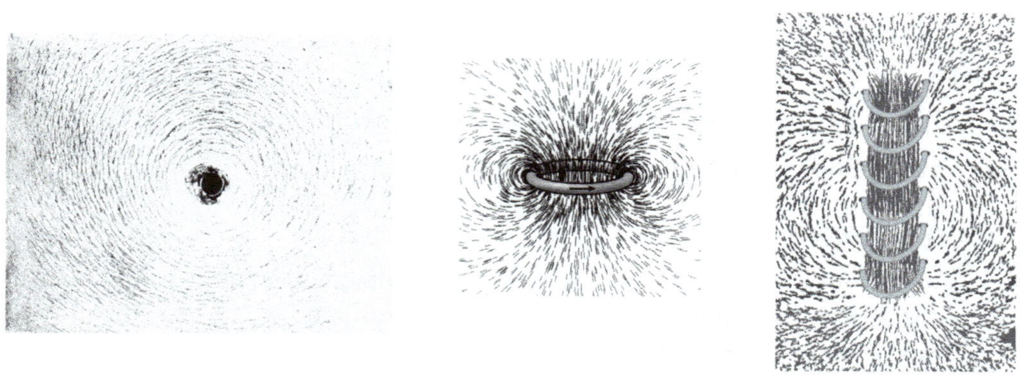

∧ Figure 10.12

Iron filings sprinkled on paper reveal the magnetic field configurations about (a) a current-carrying wire coming up out of the page, (b) a current-carrying loop, and (c) a current-carrying coil of loops.

Electromagnets

If a piece of iron is placed in a current-carrying coil of wire, the alignment of magnetic domains in the iron produces a stronger magnet. Then we have an **electromagnet**. Its strength is increased by increasing the current through the coil.

Electromagnets powerful enough to lift automobiles are a common sight in junkyards. The strength of these electromagnets is limited mainly by overheating of the current-carrying coils. The most powerful electromagnets omit the iron core and use superconducting coils through which large electrical currents easily flow.

∧ Figure 10.13

A permanent magnet levitates above a superconductor because the magnet's magnetic field cannot penetrate the superconducting material.

Superconducting Electromagnets

Ceramic superconductors have the interesting property of expelling magnetic fields. Because magnetic fields cannot penetrate the surface of a superconductor, magnets levitate above them. The reasons for this behavior are beyond the scope of this program and involve quantum mechanics. One of the hot applications of superconducting electromagnets is the levitation of high-speed trains for transportation.

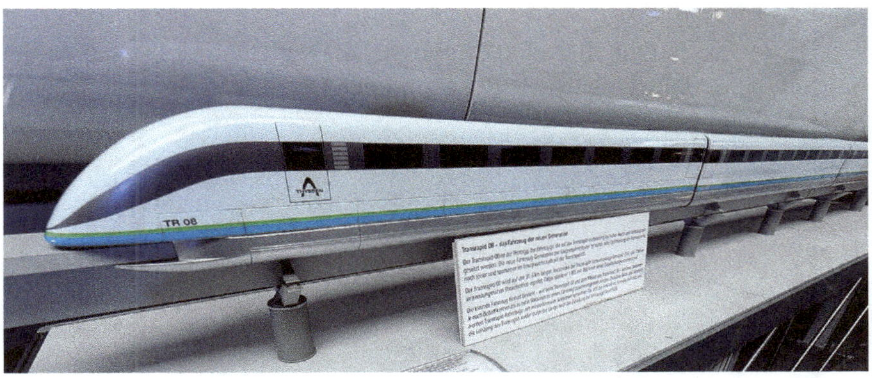

< Figure 10.14

Scale model of a prototype magnetically levitated vehicle—a magplane. Whereas conventional trains vibrate as they ride on rails at high speeds, magplanes can travel vibration-free at high speeds because they make no physical contact with the guideway they float above.

10.4 Magnetic Forces

A charged particle has to be moving to interact with a magnetic field. Charges at rest don't respond to magnets. But when moving, charged particles experience a deflecting force. The force is greatest when the particles move at right angles to the magnetic field lines. At other angles, the force is less and becomes zero when the particles move parallel to the field lines. The force is always perpendicular to the magnetic field lines and perpendicular to the velocity of the charged particle (**Figure 10.15**). So, a moving charge is deflected when it crosses through a magnetic field, but when it travels parallel to the field no deflection occurs.

This deflecting force is very different from the forces that occur in other interactions. Gravitation acts in a direction parallel to the line between masses, and electrical force acts in a parallel direction between charges. But magnetic force acts at right angles to the magnetic field and the velocity of the charged particle.

We are fortunate that charged particles are deflected by magnetic fields. This fact was useful in the first TV picture tubes where electrons were guided by magnetic fields to provide a picture. More interesting, charged

READING CHECK

In what direction of motion is the force greatest on a charged particle moving in a magnetic field? In what direction is the force weakest?

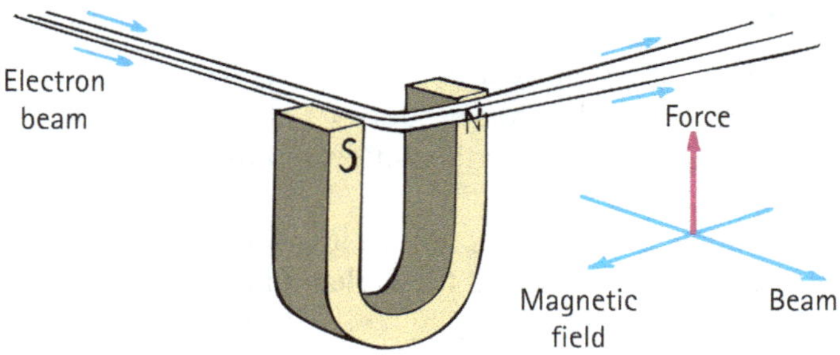

Figure 10.15 >

A beam of electrons is deflected by a magnetic field.

particles from outer space are deflected by the Earth's magnetic field. The intensity of harmful cosmic rays bombarding the Earth's surface would be more intense otherwise.

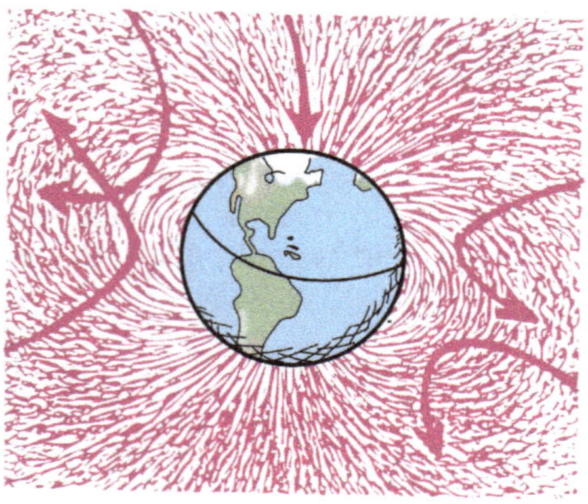

∧ Figure 10.16

The magnetic field of the Earth deflects many charged particles that make up cosmic radiation.

Magnetic Force on Current-Carrying Wires

Simple logic tells you that if a charged particle moving through a magnetic field experiences a deflecting force, then a current of charged particles moving through a magnetic field also experiences a deflecting force. If the particles are forced while moving inside a wire, the wire is also forced (**Figure 10.17**).

If we reverse the direction of current, the deflecting force acts in the opposite direction. This led to the invention of electric meters and electric motors.

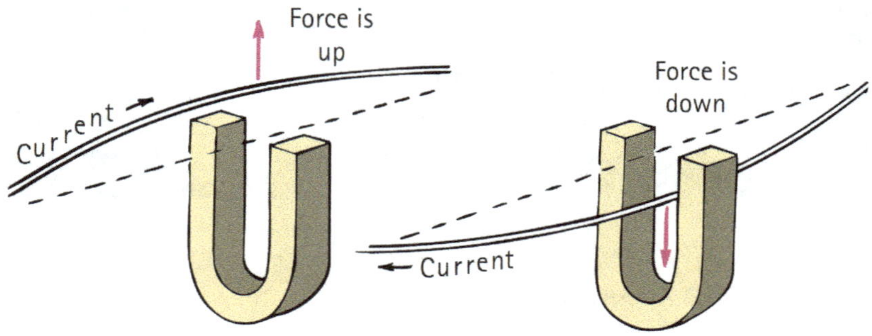

Figure 10.17 >

A current-carrying wire experiences a force in a magnetic field. (Can you see the same physics here as in Figure 10.15?).

Electric Meters

The simplest meter that can detect an electric current is a magnetic compass. The next level of complexity is a compass in a coil of wires (**Figure 10.18**). When an electric current is in the coil, each loop produces its own effect on the needle, and so a very small current can be detected. A current-indicating instrument is called a *galvanometer*.

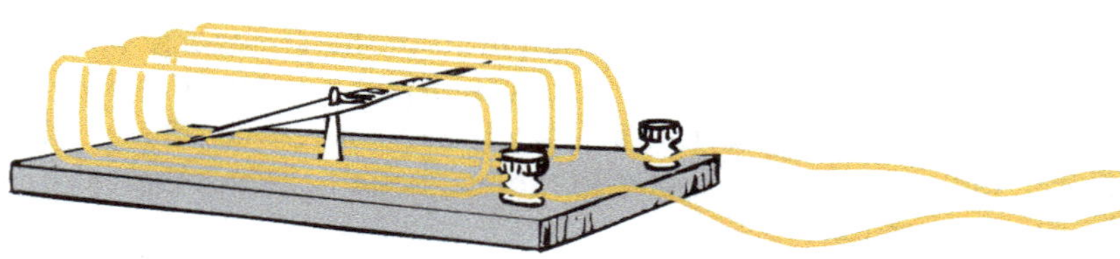

Figure 10.18 >

A very simple galvanometer.

A more common galvanometer design is shown in **Figure 10.19**. It uses more loops of wire and is therefore more sensitive. The coil is free to move, and the magnet is held stationary. The greater the current in its windings, the greater its deflection. The coil turns against a spring, with a needle fixed to show the amount of deflection. A galvanometer may be calibrated to measure current (amperes), which is then called an *ammeter*. Or it may be calibrated to measure voltage, then called a *voltmeter*.

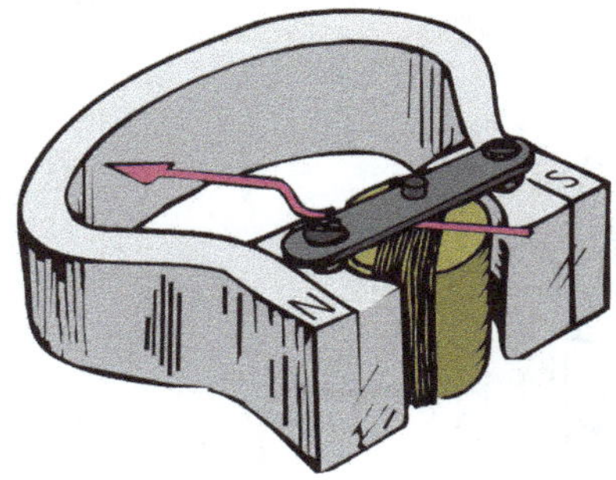

Figure 10.19

A common galvanometer design.

Electric Motors

If we change the design of the galvanometer slightly, so that deflection makes a complete turn rather than a partial rotation, we have an **electric motor**. The principal difference is that in a motor the current is made to change direction each time the coil makes a half rotation. This happens in cyclic fashion to produce continuous rotation, which has been used to run clocks, operate gadgets, electric cars, and lift heavy loads.

In **Figure 10.21**, we see the principle of the electric motor in bare outline. A permanent magnet produces a magnetic field in a region where a rectangular loop of wire is mounted to turn about the dashed axis shown. Can you see that any current in the loop has one direction in the upper side of the loop and the opposite direction in the lower side? If the upper side of the loop is forced to the left by the magnetic field, the lower side is forced to the right, as if it were a galvanometer. But unlike in a galvanometer, the current in a motor is reversed during each half revolution by stationary contacts on the shaft. The parts of the wire that rotate and brush against these contacts are called *brushes*. In this way, the current in the loop alternates so that the forces on the upper and lower regions do not change directions as the loop rotates. The rotation is continuous as long as current is supplied.

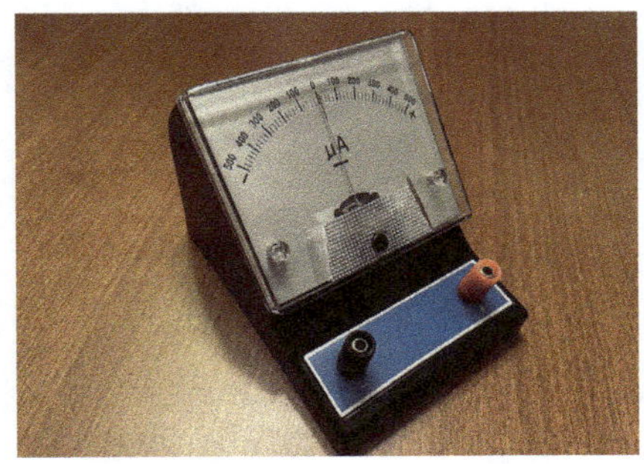

Figure 10.20

Both the ammeter (shown here) and the voltmeter are galvanometers, but calibrated differently.

We have described here only a very simple dc motor. Larger motors, dc or ac, are usually made by replacing the permanent magnet by an electromagnet that is energized by some power source. Of course, more than a single loop is used. Many loops of wire are wound about an iron cylinder, called an *armature*, which then rotates when the wire loops carry current.

The advent of electric motors brought to an end much human and animal toil in many parts of the world. To be sure, electric motors have greatly changed the way people live.

CONCEPT CHECK

What is the major similarity between a galvanometer and a simple electric motor? What is the major difference?

CHECK YOUR ANSWERS

A galvanometer and a motor are similar in both using coils positioned in a magnetic field. When a current passes through the coils, forces on the wires rotate the coils. The major difference is that the maximum coil rotation in a galvanometer is one-half turn, whereas in a motor the coil (wrapped on an armature) rotates through many complete turns. This is accomplished by alternating the direction of the current with each half turn of the armature.

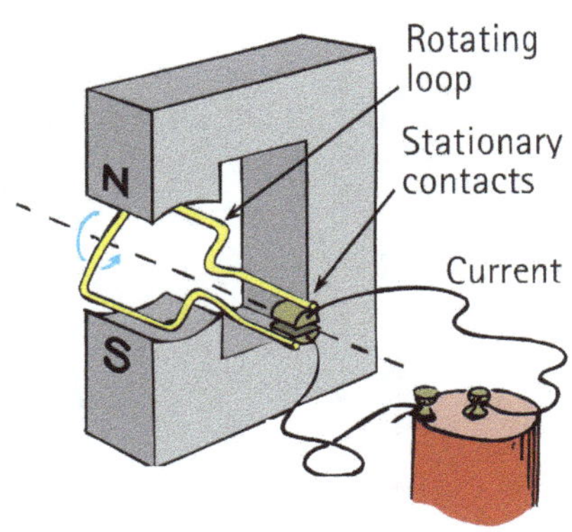

Figure 10.21

A simplified electric motor.

10.5 Electromagnetic Induction

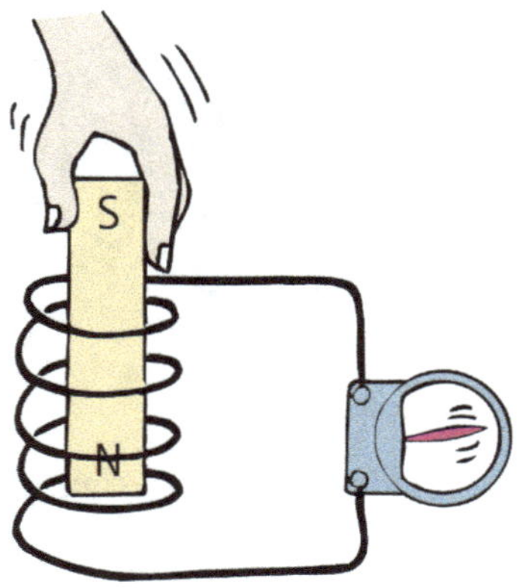

∧ **Figure 10.22**

When the magnet is plunged into the coil, charges in the coil are set in motion; voltage is induced in the coil.

In the early 1800s, the only current-producing devices were voltaic cells. These were the first batteries, which produced small currents by dissolving metals in acids. An important question asked was whether electricity could be produced from magnetism. The answer was provided in 1831 by two physicists, Michael Faraday in England and Joseph Henry in the United States—each working without knowledge of the other. Their discovery changed the world. The technology that followed made electricity commonplace, powering industries by day and lighting up cities at night.

Faraday and Henry both discovered that electric current can be produced in a wire simply by moving a magnet in or out of a coiled part of the wire (**Figure 10.22**). No battery or other voltage source is needed—only the motion of a magnet in a wire coil. They discovered that voltage is caused, or induced, by the relative motion between a wire and a magnetic field. Whether the magnetic field moves near a stationary conductor or the conductor moves in a stationary magnetic field—voltage is induced either way (**Figure 10.23**).

READING CHECK

What was the discovery of Michael Faraday and Joseph Henry in 1831?

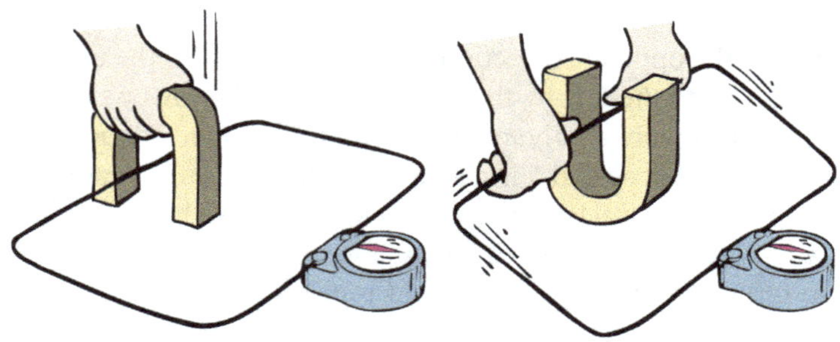

∧ **Figure 10.23**

Voltage is induced in the wire loop either when the magnetic field moves past the wire or when the wire moves through the magnetic field.

The greater the number of loops of wire moving in a magnetic field, the greater the induced voltage (**Figure 10.24**). Pushing a magnet into a coil with twice as many loops induces twice as much voltage; pushing into a coil with ten times as many loops induces ten times as much voltage; and so on. It may seem that we get something (energy) for nothing by simply increasing the number of loops in a coil of wire. But we don't: We find it is more difficult to push the magnet into more loops. This is because the induced voltage produces a current, which makes an electromagnet, which repels the magnet in our hand. So, we do more work against this "back force" to induce more voltage (**Figure 10.25**).

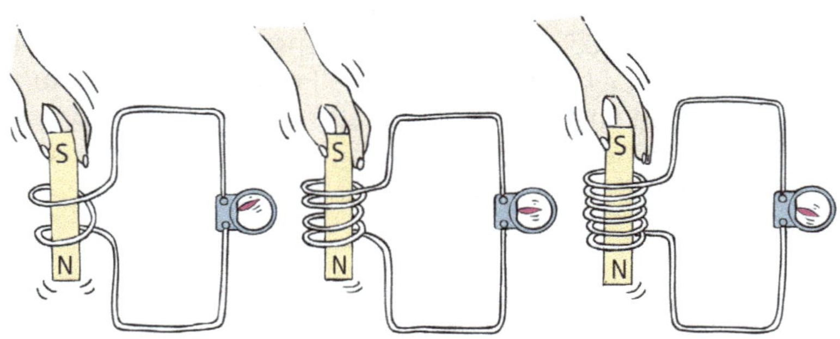

Figure 10.24 >

When a magnet is plunged into a coil of twice as many loops as another, twice as much voltage is induced. If the magnet is plunged into a coil with three times as many loops, three times as much voltage is induced.

The amount of voltage induced also depends on how fast the magnetic field lines are entering or leaving the coil. Very slow motion produces hardly any voltage at all. Quick motion induces a greater voltage. This phenomenon of inducing voltage by changing the magnetic field in a coil of wire is **electromagnetic induction**.

Figure 10.25

It is more difficult to push the magnet into a coil made up of many loops because the magnetic field of each current loop resists the motion of the magnet.

Faraday's Law

Electromagnetic induction is summarized by Faraday's law:

> **The induced voltage in a coil is proportional to the number of loops multiplied by the rate at which the magnetic field changes within those loops.**

The amount of *current* produced by electromagnetic induction depends on more than the induced voltage. It also depends on the resistance in both the coil and the circuit to which it's connected. For example, we can plunge a magnet in and out of a closed rubber loop and produce no current. Doing the same in a copper loop easily produces current. So, the current in each is quite different. The electrons in the rubber sense the same voltage as those in the copper, but their tight bonding to atoms prevents the movement of electrons that move so freely in copper.

We have mentioned two ways in which voltage can be induced in a loop of wire: by moving the loop near a magnet or by moving a magnet near the loop. There is a third way, by changing a current in a nearby loop. All three cases possess the same essential feature—a changing magnetic field in the loop.

We see electromagnetic induction all around us. On the road we see it operate when a car drives over buried coils of wire to activate a nearby traffic light. When iron parts of a car move over the buried coils, the Earth's magnetic field in the coils is changed, inducing a voltage to trigger the changing of the traffic lights. Similarly, when you walk through the upright coils in the security system at an airport, any metal you carry slightly alters the magnetic field in the coils. This change induces voltage, and sounds an alarm. When the magnetic strip on the back of a credit card is scanned, induced voltage pulses identify the card. That's why you might be asked to insert or remove your card quickly. Electromagnetic induction is everywhere. As we shall see in Chapter 12, it underlies the electromagnetic waves we call light.

fyi FOR YOUR INFORMATION

As shown by the Conceptual Academy chemistry author, guitar pickups are tiny coils with magnets inside them. The magnets magnetize the steel strings. When the strings vibrate, voltage is induced in the coils and boosted by an amplifier, and sound is produced by a speaker.

CONCEPT CHECK

If you push a magnet into a coil, as shown in Figure 10.25, you'll feel a resistance to your push. Why is this resistance greater in a coil with more loops?

CHECK YOUR ANSWER

Simply put, more work is required to provide more energy. You can also look at it this way: When you push a magnet into a coil, you cause the coil to become an electromagnet. The more loops on the coil, the stronger the electromagnet that you produce and the stronger it pushes back against you.

10.6 Power Production

A **generator** is a motor in reverse. The device is much the same, with the roles of input and output reversed. In a motor, electrical energy is the input and mechanical energy the output. In a generator, mechanical energy is the input and electric energy is the output. Both devices simply transform energy from one form to another.

A striking example of a device functioning as both motor and generator is found in hybrid and electric automobiles. When power for accelerating or hill climbing is needed, this device draws current from a rechargeable battery and acts as a motor. Braking or rolling downhill causes the device to act as a generator to recharge the battery.

Shake flashlights need no batteries. Shake the flashlight for 30 seconds or so and generate up to 5 min of bright LED illumination. This is electromagnetic induction in action! When brightness diminishes, shake again. You provide the energy to charge the device.

Fifty years after Faraday and Henry discovered electromagnetic induction, Nikola Tesla and George Westinghouse put those findings to practical use. They showed the world that electricity could be generated reliably and in sufficient quantities to light entire cities.

Tesla built generators much like those still in use but more complicated than the simple model we have discussed. His generators had armatures made up of bundles of copper wires. The armatures were forced to spin within strong magnetic fields by a turbine, which in turn was spun by the energy of either falling water or steam. The rotating loops of wire in the armature cut through the magnetic field of the surrounding electromagnets. In this way they induced alternating voltage and current.

It's important to know that generators don't produce energy—they simply convert energy from some other form to electric energy. As we discussed in Chapter 5, energy from a source, whether fossil or nuclear fuel or wind or water, is converted to mechanical energy to drive the turbine. The attached generator converts most of this mechanical energy to electrical energy. Some of your friends may think that electricity is a primary source of energy. Let them know that electricity is a carrier of energy that must have a source.

READING CHECK

Does a generator produce energy? Defend your answer.

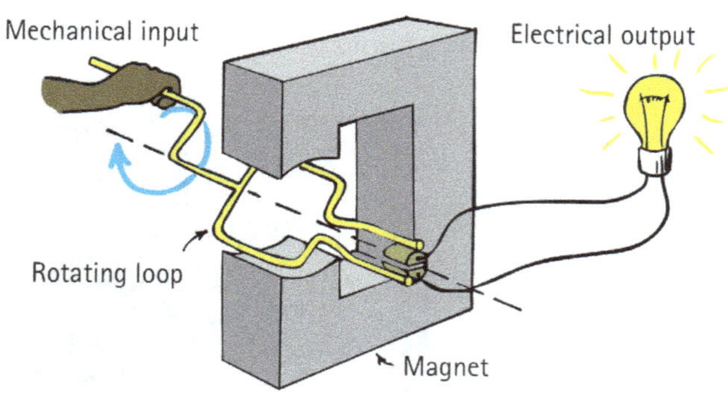

Figure 10.26

A simple generator. Voltage is induced in the loop when it is rotated in the magnetic field.

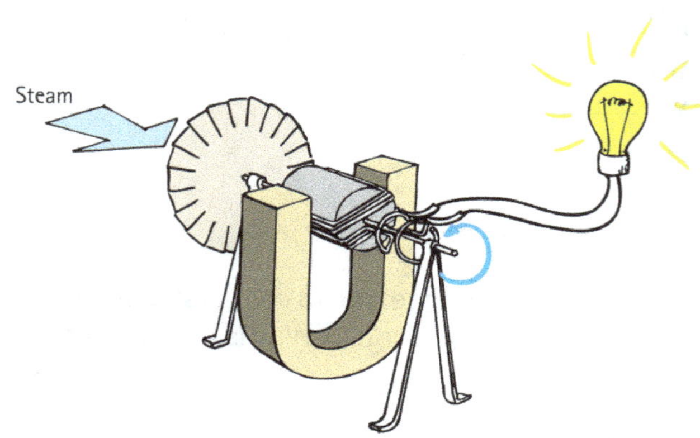

Figure 10.27

Steam drives the turbine, which is connected to the armature (gray) of the generator.

10.7 Electromagnetic Waves

The modern view of electromagnetic induction states that electric and magnetic fields are induced. These arising fields, in turn, create the voltages we have considered in this chapter. Amazingly, induction takes place whether or not a conducting wire or even any material medium is present. The induction happens through space itself. It doesn't really matter if anything is within that space. It just so happens that if a conducting material is within that space, then a notable electric current will be created. In this more general sense, Faraday's law states:

An electric field is induced in any region of space in which a magnetic field is changing with time.

There is a second effect, an extension of Faraday's law. It is the same, except that the roles of electric and magnetic fields are interchanged. It is one of nature's many symmetries. This effect was advanced by the British physicist James Clerk Maxwell in about 1860 and is known as **Maxwell's counterpart to Faraday's law**:

A magnetic field is induced in any region of space in which an electric field is changing with time.

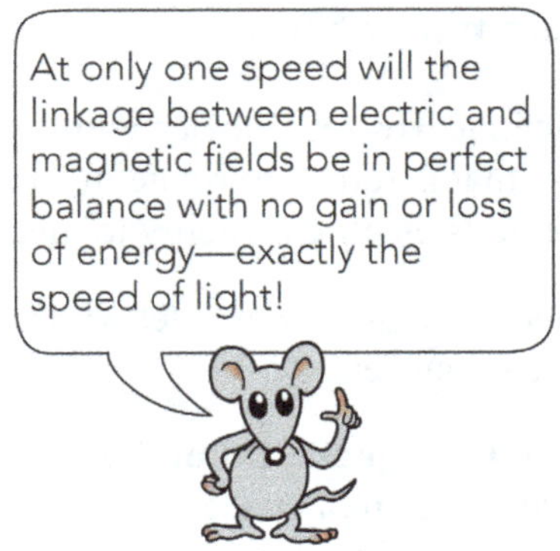

This tells us that a changing magnetic field induces an electric field, but that the reverse is also true: a changing electric field induces a magnetic field. Start one and you'll create the other, which then feeds back to recreate the former, and so on. Electric and magnetic fields within purely empty space can reinforce each other.

Maxwell recognized that these electric and magnetic fields could reinforce each other with no loss or gain of energy at some optimal speed. He calculated this speed and discovered it matched the speed of light, which had already been experimentally measured by his time. His great insight was that these electromagnetic waves and light were one and the same. If electric charges are set into vibration in the range of frequencies that match those of light, electromagnetic waves are produced that **are** light! Maxwell discovered that light is simply electromagnetic waves in the range of frequencies to which the eye is sensitive.

READING CHECK

What relationship did Maxwell discover about electromagnetic waves?

On the eve of his discovery he had a date with a young woman he was later to marry. While walking in a garden, his date remarked about the beauty and wonder of the stars. Maxwell asked how she would feel to know that she was walking with the only person in the world who knew what the starlight really was. For it was true. At that time, James Clerk Maxwell was the only person in the world to know that light of any kind is energy carried in waves of electric and magnetic fields that continually regenerate each other.

Chapter 10

Summary of Terms

Electric Motor A device employing a current-carrying coil that is forced to rotate in a magnetic field that converts electrical energy to mechanical energy.

Electromagnet A magnet whose field is produced by an electric current.

Electromagnetic Induction A magnetic field is induced in any region of space in which an electric field is changing with time.

Faraday's Law The induced voltage in a coil is proportional to the product of its number of loops, the cross-sectional area of each loop, and the rate at which the magnetic field changes witlhin those loops.

Generator An electromagnetic induction device that produces electric current by rotating a coil within a stationary magnetic field. A generator converts mechanical energy to electrical energy.

Magnetic Domains Clustered regions of aligned magnetic atoms. When these regions are aligned with one another, the substance containing them is a magnet.

Magnetic Field The region of magnetic influence around either a magnetic pole or a moving charged particle.

Magnetic Force (1) Between magnets, it is the attraction of unlike magnetic poles for each other and the repulsion between like magnetic poles. (2) Between a magnetic field and a moving charged particle, it is a deflecting force due to the motion of the particle.

Maxwell's Counterpart To Faraday's Law An electric field is induced in any region of space in which a magnetic field is changing with time.

Review Questions

1. In what way is the rule for the interaction between magnetic poles similar to the rule for the interaction between electric charges?
2. In what way are magnetic poles very different from electric charges?
3. An electric field surrounds an electric charge. What additional field surrounds a moving electric charge?
4. What two kinds of motion are exhibited by electrons in an atom?
5. Most metals are not magnetic, but iron is. Why?
6. Why will dropping an iron magnet on a hard floor results in a weaker magnet?
7. What is the shape of magnetic field lines about a current-carrying wire?
8. What happens to the direction of the magnetic field about an electric current when the direction of the current is reversed?
9. In what direction relative to a magnetic field does a charged particle move in order to experience maximum deflecting force? Minimum deflecting force?
10. Both gravitational and electrical forces act along the direction of the force fields. How is the direction of the magnetic force on a moving charge different?
11. Since a magnetic force acts on a moving charged particle, does it make sense that a magnetic force also acts on a current-carrying wire? Defend your answer.
12. What happens to the direction of the force on a wire when the current in it is reversed?

13. What is a galvanometer called when calibrated to read current? Voltage?

14. What must change in order for electromagnetic induction to occur?

15. What are the three ways that voltage can be induced in a wire?

16. What is the basic difference between a generator and an electric motor?

17. What is the basic similarity between a generator and an electric motor?

18. What commonly supplies the energy input to a turbine?

19. What is induced by the rapid alternation of a magnetic field?

20. What is induced by the rapid alternation of an electric field?

Answers to Odd-Numbered Questions

1. Just as like charges repel and unlikes attract, like poles repel and unlike poles attract.

3. Magnetic field.

5. Magnets are the pieces of iron with aligned domains. Unmagnetized pieces of iron have domains that are not aligned.

7. Magnetic lines take the shape of concentric circles about the wire.

9. Maximum force occurs when motion is perpendicular to the field. No force occurs when charged particles move parallel to field lines.

11. Yes, of course. If it acts on single charges, it acts on many moving charges.

13. Ammeter; voltmeter.

15. By moving a loop of wire near a magnet; by moving a magnet near a loop of wire, or by changing the current in the wire.

17. Both are geometrically the same. Only the roles of input and output make them different.

19. An electric field.

Chapter 11

Waves and Sound

THE MAIN IDEA

Sound is a form of energy that travels in waves that spread out though space.

The bow on this cello activates the strings into vibrations, which cause the body of the cello to resonate. The resonating cello causes the surrounding air molecules to vibrate in the same fashion creating the characteristic sound of the cello. But what exactly is the nature of sound? Can it travel in a vacuum? Can two sounds cancel each other out? Why does sound travel faster underwater? Does the reflection of sound follow the same physics rules for reflected light? Is sound related to the electromagnetic spectrum, like light? In what way do radio waves and sound waves differ? How do aircraft form shock waves that produce sonic booms? Why does the pitch of a siren change with motions of ambulances and fire engines? To answer these sorts of questions we begin by studying the nature of vibrations and the waves they produce.

11.1 Vibrations and Waves

When something moves periodically back and forth, side to side, or up and down, we say it vibrates. When a vibration carries through space and time it is a wave. A **wave** is a pattern of matter or energy that extends from one place to another. Consider a bug bobbing up and down on a still pond. Radiating outward from the bobbing bug will be waves of water. But note carefully, it's not the water itself that radiates outward from the bug. As evidenced by a floating leaf, the water simply bobs up and down in concert with the bug. What catches your eye are the energetic vibrations passing through the water—what you witness as a series of waves. So, we see that a wave is a vibration that carries through space and time.

READING CHECK
What is a wave?

Sound waves and light waves are both generated by vibrations. But they are two very different kinds of waves. Sound is the traveling vibration of matter—through solids, liquids, or gases. If there is no matter to vibrate, then no sound is possible. Sound cannot travel in a vacuum because there is no material there to vibrate. Light, however, can travel through a vacuum because light is the traveling vibration of electric and magnetic fields, and such fields can occur wherever. For example, light travels quite nicely from

FOR YOUR INFORMATION

Many things in nature vibrate—the string on a guitar, the reed in a clarinet, and your vocal cords when you speak or sing. When they vibrate in air, they make the air vibrate in the same way. When these vibrations reach your ear, they are transmitted as impulses to a part of your brain, and you hear sound.

the Sun and stars to the Earth right through the vacuum of space. So, light can pass through many materials, such as glass, although none is required.

You can see the relationship between a vibration and a wave in **Figure 11.1**. A marking pen on a bob attached to a vertical spring vibrates up and down. Notice that it traces a wave form on a sheet of paper that moves horizontally at constant speed. The wave form is a *sine curve*, with a shape like rolling hills and valleys and a precise mathematical description. As in a water wave, the high points of a sine wave are called *crests* and the low points are *troughs*. The dashed green line shows the "home" position, or midpoint of the vibration.

The term **amplitude** refers to the distance from the midpoint to the crest (or trough) of the wave. The amplitude equals the maximum displacement from the home position—from equilibrium.

The **wavelength** of a wave is the distance from one crest to the next one. Equivalently, it is the distance between any two successive identical parts of the wave. The wavelengths of waves at the beach are measured in meters, the wavelengths of ripples in a pond in centimeters, and the wavelengths of light waves in billionths of a meter (nanometers).

How frequently a vibration occurs is described by its **frequency**. The frequency of any vibrating object is the number of to-and-fro vibrations the object makes in a given time (usually one second). If a complete to-and-fro vibration (one cycle) occurs in one second, the frequency is one vibration per second. If two vibrations occur in one second, the frequency is two vibrations per second.

The unit of frequency is the **hertz** (Hz). It is named after Heinrich Hertz, who demonstrated the existence of radio waves in 1886. We call one vibration per second 1 hertz; two vibrations per second is 2 hertz, and so on. Higher frequencies are measured in kilohertz (kHz), and still higher frequencies in megahertz (MHz). A station at 101.7 MHz on your dial broadcasts radio waves that have a frequency of 101,700,000 hertz. These radio-wave frequencies are the frequencies at which electrons in the broadcasting antenna are forced to vibrate. Note: All radio waves are a form of electromagnetic radiation that travel at the speed of light from the antenna to a receiver. As we'll be discussing later, after being received, electronics are used to translate signals within those radio waves into audible sound waves. In short, you can listen to your radio, but never to a radio wave.

The **period** of a wave or vibration is the time it takes for a complete vibration. Period can be calculated from frequency, and vice versa. Suppose, for example, that a pendulum makes two vibrations in one second.

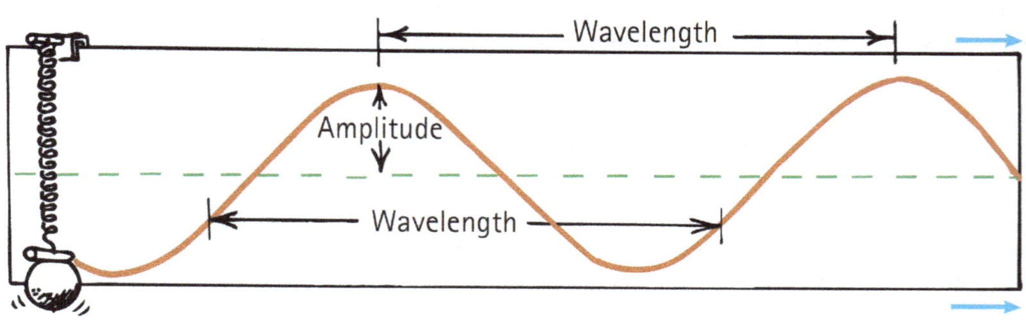

Figure 11.1 >
When the bob vibrates up and down, a marking pen traces out a sine curve on paper that moves horizontally at constant speed.

Its frequency is 2 Hz, and the time needed to complete one vibration—that is, the period—is 1/2 second. If the vibration frequency is 3 Hz, then the period is 1/3 second. Frequency and period are the inverse of each other:

$$\text{Frequency} = \frac{1}{\text{period}} \qquad \text{Period} = \frac{1}{\text{frequency}}$$

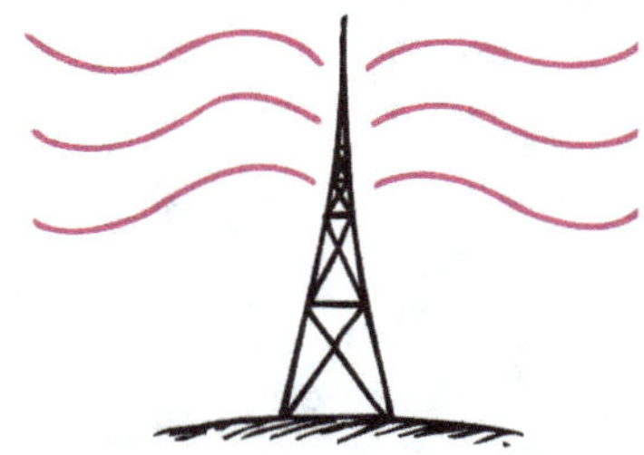

Figure 11.2
Electrons in the transmitting antenna vibrate 940,000 times each second and produce 940-kHz radio waves.

CONCEPT **CHECK**

1. An electric toothbrush completes 90 cycles every second. What are (a) its frequency and (b) its period?

2. Gusts of wind cause the Sears Building in Chicago sway back and forth, completing a cycle every ten seconds. What are (a) its frequency and (b) its period?

CHECK YOUR ANSWERS

1. (a) 90 cycles per second is 90 vibrations per second or 90 Hz; (b) 1/90 second.

2. (a) 1/10 Hz. (b) 10 s.

11.2 Wave Motion

When you drop a stone into a quiet pond, waves travel outward in expanding circles. Energy is carried by the wave, moving from place to place. The water itself goes nowhere. This can be seen by watching a leaf floating in the water. The leaf bobs up and down but doesn't travel with the waves. When you speak, the energy of your voice travels across the room at about 340 meters per second. Wave energy travels through the air. The air goes nowhere. The air itself doesn't travel at this speed. If it did, then speaking would be a windy experience.

READING CHECK

Exactly what is it that moves through the air in the case of sound? What doesn't travel in a sound wave?

Wave Speed

Wave speed is related to the frequency and wavelength of the waves. You can understand this by considering the simple case of water waves (**Figures 12.3** and **12.4**). Imagine that you fix your eyes at a stationary point on the surface of water and observe the waves passing by this point. You can measure the amount of time that passes between the arrival of one crest and the arrival of the next one (this time interval is the period). And you can also estimate the distance between crests (the wavelength). You know that speed is defined as distance divided by time. In this case, the distance is one wavelength and the time is one period. That means wave speed = wavelength/period.

For example, if the wavelength is 10 meters and the time between crests at a point on the surface is 0.5 second, the wave moves 10 meters in 0.5 seconds. So, its speed is 10 meters divided by 0.5 seconds = 20 meters per second.

Λ **Figure 11.3**
A duck creating water waves.

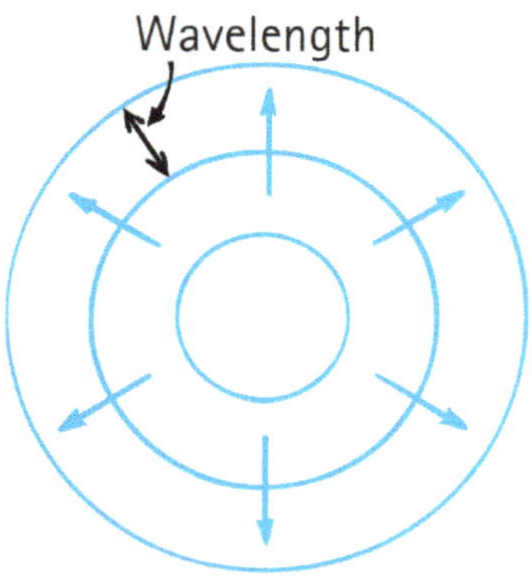

Λ **Figure 11.4**
A top view of water waves.

Since period is the inverse of frequency, the formula wave speed = wavelength/period can also be written

$$\text{Wave speed} = \text{frequency} \times \text{wavelength}$$

This relationship applies to all kinds of waves, whether they are water waves, sound waves, or light waves.

Two Types of Waves—Transverse and Longitudinal

Fasten one end of a Slinky to a wall and hold the free end in your hand. Shake it up and down and you produce vibrations that are at right angles to the direction of wave travel. The right-angled, or sideways, motion is called transverse motion. This type of wave—in which the direction of wave travel is perpendicular to the direction of the vibrating source—is called a **transverse wave**. Waves in the stretched strings of musical instruments are transverse. Electromagnetic waves are also transverse.

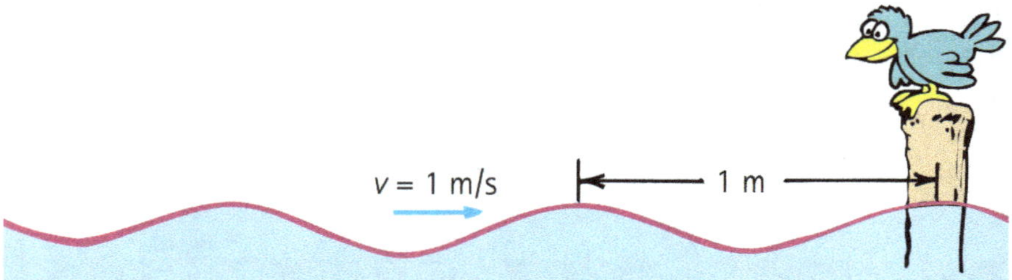

Λ **Figure 11.5**
If the wavelength is 1 m, and one wavelength per second passes the pole, then the speed of the wave is 1 m/s.

CONCEPT **CHECK**

1. If a train of freight cars, each 10 m long, rolls by you at the rate of three cars each second, what is the speed of the train?

2. If a water wave vibrates up and down three times each second and the distance between wave crests is 2 m, what are (a) the wave's frequency? (b) its wavelength? (c) its wave speed?

CHECK YOUR ANSWER

1. 30 m/s. We can see this in two ways. (1) According to the speed definition from Chapter 2, $v = d/t = (3 \times 10\text{ m})/1\text{ s} = 30\text{ m/s}$, because 30 m of train passes you in 1 s. (2) If we compare the train to wave motion, with wavelength corresponding to 10 m and frequency 3 Hz, then Speed = frequency × wavelength = 3 Hz × 10 m = 30 m/s.

2. (a) 3 Hz; (b) 2 m; (c) Wave speed = frequency × wavelength = 3/s × 2 m = 6 m/s. (Note that 3 Hz is 3 vibrations/s, and because "vibrations" has no unit, we write 3 Hz = 3/s.) It is customary to express the equation for wave speed as $v = f\lambda$, where v is wave speed, f is wave frequency, and λ (the Greek letter lambda) is wavelength.

Longitudinal waves are one in which the direction of wave travel is along the direction in which the source vibrates. You produce a longitudinal wave with your Slinky when you shake it back and forth along the Slinky's axis (**Figure 11.6a**). The vibrations are then parallel to the direction of energy transfer. Part of the Slinky is compressed, and a wave of compression travels along it. In between successive compressions is a stretched region, called a rarefaction. Both compressions and rarefactions travel in the same direction along the Slinky. Together they make up the longitudinal wave.

READING CHECK

Distinguish between a transverse wave and a longitudinal wave.

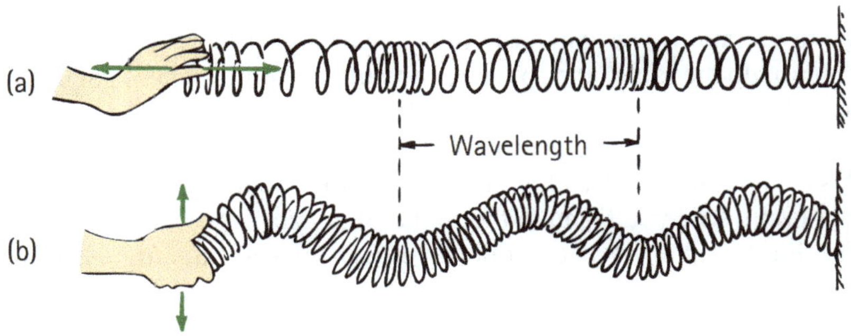

Figure 11.6
Both waves transfer energy from left to right. (a) When the Slinky is pushed and pulled rapidly along its length, a longitudinal wave is produced. (b) When the end of the Slinky is shaken up and down, a transverse wave is produced.

11.3 Sound Waves

Think of the air molecules in a room as tiny randomly-moving Ping-Pong balls. If you vibrate a ping-pong paddle in the midst of the balls, you'll set them vibrating to-and-fro. In some regions the balls are momentarily bunched up (compressions) and in other regions in between they are momentarily spread out (rarefactions). The same is done to air molecules by the vibrating prongs of a tuning fork. Vibrations made up of compressions and rarefactions spread from the tuning fork throughout the air. A sound wave is produced.

The wavelength of a sound wave is the distance between successive compressions or, equivalently, the distance between successive rarefactions. Each molecule in the air vibrates to-and-fro about some equilibrium position as the waves move by.

Our subjective impression about the frequency of sound is described as pitch. A high-pitch sound like that from a tiny bell has a high vibration frequency. Sound from a large bell has a low pitch because its vibrations are of a low frequency.

The human ear can normally hear pitches from sound in a range from about 20 to 20,000 hertz. As we age, this range shrinks Sound waves of frequencies below 20 hertz are called infrasonic waves, and those of frequencies above 20,000 hertz are called ultrasonic waves. We cannot hear infrasonic or ultrasonic sound waves. But dogs and many other animals can.

Figure 11.7
Vibrate a ping-pong paddle in the midst of a lot of ping-pong balls, and you make the balls vibrate also.

READING CHECK

How do the pitches of a tiny bell and a large bell compare?

Sound requires a medium. It can't travel in a vacuum because there's nothing to compress and stretch.

Most sound travels through air, but any elastic substance—solid, liquid, or gas—can transmit sound. An elastic substance is "springy," has resilience, and can transmit energy with little loss. Steel, for example, is elastic, whereas lead or putty is not. Many solids and liquids conduct sound better than air. You can hear the sound of a distant train clearly by placing your ear against the rail. When swimming, have a friend some distance away click two large coins together beneath the water surface while you're submerged. Observe how well water conducts the sound.

Pause to reflect on the physics of sound while you are quietly listening to your radio sometime. The radio loudspeaker is a paper cone that vibrates in rhythm with an electrical signal. Air molecules next to the vibrating cone are set into vibration. These in turn vibrate against neighboring molecules, which in turn do the same, and so on. As a result, rhythmic patterns of compressed and rarefied air emanate from the loudspeaker and vibrate the air in the whole room. This sets your eardrum into vibration, which in turn sends cascades of rhythmic electrical impulses along the cochlea nerve canal and into your brain. And you listen to the sound of music.

The loudspeaker of your radio or other sound-producing systems changes electrical signals into sound waves. The electrical signals pass through a coil wound around the neck of a paper cone. This coil acts as an electromagnet, which is located near a permanent magnet. When the current direction is one way, magnetic force pushes the electromagnet toward the permanent magnet, pulling the cone inward. When the current direction reverses, the cone is pushed outward. Vibrations in the electric signal then cause the cone to vibrate. Vibrations of the cone produce sound waves in the air.

Be clear about the distinction between *frequency* and *speed*. How frequently a wave vibrates is altogether different from how fast it moves from one location to another.

Speed of Sound

If you watch a person at a distance chopping wood or hammering, you can easily see that the blow occurs before you hear it. Likewise, you see a flash of lightning before you hear thunder. Sound takes time to travel from one location to another. The speed of sound depends on wind conditions, temperature, and humidity. But it doesn't depend on the loudness or the frequency of sound. All sounds, loud or soft, high- or low-pitched, travel at the same speed in a given medium. The speed of sound in dry air at 0°C is about 330 meters per second (nearly 1200 km/h). It travels slightly faster if

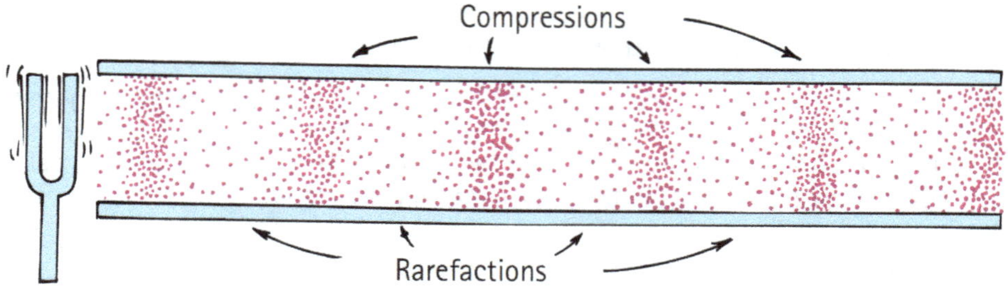

Λ Figure 11.8
Compressions and rarefactions travel (both at the same speed in the same direction) from the tuning fork through the air in the tube.

water vapor is in the air. Sound travels faster through warm air than cold air. This makes sense because faster-moving molecules in warm air bump into each other more often and transmit a pulse in less time. For each degree rise in temperature above 0°C, the speed of sound in air increases by 0.6 meter per second. So, in air at a normal 20°C room temperature, sound travels at about 340 meters per second. In water, sound speed is about 4 times its speed in air; in steel it's about 15 times.

Figure 11.9
Waves of compressed and rarefied air, produced by the vibrating cone of the loudspeaker, make up the pleasing sound of music.

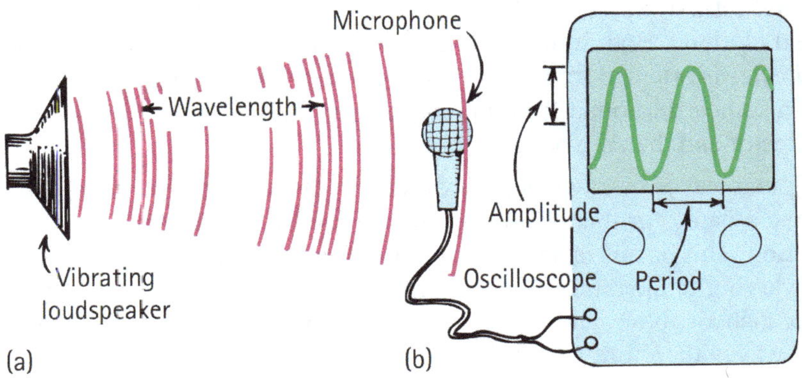

Figure 11.10
The radio loudspeaker is a paper cone that vibrates in rhythm with an electric signal. The sound produced sets up similar vibrations in the microphone, which are displayed on an oscilloscope. The shape of the waveform on the oscilloscope screen reveals information about the sound.

CONCEPT **CHECK**

1. Do compressions and rarefactions in a sound wave travel in the same direction or in opposite directions from one another?
2. What is the approximate distance of a thunderstorm when you note a 3-s delay between the flash of lightning and the sound of thunder? (Use 340 m/s for the speed of sound.)

CHECK YOUR ANSWERS

1. They travel in the same direction.
2. In 3 s the sound travels (340 m/s × 3 s) = 1020 m. Since there is no noticeable delay between the time the lightning bolt is created and the time you see it, the storm is slightly more than 1 km away.

11.4 Reflection and Refraction

We call the reflection of sound an *echo*. A large fraction of sound energy is reflected from a surface that is rigid and smooth. Less sound is reflected if the surface is soft and irregular. Sound energy that is not reflected is either transmitted or absorbed.

Sound reflects from a smooth surface the same way light does—the angle of incidence is equal to the angle of reflection (**Figure 11.11**). Reflected sound in a room makes it sound lively and full, as you have probably noticed while singing in the shower. Sometimes when the walls, ceiling, and floor of a room are too reflective, the sound becomes garbled. This is due to multiple reflections called reverberations. On the other hand, if the reflective

READING CHECK

Compared with the way in which light reflects, how does sound reflect?

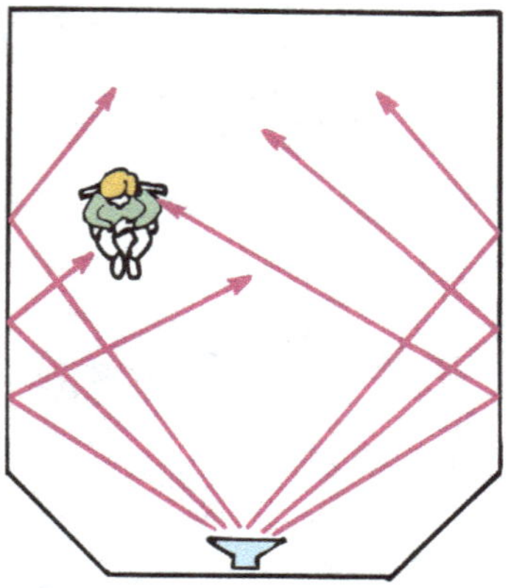

^ Figure 11.11
The angle of incident sound is equal to the angle of reflected sound.

surfaces are too absorbent, the sound level is low and the room may sound dull and lifeless. In the design of an auditorium or concert hall, a balance must be achieved between reverberation and absorption. The study of sound properties is called acoustics.

In concert halls good acoustics require highly reflective surfaces behind the stage to direct sound out to an audience. Sometimes reflecting surfaces are suspended above the stage. In some opera halls the reflective surfaces are large, shiny, plastic plates that also reflect light (**Figure 11.12**). A listener can look up at these reflectors and see the reflected images of the members of the orchestra (the plastic reflectors are somewhat curved, which increases the field of view). Both sound and light obey the same law of reflection, so if a reflector is oriented so that you can see a particular musical instrument, you'll also hear it. Sound from the instrument follows the line of sight to the reflector and then to you.

Sound waves bend when parts of the waves travel at different speeds. We have a name for this bending of sound due to differences in speed—**refraction**. This occurs in uneven winds or when sound is traveling through air of varying temperatures. On a warm day, air near the ground may be warmer than air above, and so the speed of sound near the ground is greater. Sound waves therefore tend to bend away from the ground, resulting in sound that does not seem to carry well (**Figure 11.13**).

READING CHECK

In terms of speed, what occurs when sound bends from a straight-line course?

The refraction of sound occurs under water, too, where the speed of sound varies with temperature. This poses a problem for surface vessels that chart the bottom features of an ocean by bouncing ultrasonic waves off the bottom. This is a blessing for submarines that wish to escape detection. Layers of water at different temperatures (thermal gradients) result in refraction of sound that leaves gaps, or "blind spots," in the water. This is where submarines hide. If it weren't for refraction, submarines would be easier to detect.

The multiple reflections and refractions of ultrasonic waves are used by physicians in a technique for harmlessly "seeing" inside the body without the use of X-rays. When high-frequency sound (ultrasound) enters the body,

^ Figure 11.12
The plastic plates, or acoustic shells, above the orchestra reflect both light and sound. Adjusting them is quite simple: What you see is what you hear.

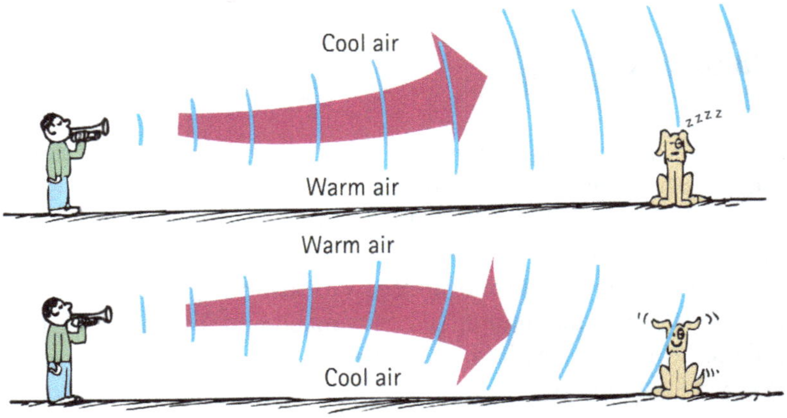

< **Figure 11.13**
Sound waves are bent in air of uneven temperatures.

it is reflected more strongly from the outside of organs than from their interior. A picture of the outline of the organs is obtained (**Figure 11.14**). This ultrasound echo technique has always been used by bats, who emit ultrasonic squeaks and locate objects by their echoes. Dolphins do this and much more.

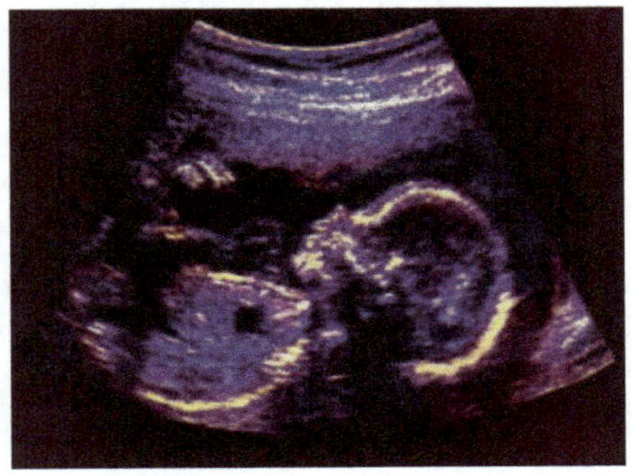

∧ **Figure 11.14**
An ultrasound image of the 14-week-old fetus of Megan Hewitt Abrams.

Dolphins and Acoustical Imaging

The primary sense of the dolphin is acoustic, because sight is not a very useful sense in the often murky and dark depths of the ocean. Whereas sound is a passive sense for us, it is an active sense for dolphins when they send out sounds and then perceive their surroundings via echoes. The ultrasonic waves emitted by a dolphin enables it to "see" through the bodies of other animals and people. Because skin, muscle, and fat are almost transparent to dolphins, they "see" only a thin outline of the body, but the bones, teeth, and gas-filled cavities are clearly apparent. Physical evidence of cancers, tumors, and heart attacks can all be "seen" by dolphins—as humans have only recently been able to do with ultrasound.

What's more interesting, a dolphin can reproduce the sonic signals that paint the mental image of its surroundings. Thus, the dolphin probably communicates its experience to other dolphins by providing the full acoustic image of what is "seen," placing the image directly in the minds of other dolphins. It needs no word or symbol for "fish," for example, but communicates an image of the real thing. It is quite possible that dolphins highlight portions of the images they send by selective filtering, as we similarly do when sharing a musical concert to others via various means of sound reproduction. Small wonder that the language of the dolphin is very unlike our own!

∧ **Figure 11.15**
A dolphin emits ultra-high-frequency sound to locate and identify objects in its environment. Distance is sensed by the time delay between sending sound and receiving its echo, and direction is sensed by differences in time for the echo to reach its two ears. A dolphin's main diet is fish and, since hearing in fish is limited to fairly low frequencies, they are not alerted to the fact they are being hunted.

CONCEPT **CHECK**

A depth-sounding vessel surveys the ocean bottom with ultrasonic sound that travels an average 1530 m/s in seawater. How deep is the water if the time delay of the echo from the ocean floor is 2 s?

CHECK YOUR ANSWER

The 2 s delay means it takes 1 s for the sound to reach the bottom (and another 1 s to return). Sound traveling at 1530 m/s for 1 s tells us the bottom is 1530 m deep.

11.5 Forced Vibrations and Resonance

If you strike an unmounted tuning fork, its sound is rather faint. Repeat but hold the fork against a table after you strike it, and the sound is louder. This is because the table is forced to vibrate, and with its larger surface it sets more air in motion. The table can be forced into vibration by a fork of any frequency. This is a case of **forced vibration**. The vibration of a factory floor caused by the running of heavy machinery is another example of forced vibration. A more pleasing example is given by the sounding boards of stringed instruments.

Drop a wrench and a baseball bat on a concrete floor, and you can tell the difference in their sounds. This is because each vibrates differently when striking the floor. They are not forced to vibrate at a particular frequency, but instead each vibrates at its own special frequency. When disturbed, any object made of an elastic material vibrates at its own special set of frequencies, which together form its special sound. We speak of an object's natural frequency, which depends on factors such as the elasticity and shape of the object. Bells and tuning forks, of course, vibrate at their own special frequencies. And interestingly, most other things, from planets to atoms and almost everything in between, have a springiness to them and vibrate at one or more natural frequencies.

READING CHECK

What besides bells and tuning forks have their own natural frequencies?

When the frequency of forced vibrations imposed on an object matches the object's natural frequency, a dramatic increase in amplitude occurs. This phenomenon is called **resonance** (which means "resounding" or "sounding again"). Putty doesn't resonate because it isn't elastic, and a dropped handkerchief is too limp. In order for something to resonate, it needs a force to pull it back to its starting position and enough energy to keep it vibrating.

A common experience illustrating resonance occurs on a swing. When pumping a swing, you pump in rhythm with the natural frequency of the swing. More important than the force with which you pump is the timing. Even small pumps or small pushes from someone else, if delivered in rhythm with the frequency of the swinging motion, produce large amplitudes.

A common classroom demonstration of resonance is illustrated with a pair of tuning forks adjusted to vibrate at the same frequency and spaced a meter or so apart. When one of the forks is struck, it sets the other fork into vibration. This is a small-scale version of pushing a friend on a swing—it's the timing that's important. When a series of sound waves impinge on the fork, each compression gives the prongs a tiny push. Because the frequency of these pushes matches the natural frequency of the fork, the succession

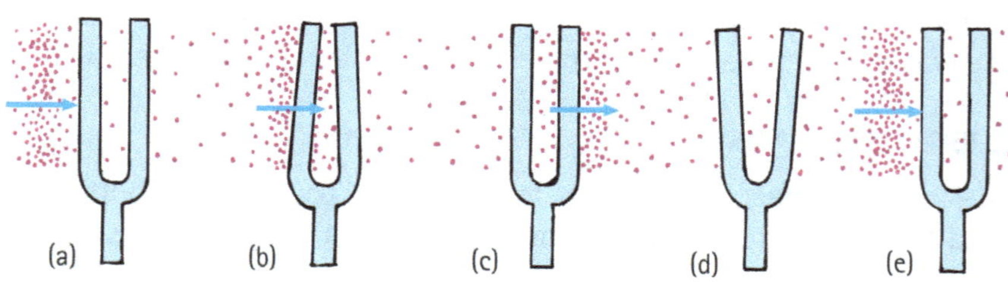

Figure 11.16 >

Stages of resonance. (a) The first compression meets the prong of the fork and gives it a tiny and momentary push; (b) the prong bends and then (c) returns to its initial position just at the time a rarefaction arrives. Still moving, the prong (d) overshoots in the opposite direction. Just when the prong restores to its initial position, (e) the next compression arrives to repeat the cycle. Now it bends farther because it is already moving.

∧ **Figure 11.17**
In 1940, four months after being completed, the Tacoma Narrows Bridge in the state of Washington was destroyed by wind-generated resonance. A mild gale produced an irregular force in resonance with the natural frequency of the bridge, steadily increasing the amplitude of vibration until the bridge collapsed.

of pushes increases the amplitude of vibration. This is because the pushes occur at the right time and in the same direction as the instantaneous motion of the fork. The motion of the second fork is called a sympathetic vibrations.

If the forks are not adjusted for matched frequencies, the timing of pushes is off and resonance does not occur. When you tune a radio or other digital device, you adjust the natural frequency of the electronics in the radio to match one of the many surrounding signals in the air. The device then resonates to one station at a time instead of playing all stations at once.

Resonance occurs whenever successive impulses are applied to a vibrating object in rhythm with its natural frequency. Cavalry troops marching across a footbridge near Manchester, England, in 1831 mistakenly caused the bridge to collapse when they marched in rhythm with the bridge's natural frequency. Since then, it is customary to order troops to "break step" when crossing bridges. A 20th-century bridge disaster was caused by wind-generated resonance (**Figure 11.17**). Today's engineers take resonance very seriously in structure design.

11.6 Wave Interference

One of the most interesting properties of all waves is **interference**. Consider transverse waves. When the crest of one wave overlaps the crest of another, the crests add together. The result is a wave of increased amplitude. This is *constructive interference*. When the crest of one wave overlaps the trough of another, the opposite occurs. The high portions of one wave simply fill in the low portions of another. This is *destructive interference* (**Figure 11.18**).

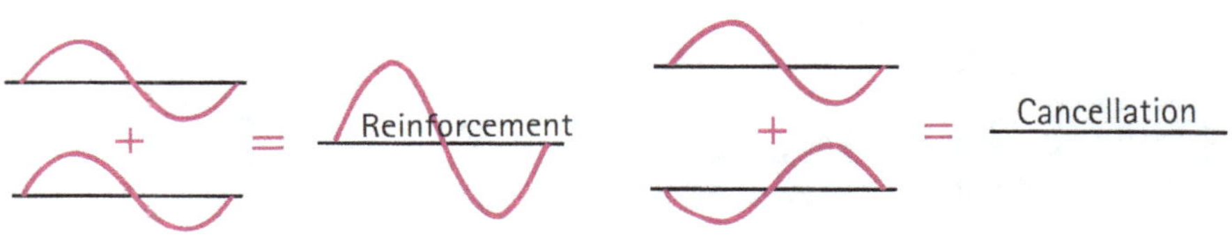

< **Figure 11.18**
Constructive and destructive interference in a transverse wave.

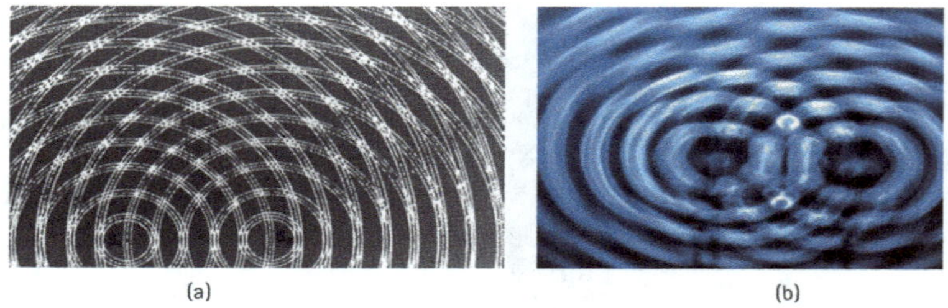

Figure 11.19 >
Two sets of overlapping water waves produce an interference pattern.

Wave interference is easiest to see in water. Study **Figure 11.19** and look at the interference pattern made when two vibrating objects touch the surface of water. Notice the regions where a crest of one wave overlaps the trough of another. This results in regions of zero amplitude. At points along these regions, the waves arrive out of step. We say they are out of phase with one another.

READING CHECK
What kind of waves exhibit the property of interference?

Interference is a property of all wave motion, whether the waves are water waves, sound waves, or light waves. We see a comparison of interference for both transverse and longitudinal waves in **Figure 11.20**. In the case of sound, the crest of a wave corresponds to a compression and the trough of a wave corresponds to a rarefaction.

Destructive sound interference is a useful property in anti-noise technology. Some noisy devices such as jackhammers are equipped with microphones that send the sound of the device to electronic microchips, which create mirror-image wave patterns of the sound signals that are fed to earphones worn by the operator. Sound compressions (or rarefactions) from the device are canceled by mirror image rarefactions (or compressions) in the earphones. The combination of signals cancels the jackhammer noise. Anti-noise devices are also common in some aircraft, which are much

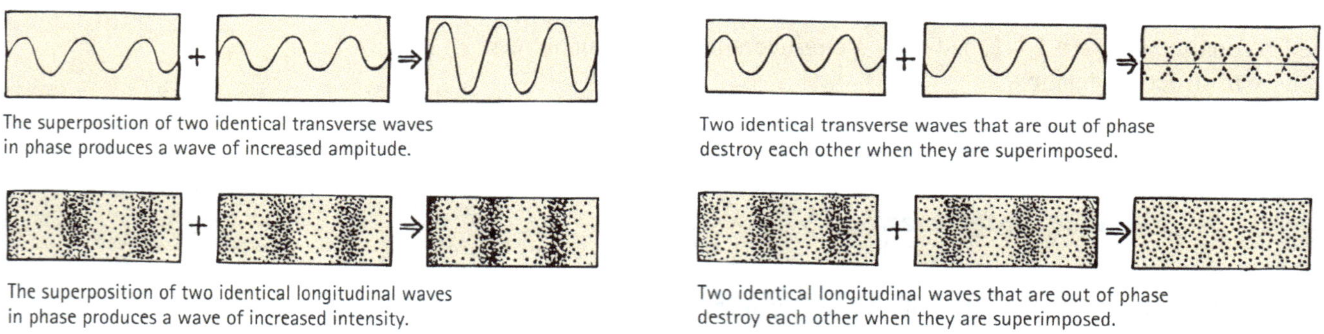

∧ Figure 11.20
Constructive (left panels) and destructive (right panels) wave interference in transverse and longitudinal waves.

∧ Figure 11.21
When a mirror image of a sound signal combines with original signal, the sound is canceled.

quieter inside than before this technology was introduced. The same occurs in hearing aids when background noise is cancelled.

Perform perhaps the most awesome personal demonstation that highlights the physics of sound! Acquire a sound system with a pair of detachable speakers as shown in Figure 11.21. Set the system for monaural sound (not stereo). Reverse the wiring on one of the speakers by switching the positive and negative wire inputs. Your speakers are then out of phase. When both speakers emit the same signal, a compression emitted by one speaker occurs at the same time a rarefaction is being emitted by the other. The resulting sound is not as full and not as loud as from speakers properly connected in phase. That's because the waves, mainly the longer ones, are being canceled by interference.

Now face the speakers toward each other. First hold them apart, with music playing, then bring them closer together. You'll hear weaker sound because shorter waves are being canceled. The sound is tinny. When the pair of speakers are brought face to face and touching each other, you'll hear very little sound! Be sure to try this and share with your friends.

Λ **Figure 11.22**
Ken Ford tows gliders in quiet comfort when he wears his noise-canceling earphones.

Beats—An Effect of Sound Interference

A tone is a sound of distinct pitch and duration. When two tones of slightly different frequency are sounded together, a fluctuation in loudness may be heard. The sound is loud, then faint, then loud, then faint, and so on. This periodic variation in loudness is called **beats**. Beats are due to interference.

Beats can occur with any kind of wave and provide a practical way to compare frequencies. For example, a piano tuner listens for beats produced between a standard tuning fork and the frequency of a particular string on the piano. When the frequencies are identical, the beats disappear. The members of an orchestra tune their instruments by listening for beats between their instruments and a standard tone produced by a piano or some other instrument.

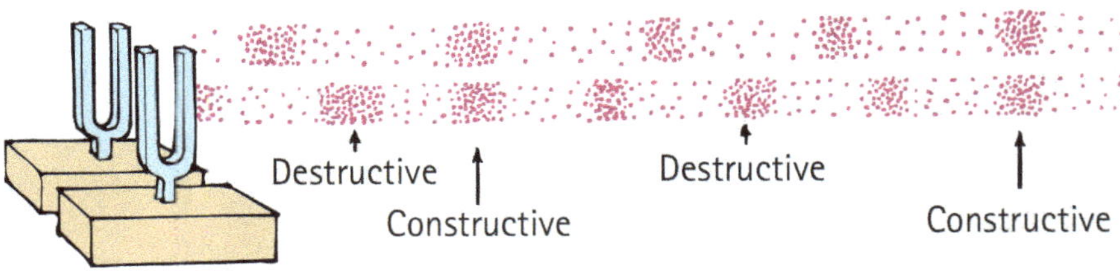

< **Figure 11.23**
The interference of two sound sources of slightly different frequencies produces beats.

Standing Waves from Waves Passing Through Each Other

Another interesting effect of interference is standing waves. Tie a rope to a wall and shake the free end up and down. (A rubber tube works even better.) Waves that hit the wall are reflected back along the rope. By shaking the rope just right, you can cause the incident and reflected waves to interfere and form a **standing wave**, where parts of the rope, called the *nodes*, are stationary. You can hold your fingers on either side of the rope at a node, and the rope will not touch them. Thus, nodes are positions of zero rope displacement. The distance between nodes is one half wavelength. Two loops make up a complete wave. The positions on a standing wave that have the largest displacements are known as *antinodes* and occur halfway between nodes.

Standing waves are produced when two sets of waves of equal amplitude and wavelength pass through each other in opposite directions. Then the waves are steadily in and out of phase with each other and produce stable regions of constructive and destructive interference (Figure 11.24).

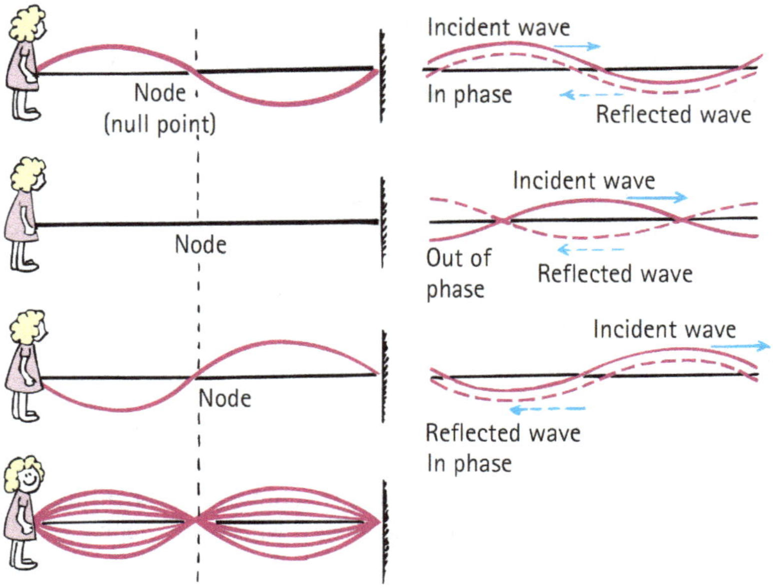

Figure 11.24 >
The incident and reflected waves interfere to produce a standing wave.

Standing waves are set up in the strings of musical instruments when plucked, bowed, or struck. Standing waves can be set up in a tub of water or a bowl of soup by sloshing it back and forth with the right frequency. They can be produced with either transverse or longitudinal vibrations.

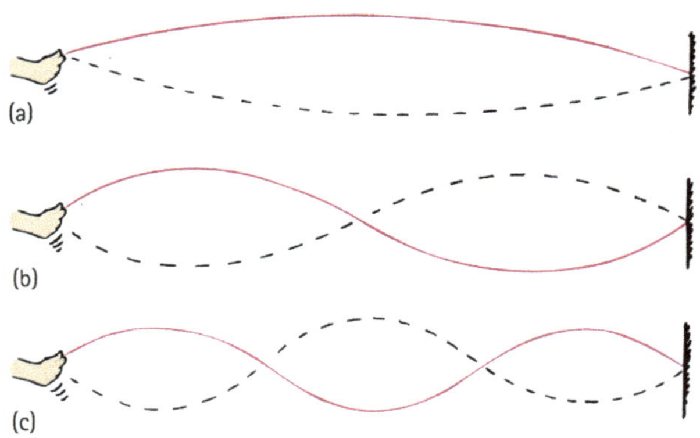

Figure 11.25 >
(a) Shake the rope until you set up a standing wave of one loop (1/2 wavelength). (b) Shake with twice the frequency and produce a wave having two loops (1 wavelength). (c) Shake with three times the frequency and produce three loops (3/2 wavelengths).

Quick Activity

Sound Board
Strike a tuning fork with a rubber hammer or the heel of your shoe. (Do not strike the tuning fork on the edge of a table.) Listen to the sound produced by the vibrating fork. Hold the base of the fork against the top of a table, or against other surfaces. How does the area of a "sounding board" affect loudness?

CONCEPT **CHECK**
Is it possible for one wave to cancel another so that no amplitude remains?

CHECK YOUR ANSWER
Yes. This is destructive interference. In a standing wave in a rope, for example, parts of the rope—the nodes—have no amplitude because of destructive interference.

11.7 The Doppler Effect

A pattern of water waves produced by a bug jiggling its legs and bobbing up and down in the middle of a quiet puddle is shown in Figure 11.26. The bug is not going anywhere but is merely treading water in a fixed position. The waves it makes are concentric circles because wave speed is the same in all directions. If the bug bobs in the water at a constant frequency, the distance between wave crests (the wavelength) is the same in all directions. Waves encounter point A as frequently as they encounter point B. This means that the frequency of wave motion is the same at points A and B, or anywhere else near the bug. This wave frequency is the same as the bug's bobbing frequency.

Suppose now that the jiggling bug begins moving to the right at a speed less than the wave speed. In effect, the bug chases part of the waves it makes. The circular waves are no longer concentric (Figure 11.27). Each one has its center where the bug was previously. The outermost wave (wave 1 in Figure 11.27) was made when the bug was at the center of that circle. The center of wave 2 was made when the bug was at the center of that circle, and so on. The centers of the circular waves move in the direction of the swimming bug. Although the bug maintains the same bobbing frequency as before, an observer at B sees the waves coming more frequently. In other words, observer B measures a higher frequency. An observer at A, on the other hand, measures a lower frequency because of the longer time between wave-crest arrivals. This change in frequency due to the motion of the source (or receiver) is called the **Doppler effect** (after the Austrian scientist Christian Doppler).

Water waves spread over the two-dimensional surface of the water. Sound and light waves, on the other hand, travel in three-dimensional space in all directions like an expanding balloon. Just as circular waves are closer together in front of the swimming bug, spherical sound or light waves ahead of a moving source are closer together and reach a receiver more frequently. The Doppler effect applies to all types of waves.

The Doppler effect is evident when you hear the changing pitch of a siren as the vehicle drives by. When it approaches, the pitch is higher than normal. The wave crests reach your ear more frequently. When the vehicle passes and moves away, you hear a drop in pitch. Then the crests of the waves hit your ear less frequently.

The Doppler effect also occurs when the receiver moves. Move toward a stationary wave source and you encounter its waves more frequently. Move away, and you encounter waves less frequently. The Doppler effect results from relative motion between a wave source and a receiver.

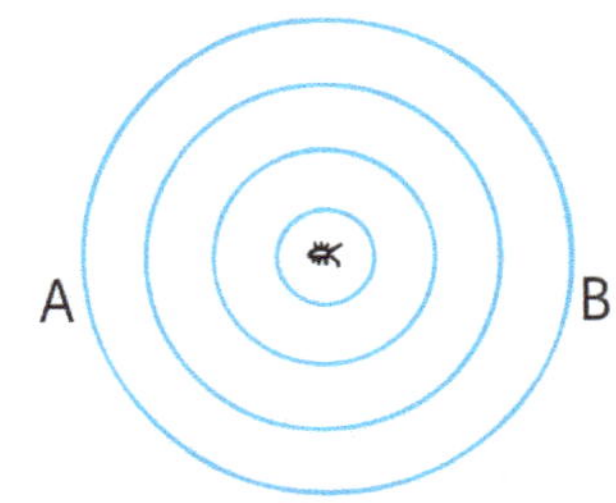

Figure 11.26
Top view of water waves made by a stationary bug jiggling in still water.

READING CHECK
What is the Doppler effect?

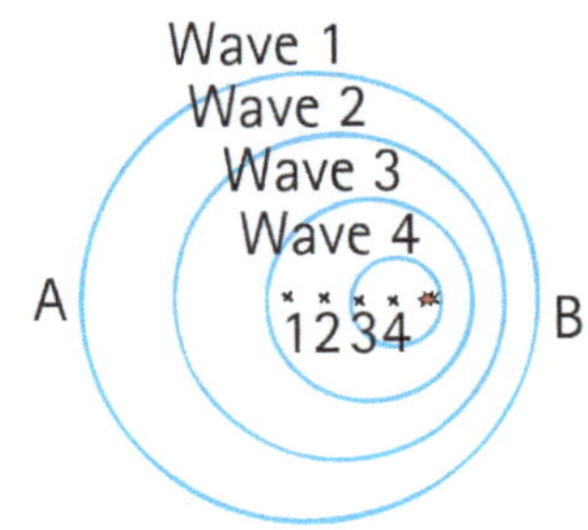

Figure 11.27
Water waves made by a bug swimming in still water toward point B.

Figure 11.28
The pitch (frequency) of sound increases when the source moves toward you and decreases when moving away.

The Doppler effect also occurs with light waves. When a source of light waves approaches, the increase in frequency is called a blue shift. That's because the increase is toward the high-frequency, blue end of the color spectrum. A decrease in frequency is called a red shift, referring to a shift toward the lower-frequency, red end of the color spectrum. The galaxies, for example, show a red shift in the light they emit. A measurement of this shift permits a calculation of the speeds at which they are receding from the Earth. A rapidly spinning star shows a red shift on the side turning away from us and a relative blue shift on the side turning toward us. This enables astronomers to calculate the star's spin rate.

CONCEPT **CHECK**

When a wave source moves toward you, do you measure an increase or decrease in wave speed?

CHECK YOUR ANSWER

Neither! It is the frequency of the waves that changes for a moving source, not the wave speed. Be clear about the distinction between frequency and speed. How frequently a wave vibrates is altogether different from how fast it moves from one place to another.

11.8 Bow and Shock Waves

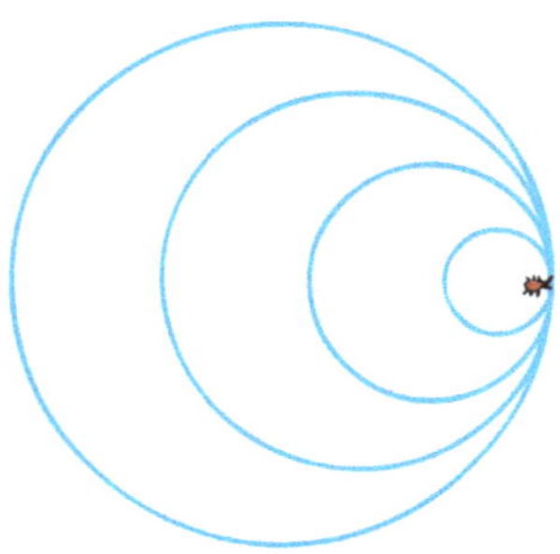

Figure 11.29
Wave pattern made by a bug swimming at wave speed.

When a moving wave source travels as fast as the waves it produces, a wave barrier is produced. Consider the bug in our previous example. When it swims as fast as the wave speed, can you see that it keeps up with the waves it produces? Instead of the waves getting ahead of the bug, they pile up and overlap directly in front of it (**Figure 11.29**). The bug encounters a wave barrier. Much effort is required of the bug to swim over this barrier—before it can swim faster than wave speed.

The same thing happens when an aircraft travels at the speed of sound. The sound waves produced by the engines overlap to produce a barrier of compressed air on the leading edges of the wings and other parts of the craft. Considerable thrust is required for the aircraft to push through this barrier. Once through, the craft can fly faster than the speed of sound without similar opposition. It is now supersonic. It is like the bug, which once over its wave barrier finds the water ahead relatively smooth and undisturbed.

When the bug swims faster than wave speed, it produces a V-shaped wave pattern like the one shown in **Figure 11.30**. It outruns the waves it produces. The V-pattern caused by the overlapping waves is called a **bow wave**, which appears to be dragging behind the bug. The familiar bow (rhymes with cow) wave generated by a speedboat knifing through the water is produced by the overlapping of many periodic circular waves.

Some wave patterns made by sources moving at various speeds are shown in **Figure 11.31**. Note that after the speed of the source exceeds wave speed, increased speeds produce a narrower and narrower V shape. Note: Bow waves generated by boats in water are more complex than indicated here. Our idealized treatment serves as an analogy for the production of the less complex shock waves in air.

READING CHECK

Compared to wave speed, how fast must an object move to produce a V-shaped wave pattern?

A speedboat knifing through the water generates a two-dimensional bow wave. A supersonic aircraft similarly generates a three-dimensional **shock wave**. Just as a bow wave is produced by overlapping circles that form a V, a shock wave is produced by overlapping spheres that form a cone. And just as the bow wave of a speedboat spreads until it reaches the shore of a lake, the conical wake generated by a supersonic craft spreads until it reaches the ground.

The bow wave of a speedboat that passes by can splash and douse you if you are at the water's edge. In a sense, you can say that you are hit by a "water boom." In the same way, when the conical shell of compressed air that sweeps behind a supersonic aircraft reaches listeners on the ground, the sharp crack they hear is described as a **sonic boom**.

We don't hear a sonic boom from slower-than-sound, or subsonic, aircraft because the sound waves reach our ears one at a time and make one continuous tone. Only when the craft moves faster than sound do the waves overlap to reach the listener in a single burst. The sudden increase in pressure is much the same in effect as the sudden expansion of air produced by an explosion. Both processes direct a burst of high-pressure air to the listener. The ear normally can't recognize a difference between an explosion and a sonic boom.

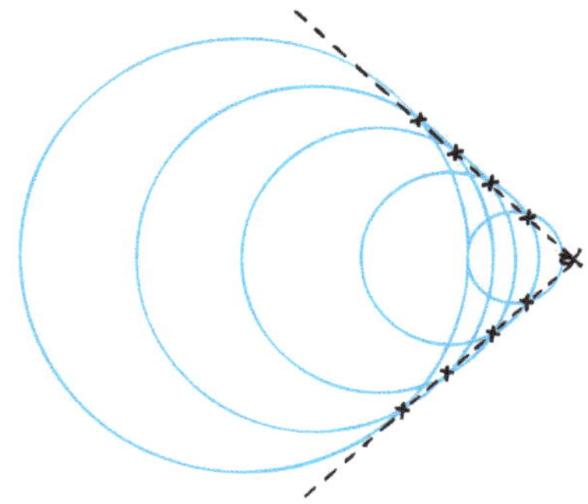

Λ Figure 11.30
A bow wave, the pattern made by a bug swimming faster than wave speed. The points at which adjacent waves overlap (X) produce the V shape.

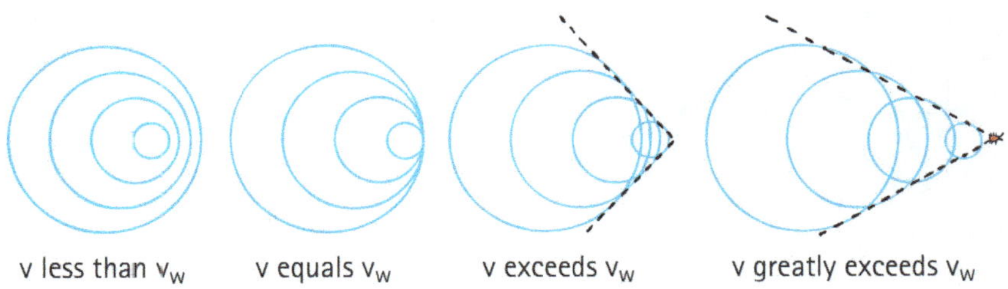

< Figure 11.31
Patterns made by a bug swimming at successively greater speeds. Overlapping at the edges occurs only when the bug swims faster than wave speed.

A water skier is familiar with the fact that next to the high hump of the bow wave is a V-shaped depression. The same is true of a shock wave. There are two cones: a high-pressure one generated at the bow of the supersonic aircraft and a low-pressure one that follows at the tail. You can see the edges of these cones in the photograph of the supersonic bullet in **Figure 11.32**. Between these two cones the air pressure rises sharply to above atmospheric pressure, then falls below atmospheric pressure before sharply returning to normal beyond the inner tail cone (**Figure 11.33**). This overpressure suddenly followed by underpressure intensifies the sonic boom.

Figure 11.32 >
Shock wave of a bullet piercing a sheet of Plexiglas. The deflection of light passing through the shock wave's compressed air makes it visible. Look carefully and see the second shock wave originating at the tail of the bullet.

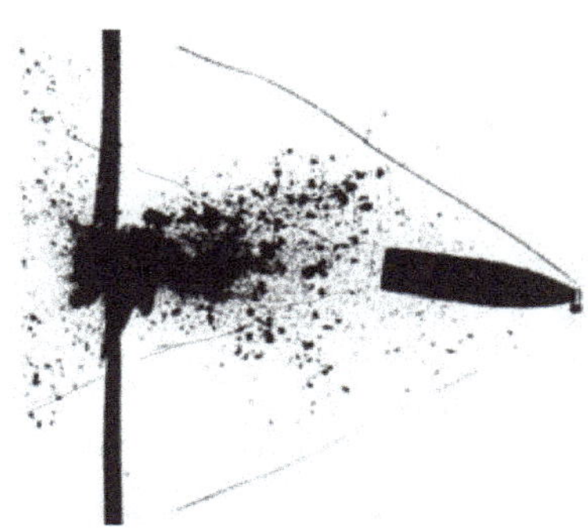

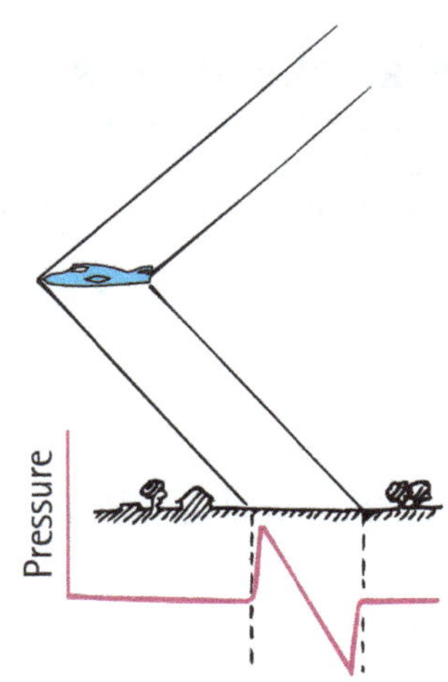

Λ Figure 11.33
A shock wave and the air pressure difference they cause.

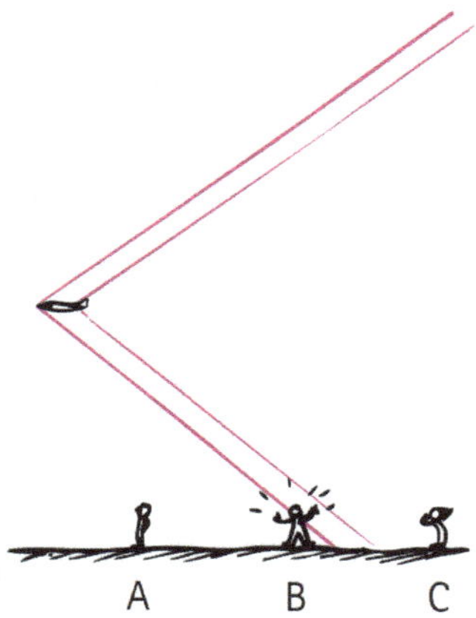

∧ Figure 11.34
The shock wave has not yet reached listener A, but is now reaching listener B and has already reached listener C.

In **Figure 11.34**, listener B is in the process of hearing a sonic boom. Listener C has already heard it, and listener A will hear it shortly. The aircraft that generated this shock wave may have broken through the sound barrier hours ago!

It is not necessary that the moving source be "noisy" to produce a shock wave. Once an object—even a silent one—is moving faster than the speed of sound, it makes sound. A supersonic bullet passing overhead produces a crack, which is a small sonic boom. If it were larger and disturbed more air in its path, the crack would be more boomlike. When a lion tamer cracks a whip, the cracking sound is a sonic boom produced by the tip when it travels faster than the speed of sound. Both the bullet and the whip are not in themselves sound sources. But when traveling at supersonic speeds they produce their own sound as they generate shock waves.

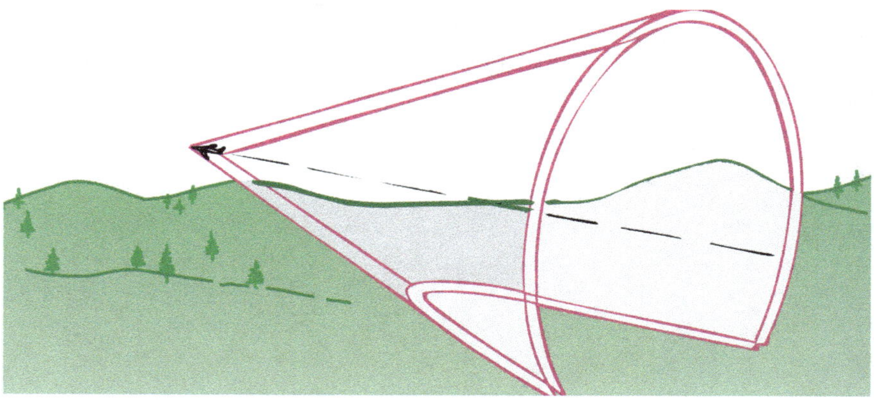

Figure 11.35
A shock wave is made up of two cones—a high-pressure cone with the apex at the bow of the aircraft and a low-pressure cone with the apex at the tail.

CONCEPT **CHECK**
Is it correct to say that a sonic boom is produced when an aircraft breaks through the sound barrier?

CHECK YOUR ANSWER
This is incorrect, and is similar to saying a boat produces a bow wave when it overtakes its own waves. A shock wave and its resulting sonic boom are swept continually behind an aircraft the entire time the craft is traveling faster than sound, just as a bow wave is swept continuously behind a speedboat.

Concept Review

Chapter 11

Summary of Terms

Amplitude For a wave or vibration, the maximum displacement on either side of the equilibrium (midpoint) position.

Beats A series of alternate reinforcements and cancellations produced by the interference of two waves of slightly different frequency, heard as a throbbing effect in sound waves.

Bow Wave The V-shaped wave produced by an object moving across a liquid surface at a speed greater than the wave speed.

Doppler Effect The change in frequency of wave motion due to the motion of the wave source or receiver.

Forced Vibration The setting up of vibrations in an object by a vibrating force.

Frequency For a vibrating body or medium, the number of vibrations per unit time. For a wave, the number of crests that pass a particular point per unit time.

Hertz The SI unit of frequency. It equals one vibration per second.

Interference The pattern formed by superposition of different sets of waves that produces mutual reinforcement in some places and cancellation in others.

Period The time in which a vibration is completed. The period of a wave equals the period of the source and is equal to 1/frequency.

Refraction The bending of a wave caused by differences in wave speed.

Resonance The response of a body when a forcing frequency matches its natural frequency.

Shock Wave The cone-shaped wave created by an object moving at supersonic speed through a fluid.

Sonic Boom The loud sound resulting from the incidence of a shock wave.

Standing Wave A stationary wave pattern formed in a medium when two sets of identical waves pass through the medium in opposite directions.

Transverse Wave A wave in which the medium vibrates in a direction perpendicular (transverse) to the direction in which the wave travels. Light is an example.

Wave A disturbance or vibration propagated from point to point in a medium or in space.

Wave Speed The speed with which waves pass a particular point:

$$\text{Wave speed} = \text{frequency} \times \text{wavelength}$$

Wavelength The distance between successive crests, troughs, or identical parts of a wave.

Review Questions

1. What is the source of all waves?
2. How do frequency and period relate to each other?
3. What is it that moves from source to receiver in wave motion?
4. What is the relationship among frequency, wavelength, and wave speed?
5. Why will sound not travel in a vacuum?
6. In what direction are longitudinal vibrations relative to the direction of wave travel?
7. How does the angle of incidence compare with the angle of reflection for sound?
8. How does the speed of sound in water compare with the speed of sound in air?
9. What causes refraction?
10. What is a reverberation?
11. Why does a struck tuning fork sound louder when it is held against a table?
12. Give three examples of forced vibration.

13. Distinguish between forced vibrations and resonance.

14. When you listen to a car radio, why are you able to hear only one station at a time rather than all stations at once?

15. What kind of waves exhibit interference?

16. Distinguish between constructive interference and destructive interference.

17. Can the Doppler effect be observed with longitudinal waves, transverse waves, or both?

18. How does the V shape of a bow wave depend on the speed of the wave source?

19. True or false: A sonic boom occurs only when an aircraft is breaking through the sound barrier.

20. Does sound tend to bend upward or downward when its speed near the ground is greater than its speed at a higher level?

Answers to Odd-Numbered Questions

1. Something vibrating.

3. Energy (not the medium itself).

5. There is no medium to vibrate.

7. Like that of light; angle of incidence equals angle of reflection

9. Slowing or speeding of one part of a sound wave relative to other parts.

11. It sets the surface of the table vibrating, which has greater surface area.

13. Forced vibration occurs at any frequency, whereas resonance occurs when vibration is at a frequency matching an object's natural frequency.

15. All waves. Interference is characteristic of all wave motion.

17. Both

19. False (aircraft may have broken through long before one encounters a sonic boom).

Chapter 12

Light and Color

THE MAIN IDEA

Let's examine the properties of light as it interacts with matter.

Light is the only thing we see, but what exactly is light? What does it have in common with sound? And why does light travel so fast? How does it travel through glass and reflect from mirrors, such as those of the James Webb Space Telescope? Does light lose speed when it passes through transparent materials like glass, or when it reflects? Why do colored lights red, green, and blue mix to form white; but red, green, and blue paints mix to a muddy brown? Does the blue of the sky have a simple explanation? And why are sunsets red and clouds white? We begin our study of light by looking at its electromagnetic nature, how it interacts with materials, and how it appears so nicely as color.

12.1 Electromagnetic Spectrum

If you shake the end of a stick back and forth in still water, you'll create waves on the water surface. If you similarly shake an electrically charged rod to-and-fro in empty space, you'll create electromagnetic waves in space, as shown in **Figure 12.1**. This is because the moving charge is in effect an electric current. Recall from Chapter 10 that a magnetic field surrounds an electric current and that the field changes as the current changes. Recall also from Chapter 10 that a changing magnetic field induces an electric field—*electromagnetic induction*. And what does the changing electric field do? It induces a changing magnetic field. The vibrating electric and magnetic fields regenerate each other to make up an *electromagnetic wave*, as shown in **Figure 12.2**. In the vacuum of space, all electromagnetic waves travel at the same speed.

∧ Figure 12.1
Shake an electrically charged object to-and-fro, and you produce an electromagnetic wave.

The classification of electromagnetic waves according to frequency is the **electromagnetic spectrum** (**Figure 12.3**). The electromagnetic spectrum is a continuous range of waves extending from radio waves to gamma rays. The descriptive names of the sections are merely a historical classification, for all these waves are the same in nature, all traveling at the same speed. They differ mainly in frequency and wavelength. Electromagnetic waves to the left of those illustrated in Figure 12.3 are long wave infrared,

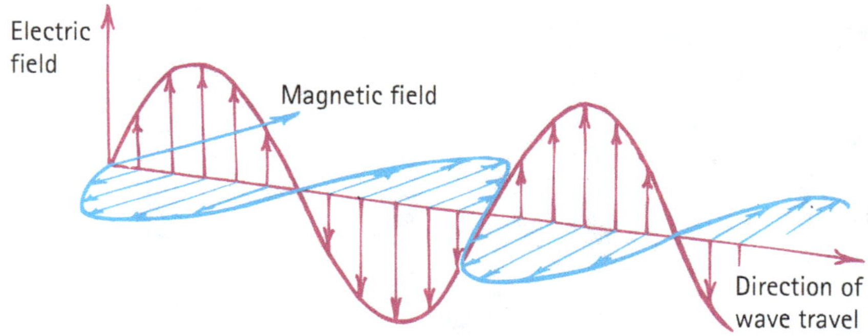

Figure 12.2 >
The electric and magnetic fields of an electromagnetic wave are perpendicular to each other and to the direction of motion of the wave.

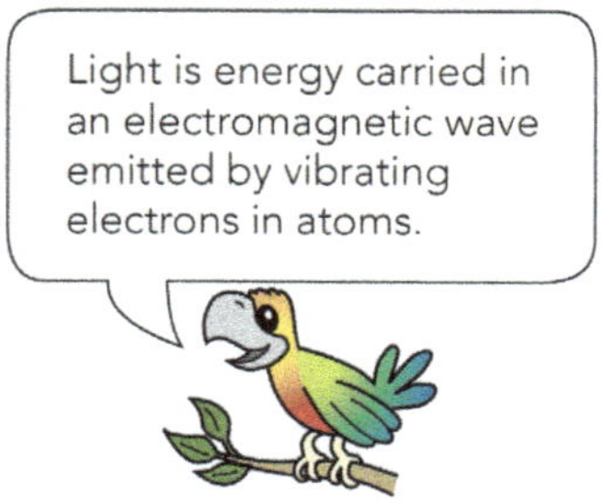

which are invisible to the human eye. To the right are the ultra-short invisible gamma waves. Surprisingly, visible light makes up less than a millionth of 1 percent of the electromagnetic spectrum.

The lowest frequency of light that our eyes can see appears red. The highest visible frequencies are nearly twice the frequency of red and appear violet. Still higher frequencies are ultraviolet, which are beyond our range of vision. These higher-frequency waves are more energetic and cause sunburns. Higher frequencies beyond ultraviolet extend into the X-ray and gamma-ray regions. There is no sharp boundary between these regions, which actually overlap each other. The spectrum is separated into regions merely for classification.

READING CHECK
How does the frequency of emitted light compare with the frequency of the vibrating electron that generates it?

The frequency of an electromagnetic wave as it vibrates through space is identical to the frequency of the vibrating electric charge that generates it. Different frequencies result in different wavelengths—low frequencies produce long wavelengths, and high frequencies produce short wavelengths. The higher the frequency of the vibrating charge, the shorter the wavelength of radiation.

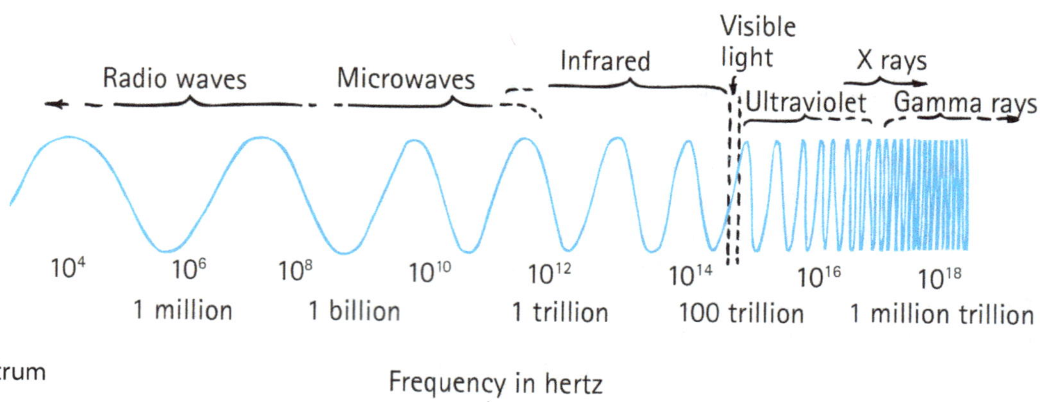

Figure 12.3 >
The electromagnetic spectrum

CONCEPT **CHECK**
Is it correct to say that a radio wave is a low-frequency light wave? Is a radio wave also a sound wave?

CHECK YOUR ANSWERS
Both a radio wave and a light wave are electromagnetic waves, and electromagnetic waves originate in the vibrations of electrons. Because radio waves have lower frequencies than light waves, a radio wave may be considered a low-frequency light wave (and a light wave a high-frequency radio wave). A sound wave, however, is a mechanical vibration of matter and is not electromagnetic. A sound wave is fundamentally different from an electromagnetic wave. So, a radio wave is definitely not a sound wave. (Don't confuse a radio wave with the sound that a loudspeaker emits.)

12.2 Transparency

When light passes through matter, some of the electrons in the matter are forced into vibration. In this way, vibrations in the emitter are transmitted to vibrations in the receiver. This is similar to the way sound is transmitted (**Figure 12.4**).

Materials such as glass and water allow light to pass through in straight lines. They are **transparent** to light. To understand how light penetrates through glass (or any transparent material), visualize the electrons in the atoms of glass as if they were connected to the nucleus by springs (**Figure 12.5**). When a light wave meets them, the electrons are set into vibration.

Figure 12.4
Just as a sound wave can force a sound receiver into vibration, a light wave can force electrons in materials into vibration.

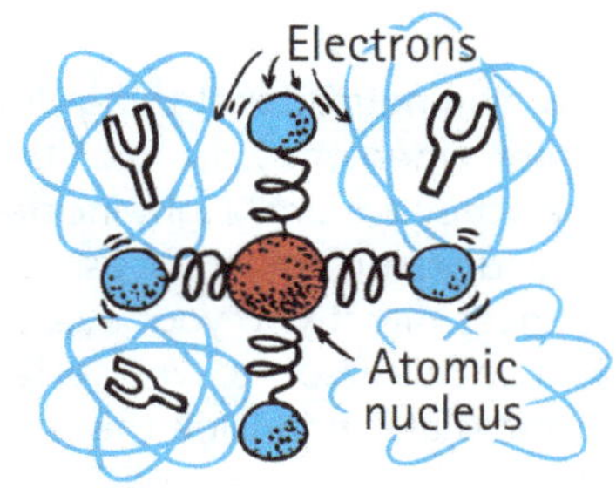

Figure 12.5
The electrons of atoms have certain natural frequencies of vibration and can be modeled as particles connected to the atomic nucleus by spring. As a result, atoms and molecules behave somewhat like optical tuning forks.

Materials that are springy (elastic) respond more to vibrations at particular frequencies than to other frequencies. Bells ring at a particular frequency, tuning forks vibrate at a particular frequency, and so do the electrons of atoms and molecules. Electrons in atoms of glass vibrate well in the ultraviolet range. So, when ultraviolet waves shine on glass, resonance occurs. The vibration of electrons builds up to large amplitudes, just as pushing someone at the resonant frequency of a swing builds to a large amplitude. The large amplitude results in a lot of chaotic bumping of neighboring atoms. This means that the ultraviolet light energy is being transformed into thermal energy. In other words, any ultraviolet light that enters into the glass is dispersed as heat—the ultraviolet light goes into the glass but does not come out the other side. We can say that the ultraviolet light is absorbed by the glass.

At lower wave frequencies, such as those of visible light, electrons in the glass atoms are forced into vibration—but at less amplitude. The atoms retain the energy for a briefer time, with less chance of collisions with neighboring atoms. This means that less energy is transformed to heat. Instead, quite nicely, the energy of vibrating electrons is re-emitted as light. Glass is transparent to light of all the *visible* frequencies. The frequency of the re-emitted light that is passed from one atom to another atom is identical to the frequency of the light that initially produced the vibration.

Interestingly, there is a slight time delay between absorption and re-emission, which explains why light travels slower through glass than through the vacuum of space. But please note: between any two atoms you'll find the vacuum of space. Light travels slower through glass because of the many pit-stops it makes along the way (**Figure 12.6**).

READING CHECK
What causes light to slow down when it passes through a transparent material?

Λ Figure 12.6
A wave of visible light incident upon a pane of glass produces vibrations in the glass atoms, which cascade to a chain of absorptions and re-emissions that pass the light energy through the material and out the other side. Because of the time delay between absorptions and re-emissions, light travels through the glass more slowly than through empty space.

FOR YOUR INFORMATION

Have you noticed that things look darker when they are wet than when they are dry? Light incident on a dry surface such as sand bounces directly to your eye. But light incident on a wet surface bounces around inside the transparent film of water. With each bounce more of the light gets absorbed by the water. So sand and other things look darker when wet.

Because of the absorption/re-emission process, light travels at different average speeds through different materials. We say average speeds because the speed of light in a vacuum, whether in interstellar space or in the space between atoms in a piece of glass, is a constant 300,000 kilometers per second, which is rounded up from the presently accepted value of 299,792 km/s. We call this speed of light c. The speed of light in the atmosphere is slightly less than in a vacuum but is usually rounded off as c. In water, light travels at 75% of its speed in a vacuum, or $0.75c$. In glass, light travels about $0.67c$, depending on the wavelength of light and the type of glass. In a diamond, light travels at less than half its speed in a vacuum, only $0.41c$. When light emerges from these materials into the air, it again travels at its original speed c.

Infrared waves, which have frequencies lower than those of visible light, vibrate entire molecules in the structure of glass and many other materials. This molecular vibration increases the thermal energy and temperature of the material, which is why infrared waves are often called heat waves. Thus, glass is not transparent to infrared light.

We use the word *transparent* to indicate when light is able to pass through a material. Glass, for example, is transparent to visible light.

Λ Figure 12.7
When the raised ball is released and hits the others, the ball that emerges from the opposite side is not the same ball that initiated the transfer of energy. Likewise, the light that emerges from a pane of glass is not the same that was incident on the glass. Both the emerging ball and emerging light are different from, though identical to, the incident ones.

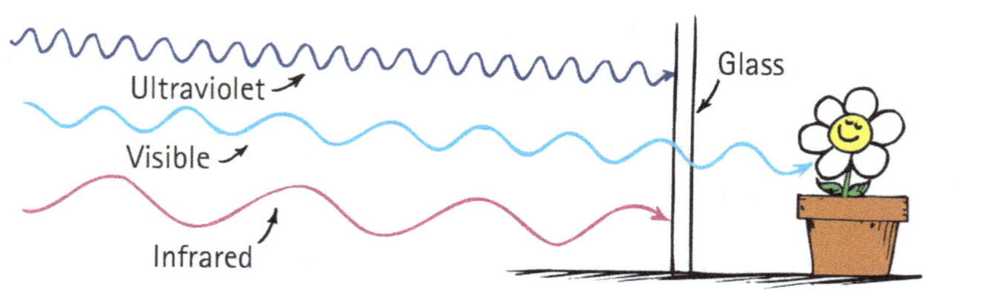

Λ Figure 12.8
Glass blocks both infrared and ultraviolet waves but is transparent to all the frequencies of visible light.

Transparent means, "you can see through it". We use the word **opaque** to indicate when light is absorbed and then transformed into random thermal energy. Opaque means "you can't see through it." A related term is the word **translucent**, which is when the material is partially transparent yet also partially opaque. For a translucent material, such as wax paper, light can pass through, but any detail is lost.

Most things around us are opaque—they absorb light without re-emitting it. Books, desks, chairs, and people are opaque. Vibrations of their atoms and molecules produced by light are turned into random kinetic energy—into thermal energy. They become slightly warmer.

Metals are opaque. As we learned in Chapter 9, the outer electrons of atoms in metals are not bound to any particular atom. They are loose and free to wander throughout the material (which is why metal conducts electricity and heat so well). When light shines on metal and sets these free electrons into vibration, their energy does not "spring" from atom to atom in the material but is instead reflected. That's why metals are also shiny.

The Earth's atmosphere is transparent to some ultraviolet light, all visible light, and some infrared light. But the atmosphere is opaque to high-frequency ultraviolet light. If all ultraviolet light got through the atmosphere we would be fried to a crisp. The small amount of ultraviolet that does get through causes sunburns. Clouds are semitransparent (translucent) to ultraviolet light, which is why you can get a sunburn on a cloudy day. Ultraviolet light is not only harmful to your skin but is also damaging to tar roofs. Now you know why tarred roofs are often covered with gravel.

Figure 12.9
Metals are shiny because light that shines on them forces free electrons into vibration, and these vibrating electrons then emit their "own" light waves as reflection.

CONCEPT **CHECK**

1. Why is glass transparent to visible light but not to ultraviolet and infrared?

2. Pretend that while you walk across a room you make several short stops along the way to greet people who are "on your wavelength." How is this analogous to visible light traveling through glass?

3. In what way is it not analogous?

CHECK YOUR ANSWERS

1. The natural vibration frequency for electrons in glass matches the frequency of ultraviolet light, so resonance occurs when ultraviolet waves shine on glass. The absorbed energy is passed on to other atoms as heat, instead of being re-emitted as light. So, glass is opaque at ultraviolet frequencies. But in the range of visible light, the forced vibrations of electrons in the glass are at smaller amplitudes. Vibrations are smaller and light is re-emitted instead of becoming thermal energy. So, the glass is transparent to visible light. Lower-frequency infrared light causes entire molecules, rather than electrons, to resonate. So, heat is generated and the glass is opaque to infrared.

2. Your average speed across the room is less than it would be in an empty room because of the time delays of your stops. In the same way, light takes a longer time to travel through glass. The result of interactions of light and atoms it encounters is lower average speed.

3. In walking across the room, it is you who begin and complete the walk. But it is different for light. The light that entered the glass is not the same light that emerges. The frequencies are the same, so they are indistinguishable.

12.3 Light Reflection

Anyone who has played pool or billiards knows that when a ball bounces from a surface, the angle of rebound is equal to the angle of incidence. Likewise for light. This is the **law of reflection**, and applies to all angles:

The angle of reflection equals the angle of incidence.

The law of reflection is illustrated with arrows that represent light rays in **Figure 12.10**. Instead of measuring angles from the reflecting surface, it is customary to measure the angles of each ray from a line perpendicular to the reflecting surface. This imaginary line is called the normal. The incident ray, the normal, and the reflected ray all lie in the same plane.

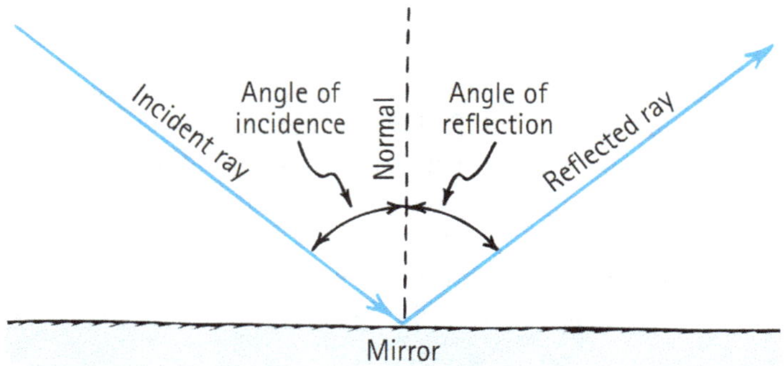

Λ Figure 12.10
The law of reflection represented on the surface of a mirror.

Look at a candle flame placed in front of a mirror. Rays of light radiate from the flame in all directions. **Figure 12.11** shows only four of the infinite number of rays leaving one point on the flame. When these rays meet the mirror, they reflect at angles equal to their angles of incidence. The rays diverge (spread out) from the flame. Notice that they diverge also from the mirror, which means they appear to emanate from behind the mirror (dashed lines) creating what we call a **virtual image**. You would see a virtual image of the candle's flame at this point behind the mirror. This image is as far behind the mirror as the flame is in front of the mirror. Both the flame and its image have the same size.

READING CHECK

When you look at something in a plane mirror, how far behind the mirror is the virtual image?

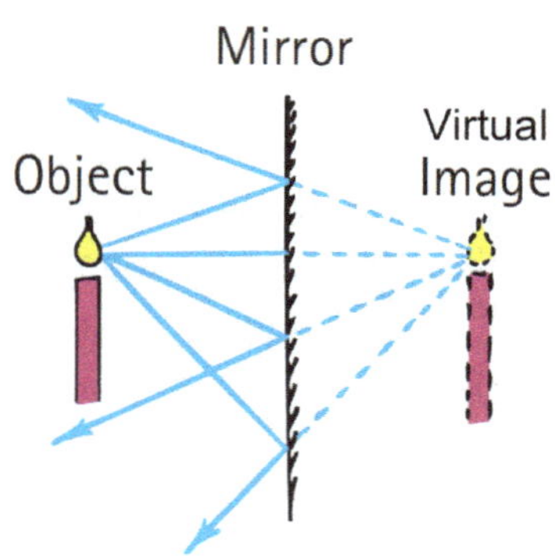

Figure 12.11 >
A virtual image is formed behind the mirror. It is located at the position where the extended reflected rays (dashed lines) converge.

When you view yourself in a mirror, for example, the size of your image is the same as the size of your twin if located as far behind the mirror as you are in front—as long as the mirror is flat. A flat mirror is called a *plane mirror.* Interestingly, as **Figure 12.12** shows, the left-right axis is not reversed, no more than reversal of your up-down axis. The axis that is reversed, is front-back.

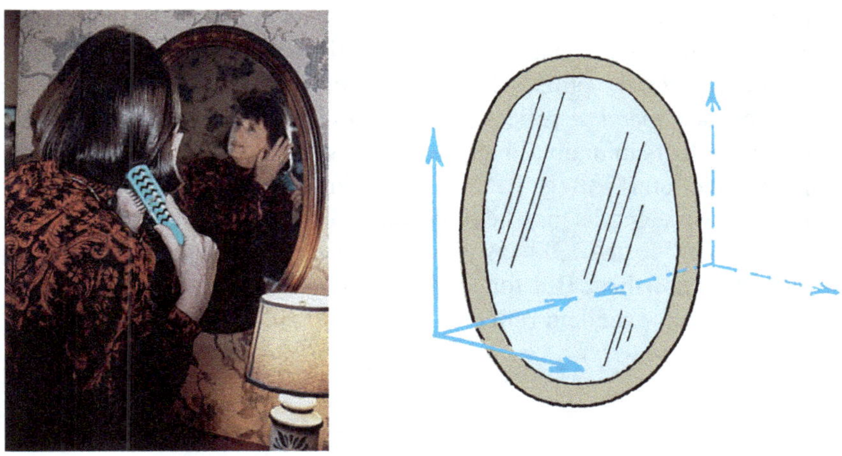

∧ Figure 12.12
Marjorie's image is as far behind the mirror as she is in front. Note that she and her image have the same color of clothing—evidence that light doesn't change frequency upon reflection.

When the mirror is curved, the sizes and distances of object and image are no longer equal. We shall not study curved mirrors in this program, except to say that the law of reflection still holds. A curved mirror behaves as a succession of flat mirrors, each tilted slightly different from the one next to it. At each point, the angle of incidence is equal to the angle of reflection (**Figure 12.13**). Note that in a curved mirror, the normals (dashed lines between the solid rays) at different points on the surface are not parallel to one another.

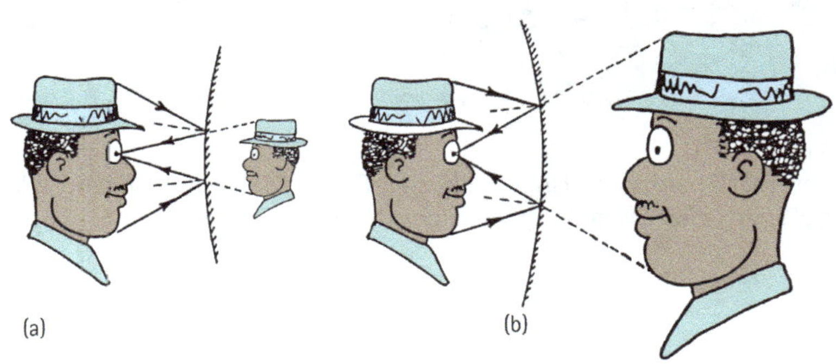

< Figure 12.13
(a) The virtual image formed by a convex mirror (a mirror that curves outward) is smaller and closer to the mirror than the object. (b) When the object is near a concave mirror (a mirror that curves inward like a "cave"), the virtual image is larger and farther away than the object. In either case the law of reflection holds for each ray.

Figure 12.14
Lillian Hewitt plays with multiple mirrors set up within an elevator. Note how successive reflections grow darker. Do you know why?

Whether a mirror is flat or curved, the human eye-brain system cannot ordinarily tell the difference between an object and its reflected image. So, we see the illusion that an object exists behind a mirror (or in some cases in front of a concave mirror). Light reaches our eye in the same manner whether it comes from an object or a reflection of the object.

Only part of the light that strikes a surface is reflected. On a pane of clear glass, for example, light perpendicular to the surface reflects only about 4 percent from each surface of the pane. So, the image you see reflected in a glass window is rather dim. By contrast, on a polished aluminum or silver mirror surface about 90 percent of incident light is reflected. So, an image you see in a mirror is about 90 percent as bright. Position two plane mirrors in parallel facing each other and you'll find "infinite" reflections, as shown in **Figure 12.14**. Because each reflection is not as bright, the reflections gradually go darker. Mirrors are fun physics!

Diffuse Reflection

If a surface is so smooth that the distances between microscopic irregularities on the surface are less than about one-eighth the wavelength of the light, the surface is said to be polished. We call reflection from such a smooth or polished surface *specular* reflection. From a less smooth surface, such as paper, light reflects in many directions. This is not specular reflection, but is *diffuse* reflection (**Figure 12.15**). Interestingly, an irregular surface for one range of wavelengths may be polished for other ranges. The wire-mesh "dish" shown in **Figure 12.16** is very rough for light waves. We certainly don't see it as polished. But for long-wavelength radio waves it is quite polished and is an excellent reflector.

READING CHECK
In terms of the wavelength of light, what does it mean to say a reflecting surface is polished?

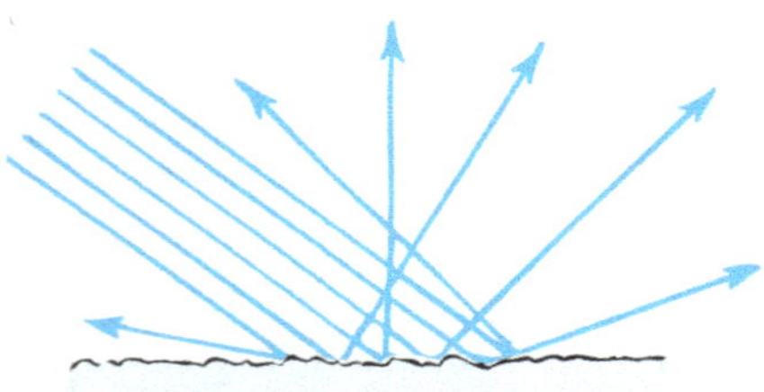

Figure 12.15
Diffuse reflection. Although each ray obeys the law of reflection, the many different surface angles that light rays encounter in striking a rough surface cause reflection in many directions.

Figure 12.16
The open-mesh parabolic dish is a diffuse reflector for short-wavelength light but a polished reflector for long-wavelength radio waves

Light reflecting from paper diffuses. The paper may be smooth to a radio wave, but to a light wave it is rough, as **Figure 12.17** clearly indicates. Rays of light striking the paper encounter millions of tiny flat surfaces facing in all directions. The incident light therefore reflects in all directions. This is good because it lets us see the page without glare. We don't have to hold our head in any special position to read the page. Similarly, for other objects. While you are in an automobile you can see the road ahead at night because of diffuse reflection by the road surface. When the road is wet, however, water provides a more mirrored surface and it is harder to see. Most of our environment is seen by diffuse reflection.

Λ Figure 12.17
A magnified view of the surface of ordinary paper.

CONCEPT **CHECK**
Why is it more dangerous to drive a car on a rainy night?

CHECK YOUR ANSWER
In addition to the less diffuse road surface when wet, as described above, there's another reason. Headlights from oncoming cars reflect from the more mirrored surface full force into your eyes. Glare is more intense on a rainy night.

Quick Activity

Twisting Mirrors

1. Stand a pair of pocket-size mirrors on edge with the faces parallel to each other. Place an object such as a coin between the mirrors and look at the reflections in each mirror. Nice?
2. Now set up the mirrors at right angles and place a coin between them. You'll see four coins.
3. Change the angle of the mirrors and see how many images of the coin you can see. Look at yourself in the pair of mirrors when they're at right angles to each other. Wink. Notice that you see yourself as others see you.
4. Rotate the mirrors, still at right angles to each other. Does your image rotate also?
5. Now place the mirrors 60° apart so you again see your face. Again, rotate the mirrors and see if your image rotates also. Amazing?

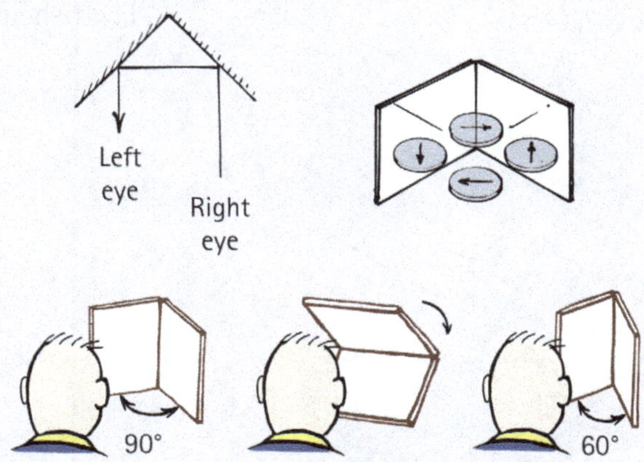

12.4 Light Refraction

In section 12.2 we learned that light slows down when it enters glass and other transparent materials. When light passes from one medium to another, we call the process **refraction**. When light is not perpendicular to the surface of penetration, bending occurs.

To better understand the bending of light in refraction look at the pair of toy cart wheels in **Figure 12.18**. The wheels roll from a smooth sidewalk onto a grass lawn. If the wheels meet the grass at an angle, as the figure shows, they are deflected from their straight-line course. Note that the left wheel slows first when it interacts with the grass on the lawn. The right wheel maintains its higher speed while on the sidewalk. It pivots about the slower-moving left wheel because it travels farther in the same time. So, the direction of the rolling wheels is bent toward the "normal," the black dashed line perpendicular to the grass-sidewalk border in Figure 12.18.

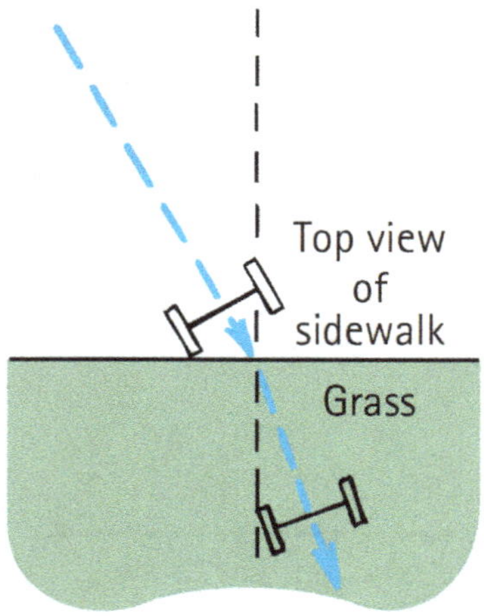

Figure 12.18
The direction of the rolling wheels changes when one wheel slows down before the other does.

Figures 12.19 and **12.20** show how a light wave bends in a similar way. Note the direction of light, indicated by the blue arrow (the light ray), and also note the wavefronts (red) drawn at right angles to the ray. In the figure the wave meets the water surface at an angle. This means that the left portion of the wave slows down in the water while the remainder in the air travels at speed c. The light ray remains perpendicular to the wavefront and therefore bends at the surface. It bends like the wheels bend when they roll from the sidewalk into the grass. In both cases the bending is caused by a change in speed.

READING CHECK
What causes light to bend?

Figure 12.21 shows a beam of light entering water at the left and exiting at the right. The path would be the same if the light entered from the right and exited at the left. Light paths are reversible for both reflection and refraction. If you see someone's eyes by means of reflection or refraction, such as with a mirror or a prism, then that person can see you also.

Refraction causes many illusions. One of them is the apparent bending of a stick that is partially in water. The submerged part appears closer to the surface than it actually is. Likewise, when you look at a fish in water. The fish appears nearer to the surface and closer than it really is (**Figure**

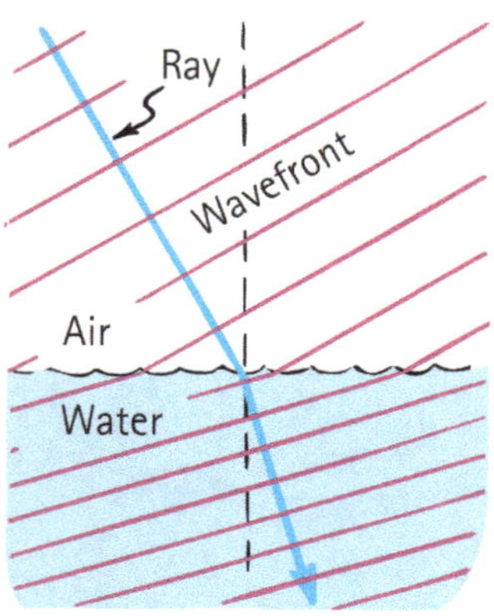

Figure 12.19
Refraction is indicated by the bending of the light ray entering water. Notice the wavelength is shorter in the water.

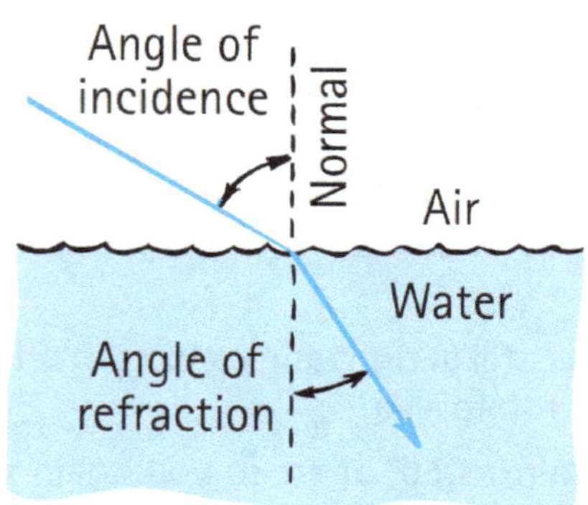

Figure 12.20
Angles of refraction above and below the water surface are expressed from the normal.

12.22). If we look straight down into water, an object submerged 4 meters beneath the surface appears to be only 3 meters deep. Because of refraction, submerged objects appear to be magnified.

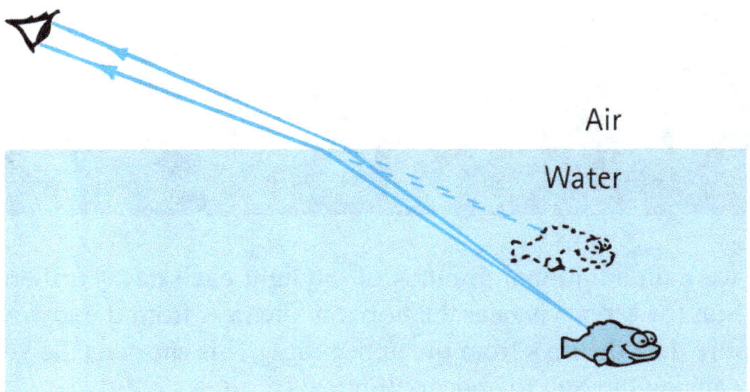

∧ **Figure 12.22**
Because of refraction, a submerged object appears to be nearer to the surface than it actually is.

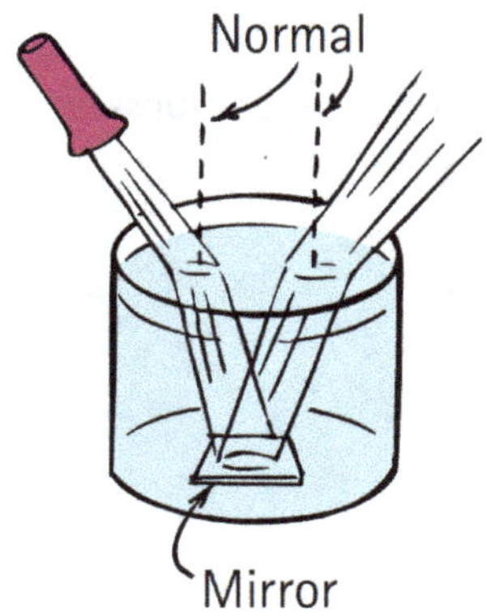

∧ **Figure 12.21**
Light refracts into water, reflects off a submerged mirror, and then refracts out of the water.

CONCEPT **CHECK**
If the speed of light were the same in all media, would refraction still occur when light passes from one medium to another?

CHECK YOUR ANSWER
No, refraction occurs when light changes speed.

Illusions and Mirages are Caused by Atmospheric Refraction

Refraction occurs in Earth's atmosphere. Whenever we watch a sunset, we see the Sun for several minutes after it has actually sunk below the horizon (**Figure 12.23**). The Earth's atmosphere is thin at the top and dense at the bottom. Because light travels faster in thin air than in dense air, parts of the wavefronts of sunlight higher up travel faster than parts of the wavefronts closer to the ground. Light rays bend. The density of the atmosphere changes gradually, so the bending of light is gradual and follows a curved

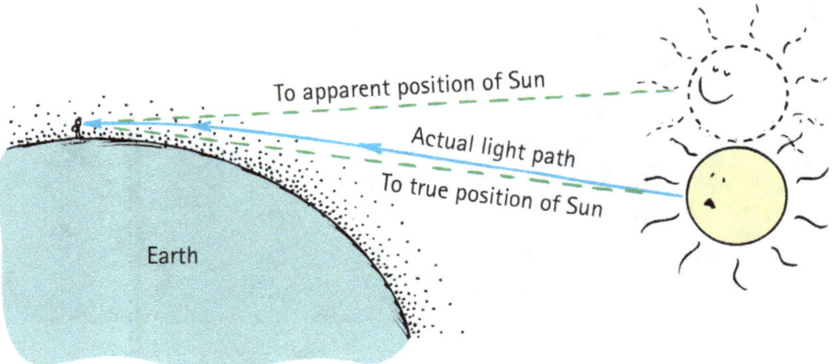

< **Figure 12.23**
Because of atmospheric refraction, when the Sun is near the horizon it appears higher in the sky.

Figure 12.24 >
The Sun's round shape at sunset or sunrise appears flattened by differential refraction.

path. So, we gain additional minutes of daylight each day. Furthermore, when the Sun (or Moon) is near the horizon, the rays from the lower edge are bent more than the rays from the upper edge. This shortens the vertical diameter, causing the Sun to appear elliptical (**Figure 12.24**).

Mirages are a common sight on a desert. The sky appears to be reflected from water on the distant sand. But when we get there, the sand is dry. Why is this so? The air is very hot just above the surface and cooler above. Light travels faster through the thinner hot air below than through the denser cool air above. So, wavefronts near the ground travel faster than above. The result is upward bending (**Figure 12.25**). So, we see an upside-down view as if reflection were occurring from a water surface. We see a mirage, which is formed by real light and can be photographed (**Figure 12.26**). A mirage is not, as many people think, a trick of the mind.

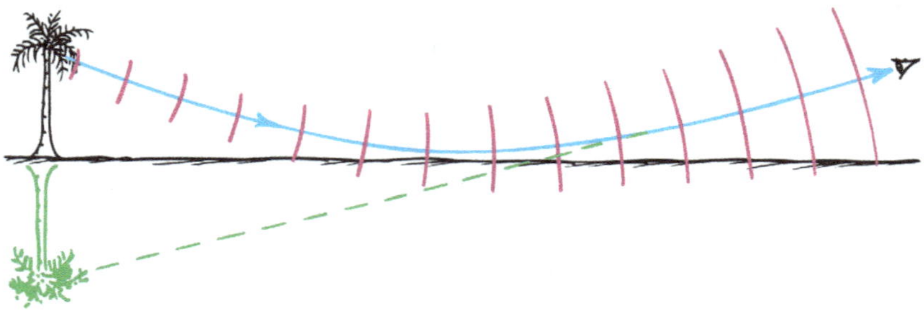

Figure 12.25 >
Light from the top of the tree gains speed in the warm and less dense air near the ground. When the light grazes the surface and bends upward the observer sees a mirage.

Figure 12.26 >
A mirage. The apparent wetness of the road is not reflection by water but rather refraction through the warmer and less-dense air near the road surface.

When we look at an object over a hot stove or over a hot pavement, we see a wavy, shimmering effect. This is due to varying densities of air caused by changes in temperature. The twinkling of stars results from similar density variations in the sky, where light passes through moving layers in the atmosphere.

READING CHECK
Why do stars twinkle?

CONCEPT CHECK
If the speed of light were the same in air of various temperatures and densities, would there still be slightly longer daytimes, twinkling stars at night, mirages, and a slightly squashed Sun at sunset?

CHECK YOUR ANSWER
No.

12.5 Color

To the scientist, color is not in a material. Color is a physiological experience and is in the eye of the beholder. So, when we say that light from a rose is red, in a strict sense we mean that it appears red. Many organisms, including people with deficient color vision, do not see the rose as red at all.

The colors we see depend on the frequency of the light we see. Light of different frequencies is perceived as having different colors; the lowest frequency we detect appears to most people as the color red. The highest frequency appears as violet. In between is the infinite number of hues that make up the color spectrum of the rainbow. By convention these hues are grouped into the seven colors; red, orange, yellow, green, blue, indigo, and violet. You'll notice that the color "white" is not listed as one of these hues. Physiologically, white is what the brain perceives when all the colors are received by the eye at the same time. Black is what the brain perceives when no light is received.

∧ Figure 12.27
Sunlight passing through a prism separates into a color spectrum. The colors of things depend on the colors of the light that illuminate them.

< Figure 12.28
When printed on white paper, the square on the left reflects all the colors illuminating it. In sunlight it is white. When illuminated with blue light, it is blue. The square on the right, when printed, absorbs all the colors illuminating it. In sunlight it becomes warmer than the white square.

Selective Reflection

Except for sources of light such as lamps, stars, and our Sun, most things around us reflect rather than emit light. But as an object reflects light, it might not reflect every frequency. A banana, for example, only reflects yellow while absorbing all the other colors. That's why a banana is yellow. Similarly, a leaf reflects green light while absorbing all the other colors. That's why the leaf is green. So, the color of an object or any material is determined by the light it selectively reflects.

Look toward a rose in a pitch-dark room and you'll see the rose itself does not emit light. Turn on a red light and you'll see that the flower petals are red (**Figure 12.29**). The reason for this is because those petals are able to reflect red light. Interestingly, you'll also see that the leaves coming from the stem are dark. This is because the leaves absorb red light—the red light goes in, but doesn't come back out. Now switch the red light for a green light. The leaves will now appear green because green is the color they selectively reflect. But what of the flower petals that appear dark under the green light. Why? Because they absorb green light—the green light goes in, but doesn't come back out. Looking at a "green" object with red light, or a "red" object with green light is a mind-opening experience. You'll discover that the color of a non-emitting surface depends upon the light source. Lastly, turn on a full spectrum light or bring the rose into daylight. Here, all frequencies of visible light are available so the flower petals appear red and the leaves green.

Figure 12.29 >
A rose under (a) red light, (b) green light, and (c) daylight.

The reflected colors of most objects are not pure single-frequency colors, but are a mixture of frequencies. For example, yellow daffodils reflect red and green as well as yellow. So, yellow daffodils selectively reflect a broad band of frequencies.

An object can reflect only those frequencies present in the illuminating light. The appearance of a colored object therefore depends on the kind of light that illuminates it. An incandescent lamp of old, an LED, or a fluorescent lamp will show various hues of color. Fluorescent lamps are richer in the higher frequencies, and so blues are enhanced under them. For this reason, it is difficult to tell the daylight color of objects viewed in artificial light. What a color looks like depends on the light source (**Figure 12.31**).

∧ Figure 12.30
The rabbit's dark fur absorbs all the radiant energy in incident sunlight and therefore is black. Light fur on other parts of the body reflects light of all frequencies and therefore is white.

∧ Figure 12.31
Color depends on the light source.

Selective Transmission

The color of a transparent object depends on the color of the light it transmits. A red piece of glass appears red because it absorbs all the colors of white light, except red. So red is transmitted. Similarly, a blue piece of glass appears blue because it transmits primarily blue and absorbs the other colors that illuminate it. These pieces of glass contain dyes or pigments—fine particles that selectively absorb light of particular frequencies and selectively transmit others. From an atomic point of view, electrons in the pigment molecules are set into vibration by the illuminating light. Light of some of the frequencies is absorbed by the pigments. The rest is re-emitted from atom to atom in the glass. The energy of the absorbed light increases the kinetic energy of the atoms and the glass is warmed. Ordinary window glass doesn't have a color because it transmits light of all visible frequencies equally well. Though a bunch of "clear" glass panes stacked upon each other will appear somewhat green, which indicates that the common glass has a slight tendency to selectively transmit green.

READING CHECK
Why does red glass appear red?

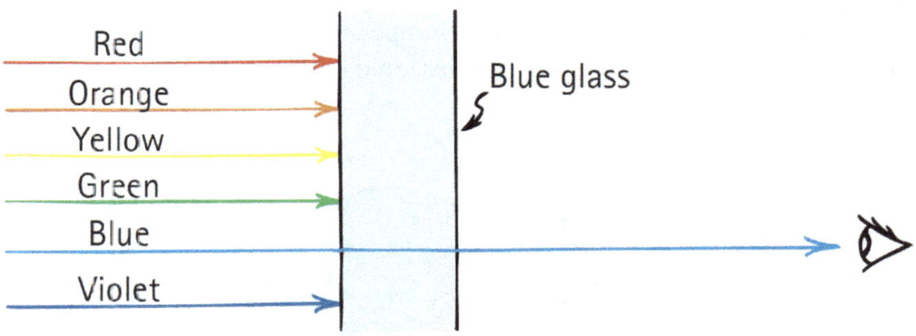

< Figure 12.32
Only energy having the frequency of blue light is transmitted; energy of the other frequencies is absorbed and warms the glass.

CONCEPT **CHECK**

1. When red light shines on a red rose, why do the leaves become warmer than the petals?
2. When green light shines on a red rose, why do the petals look black?
3. If you hold any small source of white light between you and a piece of red glass, you'll see two reflections from the glass: One from the front surface and one from the back surface. What color is each reflection?

CHECK YOUR ANSWERS

1. The leaves don't reflect red light, but absorb it. So, the leaves become warmer.
2. The petals don't reflect the green light, but absorb it. Because green is the only color illuminating the rose and because green contains no red to be reflected, the rose reflects no color and appears black.
3. The reflection from the front surface is white because the light doesn't reach far enough into the colored glass to allow absorption of non-red light. Only red light reaches the back surface because the pigments in the glass absorb all the other colors, and so reflection from the back is red.

12.6 Mixing Colors

You can see that white light from the Sun is composed of all the visible frequencies when you pass sunlight through a prism. The white light is dispersed into a rainbow-colored spectrum. The distribution of solar frequencies (**Figure 12.33**) is uneven, and is most intense in the yellow-green part of the spectrum. How fascinating that our eyes have evolved to have maximum sensitivity in this range. That's why fire engines are painted yellow-green, particularly at airports, where visibility is vital. Our sensitivity to yellow-green light is also why we see better under the illumination of yellow sodium-vapor lamps at night than under lamps that emit white light of the same brightness.

All the colors combined make white. Interestingly, we see white also from the combination of only red, green, and blue light. We can understand this by dividing the solar radiation curve into three regions, as in **Figure 12.34**. Three types of cone-shaped receptors in our eyes perceive color. Each is stimulated only by certain frequencies of light. Light of lower visible frequencies stimulates the cones sensitive to low frequencies and appears red. Light of middle frequencies stimulates the mid-frequency-sensitive cones and appears green. Light of higher frequencies stimulates the higher-frequency-sensitive cones and appears blue. When all three types of cones are stimulated equally, we perceive white.

READING CHECK

What three colors are the cones of most human eyes sensitive to?

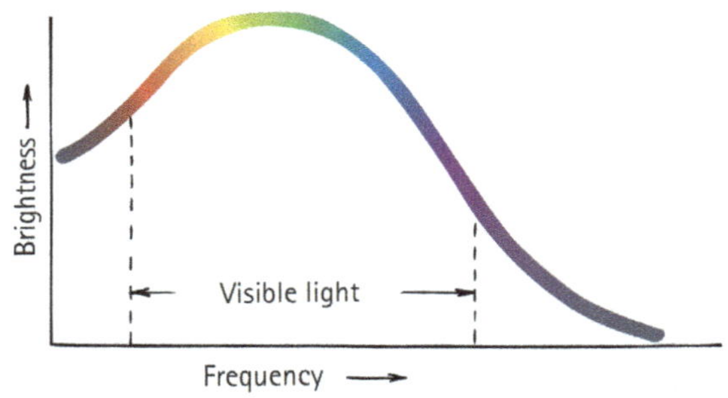

Figure 12.33
The radiation curve of sunlight is a graph of brightness versus frequency. Sunlight is brightest in the yellow-green region nearly in the middle of the visible range.

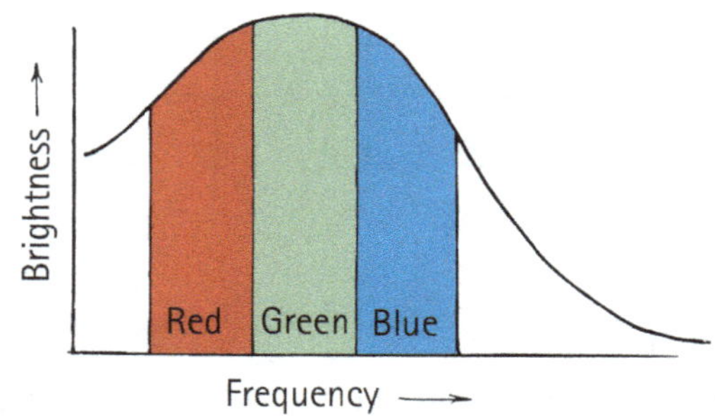

Figure 12.34
Radiation curve of sunlight divided into three regions—red, green, and blue. These are the additive primary colors.

Additive Primary Colors—Red, Green, and Blue

When red, green, and blue lights are projected on a screen, they overlap to produce white. In the language of scientists, colored lights that overlap add to each other. So, we say that red, green, and blue light add to produce white light. Any two of these colors of light add to produce another color (**Figure 12.35**). Various amounts of red, green, and blue, produce any color in the spectrum. For this reason, red, green, and blue are called the **additive primary colors.** A close examination of the picture on a TV screen reveals that the picture is a mixture of tiny spots, each less than a millimeter across. When the screen is lit, some of the spots are red, some green, and some blue. The mixtures of these primary colors when viewed at a distance provide a complete range of colors, plus white.

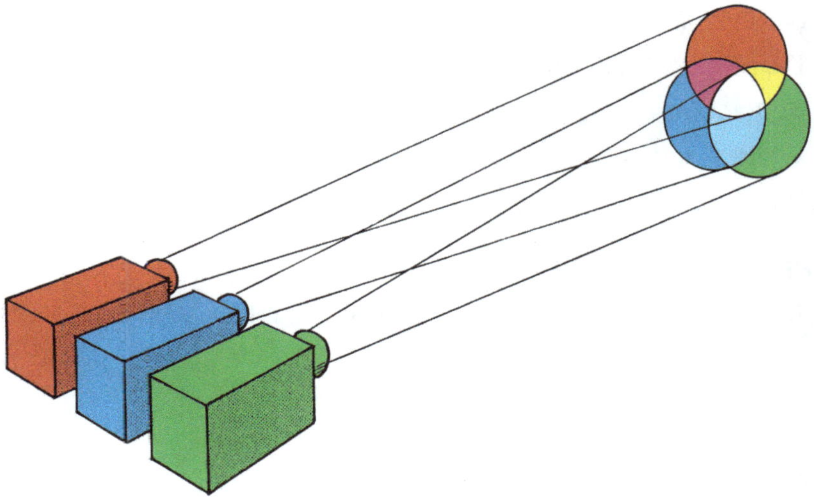

< **Figure 12.35**
Color addition by the mixing of colored lights. When three projectors shine red, green, and blue light on a white screen, the overlapping parts produce different colors. White is produced where all three overlap.

FOR YOUR INFORMATION

It's interesting to note that the "black" you see on the darkest scenes on a computer screen is simply the color of the screen itself, which is more a light gray than black. Because our eyes are sensitive to the contrast with the illuminated parts of the screen, we see this gray as black.

Complementary Colors

Here's what happens when two of the three additive primary colors are combined:

Red + Blue = Magenta

Red + Green = Yellow

Blue + Green = Cyan

We say that magenta is the opposite of green; cyan is the opposite of red; and yellow is the opposite of blue. When you add any color and its opposite color, you get white.

Magenta + Green = White (= Red + Blue + Green)

Yellow + Blue = White (= Red + Green + Blue)

Cyan + Red = White (= Blue + Green + Red)

When two colors are added together to produce white, they are called **complementary colors**. Every hue has some complementary color that when added to it makes white.

The fact that a color and its complement combine to produce white light is nicely used in lighting stage performances. Blue and yellow lights shining on performers, for example, produce the effect of white light—except where one of the two colors is absent, as in the shadows. The shadow of one lamp, say the blue, is illuminated by the yellow lamp and appears yellow. Similarly, the shadow cast by the yellow lamp appears blue. This is a most interesting effect.

We see this effect in Figure 12.36, where red, green, and blue light shine on the golf ball. Note the shadows cast by the ball. The middle shadow is cast by the green spotlight and is not dark because it is illuminated by the red and blue lights, which make magenta. The shadow cast by the blue light appears yellow because it is illuminated by red and green light. Can you see why the shadow cast by the red light appears cyan?

∧ **Figure 12.36**
The white golf ball appears white when illuminated with red, green, and blue lights of equal intensities. Why are the shadows of the ball cyan, magenta, and yellow?

CONCEPT CHECK

1. From Figure 12.35, find the complements of cyan, yellow, and red.
2. Red + blue = _______.
3. White − red = _______.
4. White − blue = _______.

CHECK YOUR ANSWERS

1. Red, blue, cyan. 2. Magenta. 3. Cyan. 4. Yellow.

Figure 12.37

Only four colors of ink are used to print color illustrations and photographs— (a) magenta, (b) yellow, (c) cyan, and (e) black. When magenta, yellow, and cyan are combined, they produce (d). Addition of black (e) produces the finished result (f).

Mixing Colored Pigments

Every artist knows that if you mix red, green, and blue paint, the result is not white but rather a muddy dark brown. Red and green paint certainly do not mix to form yellow, as is the rule for combining colored lights.

Mixing pigments in paints and dyes is entirely different from mixing lights. Pigments are tiny particles that absorb specific colors. Pigments that produce a particular color absorb other colors that blend to the complementary color. For example, something red absorbs cyan, which is why it reflects red. In effect, cyan has been subtracted from white light. Something painted blue absorbs yellow, and so reflects all the colors except yellow. Take yellow away from white and you've got blue. The colors magenta, cyan, and yellow are the subtractive primaries.

READING CHECK

What color is seen by pigments that absorb cyan?

The variety of colors you see in the colored photographs in any printed book are the result of magenta, cyan, and yellow dots. Light illuminates the book, and light of some frequencies is subtracted from the reflected light. The rules of color subtraction differ from the rules of light addition. We'll not go deeper into this topic. Much information is available elsewhere on the internet.

Λ Figure 12.38
Look through a magnifying glass and you will see that the color green on a printed page is composed of cyan and yellow dots.

Λ Figure 12.39
Dyes or pigments, as in the three transparencies shown, absorb (subtract) light of some frequencies and transmit only part of the spectrum. When white light passes through overlapping sheets of these colors, all light is blocked (subtracted). Then we have black. Where only yellow and cyan overlap, only green is not subtracted. Various proportions of yellow, cyan, and magenta dyes will produce nearly any color in the spectrum.

Λ Figure 12.40
The rich colors of Sneezlee represent many frequencies of light. The photo, however, is a mixture of only yellow, magenta, cyan, and black.

12.7 Colors of the Sky

Not all colors are the result of the addition or subtraction of light. Some colors, like the blue of the sky, are the result of *selective scattering*. (**Figure 12.41**). Air molecules, such as nitrogen, N_2, and oxygen, O_2, behave much like tiny little tuning forks. They naturally vibrate strongly at frequencies of around 6.5×10^{14} Hz and above, which corresponds to the blue end of the light spectrum. When they get hit by the Sun's radiant energy, they start vibrating at their natural blue frequencies.

Just as a tuning fork emits the sound of its natural frequency, Sun hit air molecules emit the light of their natural frequency, which is blue. As these air molecules vibrate, they send out blue light in all directions. That's why the clear sky is blue in all directions. This includes when you look up, but also when you look downward from a high-flying airplane where you'll see all below has a bluish hue because of the atmosphere between you and the ground. But the sky is only blue in the daytime. At night, there is no solar energy to cause air molecules to vibrate their blue frequencies, which is why the sky is dark in the nighttime.

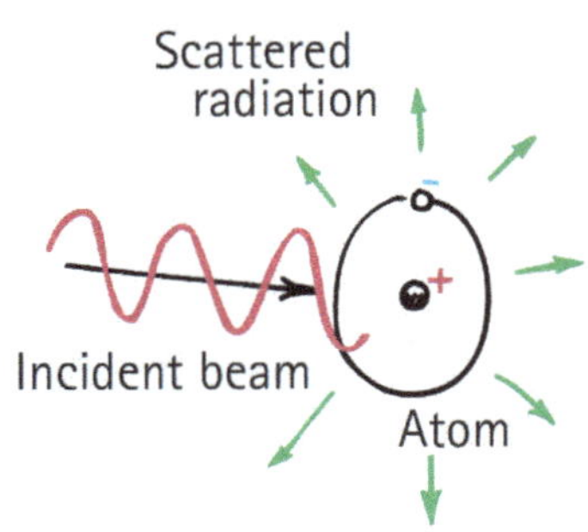

Λ Figure 12.41
A beam of light falls on an atom and increases the vibrational motion of electrons in the atom. The vibrating electrons re-emit the light in various directions. Light is scattered.

FOR YOUR INFORMATION

There is no atmosphere on the Moon, which is why the day time sky is black.

Of the visible frequencies of sunlight, violet is scattered the most by the nitrogen and oxygen in Earth's atmosphere—even more so than blue. The other colors are scattered to an even lesser extent in the following order: green, yellow, orange, and red, where red is scattered only a tenth as much as violet. Although violet light is scattered more than blue, our eyes are not very sensitive to violet light. Therefore, the blue scattered light is what predominates in our vision, so we see a blue sky! Insects are more sensitive to violet and they also can see ultraviolet. So, while we humans see the sky as blue, insects see it otherwise. Beauty is in the eye of the beholder.

The blueness of the sky varies in different locations under various conditions. A main factor is the content of water vapor in the atmosphere. On clear dry days the sky is a much deeper blue than on clear humid days. Places where the upper air is exceptionally dry, such as Italy and Greece, have beautiful blue skies that have inspired painters for centuries. Where the atmosphere contains a lot of dust particles and other particles larger than oxygen and nitrogen molecules, light of the lower frequencies is also strongly scattered. This causes the sky to appear less blue, with a whitish appearance. After a heavy rainstorm when the airborne particles have been washed away, the sky becomes a deeper blue.

READING CHECK

Why does the sky sometimes have a whitish appearance, rather than blue?

(a) (b)

Figure 12.42

(a) In clean air the scattering of high-frequency light gives us a blue sky. (b) When the air is full of particles larger than oxygen and nitrogen molecules, lower-frequency light is also scattered, which adds to the blue to give a whitish sky.

Selective scattering is not limited to particles in the air. The intriguing vivid blue of lakes in the Canadian Rocky Mountains is due to scattering. The lakes are fed by runoff from melting glaciers that contain fine particles of silt, called rock flour, which remain suspended in the water. Light scatters from these tiny particles and gives the water its eerily vivid turquoise color (**Figure 12.43**).

Figure 12.43 >

The extraordinary blue of Canadian Rocky Mountain lakes is produced by scattering from extremely fine particles of glacial silt suspended in the water.

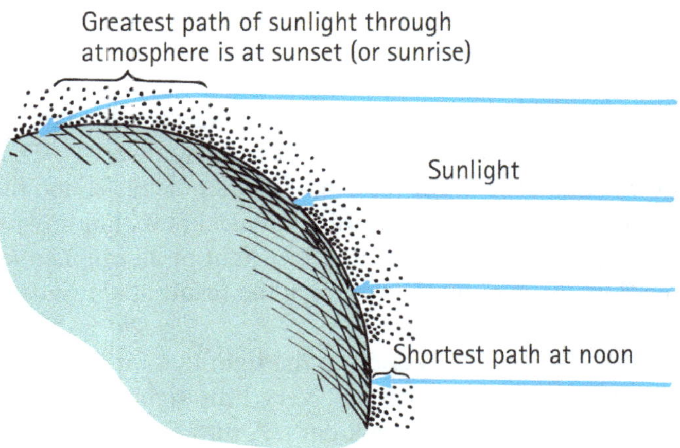

From Earth

Figure 12.44
A sunbeam must travel through more atmosphere at sunset than at noon. As a result, more blue is scattered from the beam at sunset than at noon. By the time a beam of initially white light reaches the ground, only light of the lower frequencies survives to produce a red sunset.

Why Sunsets Are Red

Light that isn't scattered is light that is transmitted. Because red, orange, and yellow light are the least scattered by the atmosphere, light of these low frequencies is better transmitted through the air. Red is scattered the least and passes through the atmosphere better than any other color.

At noon, sunlight travels through the least amount of atmosphere to reach the Earth's surface (**Figure 12.44**). Only a small amount of high-frequency violet and blue light is scattered from the sunlight, enough to make the Sun look yellowish. As the day progresses and the Sun descends lower in the sky, the path through the atmosphere is longer, and more violet and blue are scattered from the sunlight. The removal of violet and blue leaves the transmitted light redder. So, the Sun becomes progressively redder, going from yellow to orange and finally to a red-orange at sunset.

READING CHECK
What color results when violet and blue are removed from sunlight?

CONCEPT CHECK

If molecules in the sky scattered low-frequency light more than high-frequency light, what color would the sky be? What color would sunsets be?

CHECK YOUR ANSWER

If low-frequency light were scattered, the noontime sky would appear reddish-orange. At sunset more reds would be scattered by the longer path of the sunlight, and the sunlight would be predominantly blue and violet. So, sunsets would appear blue! This is precisely what happens on Mars where the atmosphere lacks blue-scattering nitrogen and oxygen molecules, but is rich in red-scattering dust particles—scatter the red away from sunlight and what remains is blue. Sunsets on Mars are blue!

From Mars

Why Clouds Are White

As moist air rises, it cools. Within that cool moist air, water molecules condense to form droplets so tiny they can be suspended by updrafts. These water droplets are considerably larger than individual molecules, and they behave differently. In contrast to the scattering that occurs within tiny particles, water droplets primarily refract and reflect light of all colors about equally. With all colors thrown in all directions, the result near midday is a white cloud.

Each droplet absorbs a small fraction of the light that hits it, so with enough droplets, a lot of absorption occurs. Very little light penetrates a really big cloud, which is why it may look dark. A further increase in the size of the droplets causes them to fall as raindrops, and we have rain.

Λ Figure 12.45

In clouds, water molecules are clumped close together, and reflection produces their whiteness. You know that color depends upon the light source. So why are clouds more reddish when the Sun is low in the sky?

Λ Figure 12.46

Water is cyan because it absorbs red light. The froth in the waves is white because, like clouds, it is composed of a variety of tiny clusters of water droplets that scatter all the visible frequencies.

Concept Review

Chapter 12

Summary of Terms

Additive Primary Colors The three colors—red, blue, and green—that when added in certain proportions produce any other color in the visible-light part of the electromagnetic spectrum.

Complementary Colors Any two colors that when added produce white light.

Electromagnetic Spectrum The range of electromagnetic waves extending in frequency from radio waves to gamma rays.

Law of Reflection The angle of a reflection equals the angle of incidence. The reflected and incident rays lie in a plane that is normal to the reflecting surface.

Opaque The term applied to materials through which light cannot pass.

Reflection The return of light into the medium from which it came.

Refraction The bending of an oblique ray of light when it passes from one transparent medium to another.

Subtractive Primary Colors The three colors of absorbing pigments—magenta, yellow, and cyan—that when mixed in certain proportions reflect any other color in the visible-light part of the electromagnetic spectrum.

Translucent The term applied to a material that is partially transparent yet also partially opaque. Light can pass through, but details are obscured.

Transparent The term applied to materials through which light can pass in straight lines.

Review Questions

1. How do the speeds of various electromagnetic waves compare to one another?
2. In terms of frequency, how does a radio wave compare with the frequency of the vibrating electrons that produces it?
3. In what region of the electromagnetic spectrum is the resonant frequency of electrons in glass?
4. What is the fate of the energy in ultraviolet light incident on glass?
5. How does the average speed of light in glass compare with its speed in a vacuum?
6. Relative to the distance of an object in front of a plane mirror, how far behind the mirror is the image?
7. When a wheel rolls from a smooth sidewalk onto grass, the interaction of the wheel with the blades of grass slows the wheel. What slows light when it passes from air into glass or water?
8. What is the angle between a ray and its wavefront?
9. Why does a setting Sun often appear elliptical instead of round?
10. What produces the wavy effect in air over a hot stove?
11. What is the appearance of a material that absorbs all the light that shines on it and reflects none?
12. What is the color of the peak frequency of solar radiation?
13. What color light are our eyes most sensitive to?
14. Why are red, green, and blue called the additive primary colors?
15. What are the subtractive primary colors? Why are they so called?
16. Why are red and cyan called complementary colors?

17. Is the blueness of the sky due to reflection, refraction, or scattering?

18. Why does the sky sometimes appear whitish?

19. Why does the Sun look reddish at sunrise and sunset but not at noon?

20. What is the evidence for a variety of particle sizes in a cloud?

Answers to Odd-Numbered Questions

1. All the same speed.

3. Ultraviolet.

5. Slower speed in glass, fastest speed in vacuum.

7. Interaction of light with atoms in the medium.

9. Light from its upper surface travels faster through slightly less dense air and bends downward, shrinking the vertical diameter.

11. The appearance is black.

13. Our eyes are most sensitve to yellow-green.

15. Magenta, cyan, yellow. When mixed, the color seen is the color that survives subtraction.

17. Blueness of the sky is the result of scattering.

19. The Sun is low in the sky at Sunrise and Sunset, resulting in a longer light path through the air, which scatters away blue leaving primarily lower-frequency light to reach an observer.

Chapter 13

Properties of Light

THE MAIN IDEA

Light reveals its nature through phenomena.

13.1 Rainbows
13.2 Lenses
13.3 Image Formation
13.4 Light Diffraction
13.5 Light Interference
13.6 Light Polarization
13.7 Wave-Particle

What causes the colors of a rainbow and why are they bow-shaped? Why are the colors reversed in the outer secondary rainbow? Why is the Sun always directly behind you when looking at a rainbow? And is it true that at the center of each rainbow you'd find your very own shadow? Beyond rainbows, what causes the iridescent colors in soap bubbles? Can light waves cancel other light waves? How do lenses redirect light to form images? What's the difference between regular sunglasses and Polaroid sunglasses? Is light a wave or is light a stream of particles, or both? We'll answer these sorts of fascinating questions and more in this chapter. Onward!

13.1 Rainbows

To understand rainbows we need to know about **dispersion**, the separation of light into colors arranged according to their frequency, as was nicely illustrated by Newton's experiment with a glass prism held in sunlight centuries ago. All the component colors of white sunlight separate in the glass prism, though only red, green, and blue are shown in **Figure 13.1**. The red part of the white beam slows and bends as it enters the glass, green bends a bit more, and blue is slowed and bent even more. When emerging, light undergoes a second refraction and the separation is further increased. Light is dispersed by a prism.

Newton further found that when he held a pair of identical prisms in sunlight, inverting the second one in such a way that the two opposite faces of the prism pair were parallel, the dispersed colors recombined to make white light again (**Figure 13.2**). This was confirmation that white light indeed is the combination of all the rainbow colors.

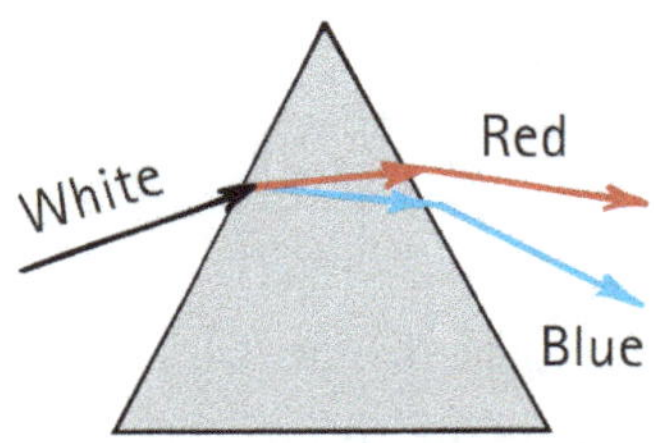

Λ Figure 13.1
Dispersion by a prism makes the color components of white light visible.

^ Figure 13.2
Isaac Newton used a glass prism to show that sunlight dispersed into a spectrum of colors, from red to violet. To human eyes, the colors seem to separate into bands, which Newton identified as red, orange, yellow, green, blue, indigo, and violet.

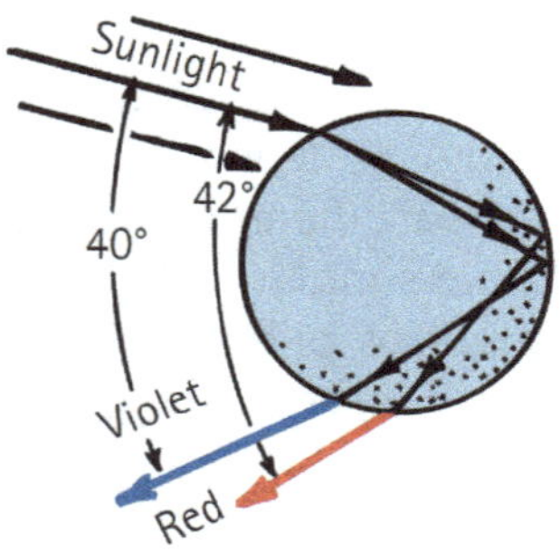

^ Figure 13.3
Dispersion of sunlight by a single raindrop.

Raindrops behave as prisms. You can see how a raindrop disperses light in **Figure 13.3**. Follow the ray of sunlight as it enters the drop near its top surface. Some of the light here is reflected (not shown), and the remainder is refracted into the water. At this first refraction, the light is dispersed into its spectrum colors, red being deviated the least and violet the most. When the light reaches the opposite side of the drop, each color is partly refracted out into the air (not shown) and partly reflected back into the water (shown). Arriving at the lower surface of the drop, each color is again partly reflected (not shown) and partly refracted into the air (shown). This refraction at the second surface, like that in a prism, increases the dispersion already produced at the first surface.

Although each drop disperses a full spectrum of colors, an observer is in a position to see only a single color from any one drop (**Figure 13.4**). If violet light from a single drop reaches an observer's eye, red light from the same drop travels a lower path (toward the observer's feet). To see red light, you should look to a higher drop in the sky. The color red is seen when the angle between a beam of sunlight and the dispersed light is about 42°. The color violet is seen when the angle between the sunbeams and dispersed light is about 40°.

A rainbow is not the flat, two-dimensional arc it appears to be. It appears flat for the same reason a spherical burst of fireworks high in the sky look flat—because they're far away. A rainbow is actually part of a

READING CHECK
Why do different colors of light bend differently in a transparent material?

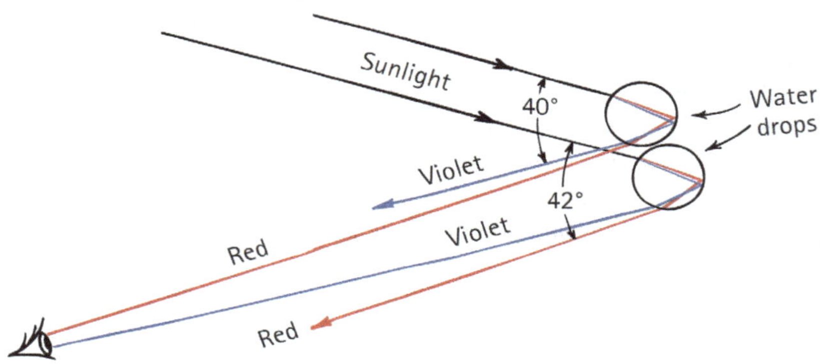

Figure 13.4 >
Sunlight incident on two raindrops as shown emerges as dispersed light. An observer sees the red light from the upper drop and the violet light coming from the lower drop. Millions of drops produce the whole spectrum of visible light.

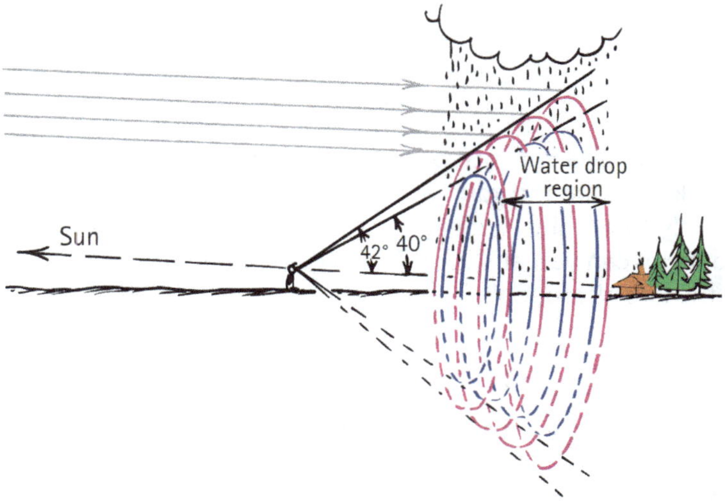

< Figure 13.5
When your eye is located between the Sun (not shown off to the left) and a water-drop region, the rainbow you see is the edge of a three-dimensional cone that extends through the water-drop region. (Innumerable layers of drops form innumerable two-dimensional arcs like the four suggested here.)

three-dimensional cone. The tip of the cone is at your eye. Think of a glass cone. If you held the pointed tip very close to your eye, what would you see? You'd see the glass as a circle. Likewise with a rainbow. All the drops that disperse the rainbow's light toward you lie in the shape of a cone—a cone of different layers. Four of the innumerable layers are shown in **Figure 13.5**. Drops that deflect red to your eye are on the outside, orange is beneath the red, yellow is beneath the orange, and so on all the way to violet on the inner conical surface. The thicker the region containing water drops, the thicker the conical edge you see through, and the more vivid the rainbow.

Your cone of vision intersects the cloud of drops and creates your rainbow. It is ever so slightly different from the rainbow seen by a person nearby. So, when a friend says, "Look at the pretty rainbow," you can reply, "Okay, move aside so I can see it, too." Everybody sees his or her own personal rainbow.

Another fact about rainbows: When you move, your rainbow moves with you. This is most apparent when observing a rainbow from a moving car—you'll see the rainbow traveling with you. This happens because the rainbow results from a *reflection* off the inner surface of the raindrops. Consider your reflection in a mirror. If you move, your reflection will move with you. This is the nature of any reflection, including those within a raindrop. Your personal rainbow will always travel with you.

Often a larger, secondary rainbow can be seen arcing at a greater angle around the primary bow. We won't treat this secondary rainbow except to say that it is formed by similar circumstances and is a result of double reflection within the raindrops (**Figure 13.7**). Because of this extra reflection (and extra refraction loss), the secondary rainbow is much dimmer and its colors are reversed, as shown in **Figure 13.8**.

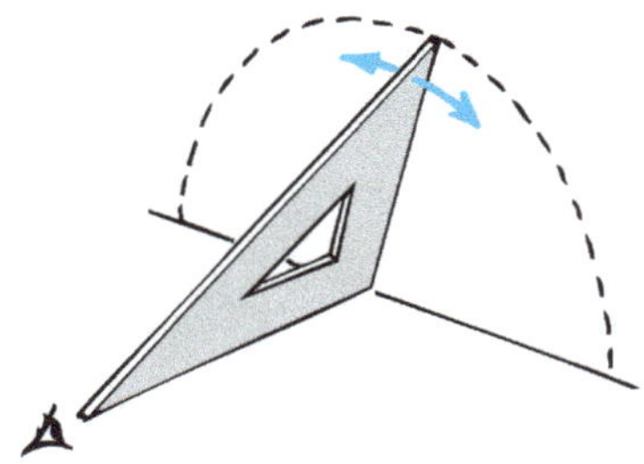

Λ Figure 13.6
Only raindrops along the dashed line disperse red light to the observer at about a 42° angle; hence, the light forms a bow.

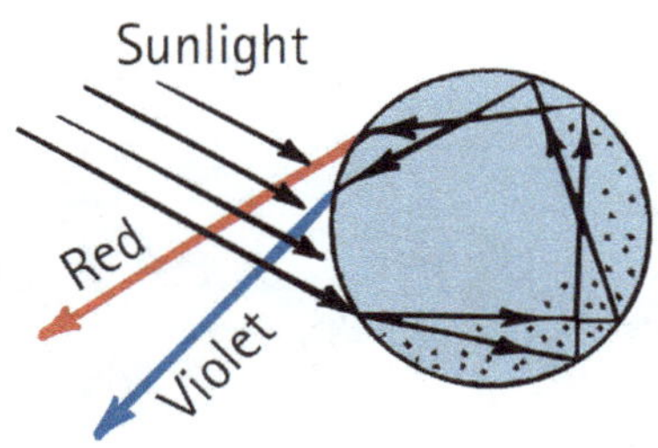

Λ Figure 13.7
Double reflection in a drop produces a secondary bow.

< Figure 13.8
The outer secondary rainbow is dimmer and its colors reversed. Just outside the inner primary rainbow the sky is much darker because, aside from the secondary rainbow, no light leaves the water droplet at angles greater than 42°.

Quick Activity

Your Own Personal Rainbow

On a bright sunny day with the Sun high in the sky, spray a fine mist of water from a garden hose. Look downward for a full circle rainbow within the mist. Look carefully and you'll actually see two circular rainbows—one for each eye. Close and open each eye separately to see this. We each see our own personal rainbow. That includes one for each eye. Having done this activity can you answer this: Why are rainbows only seen when the Sun is low in the sky, such as mornings or the late afternoon?

Unusual Rainbows

There are many ways that light can interact with moisture in the atmosphere resulting in a rainbow phenomenon. **Figure 13.9a** shows a "glory rainbow," which can be seen when flying above a cloud bank. Look carefully and you'll see the shadow of the airplane at the exact center. This tells you that the Sun is directly behind the high-flying photographer. **Figure 13.9b** shows an extremely rare "twinned rainbow," which results upon the merger of two different rainstorms with raindrops of different shapes and sizes. **Figure 13.9c** shows a "solar halo" that results when sunlight passes through ice crystals high in the atmosphere. A similar lunar halo is shown in **Figure 13.9f**. Notice how for both solar and lunar halos you are looking toward (not away from) the light source. This happens because the geometry of the ice crystals impacts the angles of reflection.

Under the right conditions, moonlight can also result in a rainbow, also known as a "moon bow," as shown in the time-lapse photo of **Figure 13.9d**. And recall from the previous chapter how clouds are white because the tiny water droplets within them refract and reflect all colors about equally. Sunlight reflecting off these droplets within a fog can thus result in a "white rainbow" as shown in **Figure 13.9e**.

Figure 13.9

(a) A glory rainbow with airplane shadow. (b) A twinned rainbow resulting from the convergence of two rain storms. (c) A solar halo from ice crystals high in the atmosphere. (d) A moonbow formed from the moonlight. (e) A white rainbow within a fog. (f) A lunar halo from ice crystals high in the atmosphere.

CONCEPT CHECK

1. Suppose you point to a wall with your arm extended. Then you sweep your arm around, making an angle of about 42° to the wall. If you rotate your arm in a full circle while keeping the same angle, what shape does your arm describe? What shape on the wall does your finger sweep out?

2. If light traveled at the same speed in raindrops as it does in air, would we have rainbows?

CHECK YOUR ANSWERS

1. Your arm describes a cone, and your finger sweeps out a circle. Likewise with rainbows.

2. No.

13.2 Lenses

Consider the many glass prisms arranged as shown in **Figure 13.10**. They refract incoming parallel light rays so that they converge to (or diverge from) a point. The arrangement shown in **Figure 13.10a** converges the light, and we have a **converging lens**. Notice that it is thicker in the middle. In **Figure 13.10b**, the middle is thinner than the edges, and makes a lens that diverges the light—we have a **diverging lens**. Note that the prisms in (b) diverge incident rays that make them appear to originate from a single point in front of the lens.

In both lenses, the greatest deviation of rays occurs at the outermost prisms because they have the greatest angle between the two refracting surfaces. No deviation occurs exactly in the middle, for in that region the two surfaces of the glass are parallel to each other. Real lenses are not made of prisms, of course. They are made of a solid piece of glass with surfaces ground usually to a circular curve. In **Figure 13.11** we see how smooth lenses refract waves. We find that the convergence or divergence of the light waves is a function of the surface angles as well as the thickness of the glass.

READING CHECK

How much deviation of light rays occurs in the exact middle of a lens?

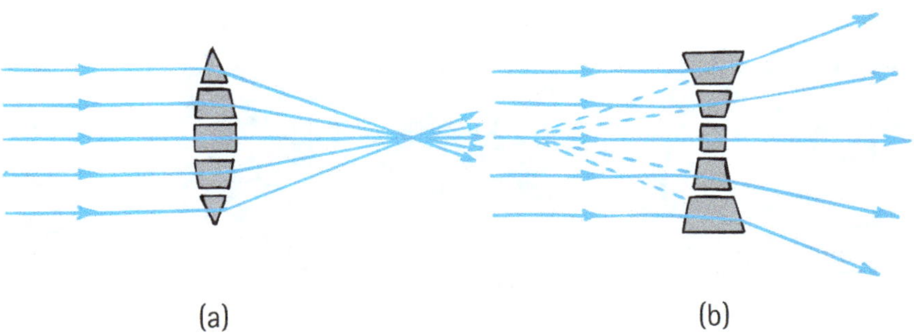

< Figure 13.10
Imagine a lens as a set of blocks and prisms.

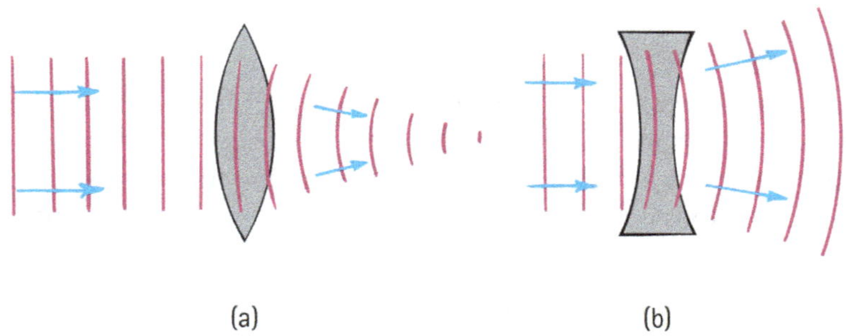

Figure 13.11 >
Wavefronts travel more slowly in glass than in air. In (a), the waves have farther to travel through the center resulting in a convergence. In (b), the waves have farther to travel through the edges resulting in a divergence.

Some key features for a converging lens are seen in **Figure 13.13**. The *centers of curvature* are where the opposite sides of the lens are parallel. The *principal axis* runs straight through the middle of the lens connecting the two centers of curvature. The focal point is the point of convergence for light parallel to the principal axis. This includes light moving from left to right or from right to left. Because a lens has two surfaces, it has two focal points. The *focal length* of the lens is the distance between the center of the lens and either focal point.

Quick Activity

Playing Around with Lenses

Determine the magnifying power of a lens, from your's or a friend's reading glasses, by focusing on the lines of a ruled piece of paper. Compare the number of line spaces within the non-magnified image to those within the magnified image. For example, three line spaces fitting into one magnified line space gives a magnification of 3:1, or 3x. Does the distance between the lens and the paper make a difference? How about the distance between the lens and your eye?

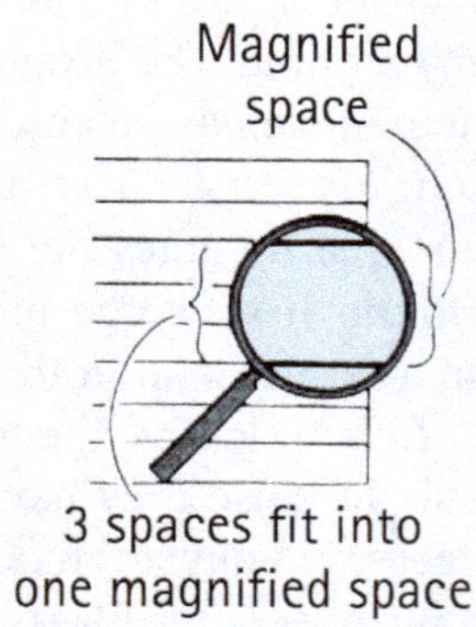

Figure 13.12 >
The moving patterns of bright and dark areas at the bottom of the pool result from the uneven surface of the water, which behaves like a blanket of undulating lenses. Just as we see the pool bottom shimmering, a fish looking upward at the Sun would see it shimmering too. Because of similar irregularities in the atmosphere, we see the stars twinkle.

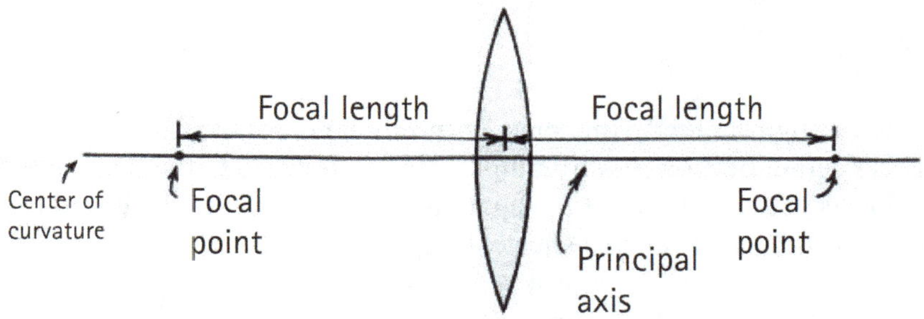

< **Figure 13.13**
Key features of a converging lens.

The Human Eye

With all of today's technology, the most remarkable optical instrument known is the eye (**Figure 13.14**). Light enters through your *cornea*, which does about 70 percent of the necessary bending of the light before it passes through an opening in the iris (colored part of the eye). The opening is called the *pupil*. Light then passes through your *lens*, which provides the extra bending power needed to focus images of nearby objects on your extremely sensitive *retina*. An image of the visual field outside your eye is spread over the retina. The retina is not uniform. There is a spot in the center of your field of view called the *fovea*—region of most distinct vision. You see greater detail here than at the side parts of your eye. There is also a spot in your retina where the nerves carrying all the information exit; this is your *blind spot*.

You can demonstrate that you have a blind spot in each eye. Close your left eye, and look at the round dot and X below with your right eye only. You can see both the dot and the X at a distance. Now move your face slowly toward the image, with your right eye fixed upon the dot, and you'll reach a position about 20–25 centimeters from your eye where the X disappears. When both eyes are open, one eye "fills in" the part to which your other eye is blind. Now repeat with only the left eye open, looking this time at the X, and the dot will disappear. But note that your brain fills in the two intersecting lines. Your brain fills in the "expected" view even with one eye closed. Instead of seeing nothing, your brain graciously fills in the appropriate background. Repeat this for small objects on various backgrounds. You not only see what's there—you see what's not there!

FOR YOUR INFORMATION

For a converging lens, the focal point is on the opposite side of the light source. For a diverging lens, the focal point is on the same side of the light source, in Figure 13.10b.

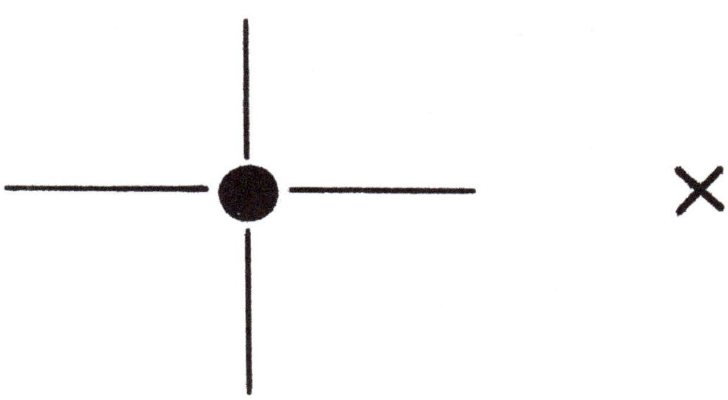

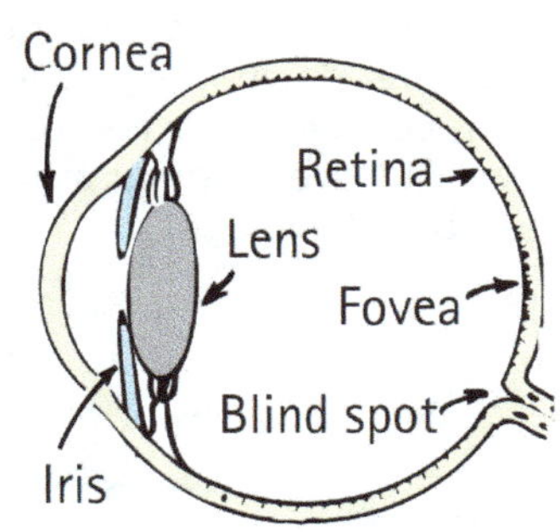

∧ **Figure 13.14**
The human eye

The light receptors in your retina do not connect directly to your optic nerve but are instead interconnected to many other cells. Through these interconnections a certain amount of information is combined and "digested" in your retina. In this way the light signal is "thought about" before it goes to the optic nerve and then to the main body of your brain. So, some brain functioning occurs in your eye. Amazingly, your eye does some of your "thinking."

13.3 Image Formation

If there is light around you, such as in the daytime, then there are light rays reflecting off you. It's from this reflected light that others are able to see you as you. Hold a sheet of paper up to your face. Light reflected from your face falls all over this sheet of paper. Pick any point on that sheet of paper and you can be sure it receives light from your forehead, your nose, mouth, chin, and all other parts at the same time. You don't see your face in the paper because the microscopically rough surface of the paper reflects all that light from your face in all directions.

Now place a card directly between your face and the paper. The light from your face will no longer reach the paper because the card is blocking that light. But here's a clever trick: Poke a tiny hole within that card. This pinhole allows light from one part of your face to fall only upon one part of the paper, rather than the entire paper. For example, light from your forehead will only be able reach the lower part of the paper. Likewise, light from your chin will only be able to reach the upper part of the paper. Each bit of the paper will then be reflecting light from only one part of your face. The result will be that the image of your face is shown on the paper. This is called *pinhole optics*. Because the amount of light coming from your face is not very much, the pinhole image on the paper will be very faint. But this image will be visible from all directions because the paper is reflecting that light in all directions. The same holds true of a vase and flowers in **Figure 13.15b**.

The first cameras had no lenses and admitted light through a small pinhole. Long exposure times were required because of the small amount of light admitted by the pinhole. That meant subjects being photographed had to remain still, which meant people rarely smiled. Motion would produce a blur. If the hole were a bit larger, exposure time would be shorter, but overlapping rays would produce a blurry image—too large a hole would allow too much overlapping resulting in no clear image. That's where a converging lens comes in (**Figure 13.15c**). The lens converges light onto the film without any overlapping of rays. Moving objects can be taken with the lens camera because of the short exposure time. That's why early photographs taken with lens cameras were called snapshots—and with short exposure times, people felt more free to smile. Furthermore, with short exposure times, a series of pictures could be captured onto a reel of film, which when developed and projected onto a screen allowed for a "motion picture," also known as a movie.

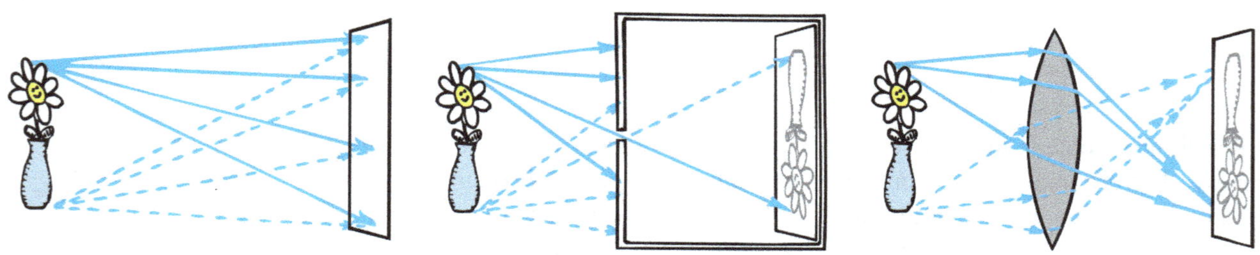

(a) (b) (c)

V Figure 13.15
Image formation. (a) No image appears on the wall because rays from all parts of the object overlap all portions of the wall. (b) A single small opening (pinhole) in a barrier prevents overlapping rays from reaching the wall; a dim upside-down image is formed. (c) A lens converges the rays upon the wall without overlapping; more light makes a brighter image.

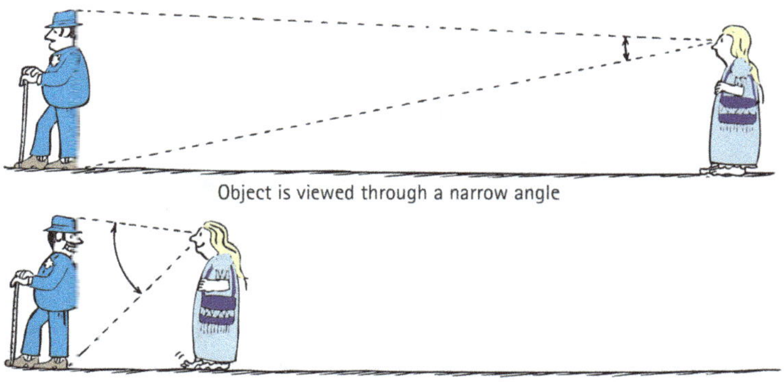

< Figure 13.16
Viewing through a narrow and wide angle.

The simplest use of a converging lens is a magnifying glass. To understand how it works, think about how you examine objects near and far. With unaided vision, you see a distant object through a relatively narrow angle of view, and a close object through a wider angle of view (**Figure 13.16**). To see the details of a small object, you want to get as close to it as possible for the widest-angle view. But your eye can't focus when it's too close to the object. That's where the magnifying glass plays a role. When close to the object, the magnifying glass gives you a clear image that would be blurry otherwise.

Quick Activity

Making a Pinhole Camera

Make your own historic first camera. Cut off one end of a small cardboard box, shoe boxes are fine, and cover it with translucent sheet, such as wax paper. Poke a clean-cut small hole with a sharp pencil at the other end. Aim the "camera" at a bright object and if lighting is right you'll see an upside-down image of the object displayed on the translucent sheet. The tinier the pinhole, the dimmer but sharper the image.

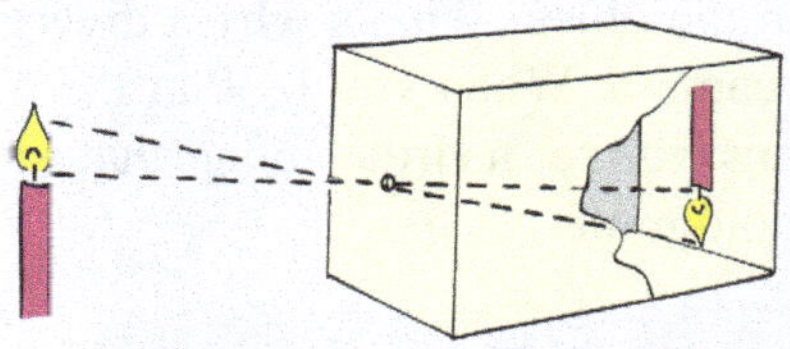

Or simply look at the spots of light on the ground beneath a sunlit tree. It so happens that when the openings between leaves in a tree are small compared to their distance from the ground, the openings behave as pinholes. The greater the distance between the tree openings and the ground, the larger the sunballs. Siu Bil Lee sits next to sunballs hardly noticed by people walking by. People don't notice them until pointed out. Want to impress your friends? When walking with them beneath sunlit trees, point out the sunball images of the Sun that may have eluded their notice over a lifetime. And as pointed out by Tracy, during a solar eclipse these sunballs tranform into sun crescents!

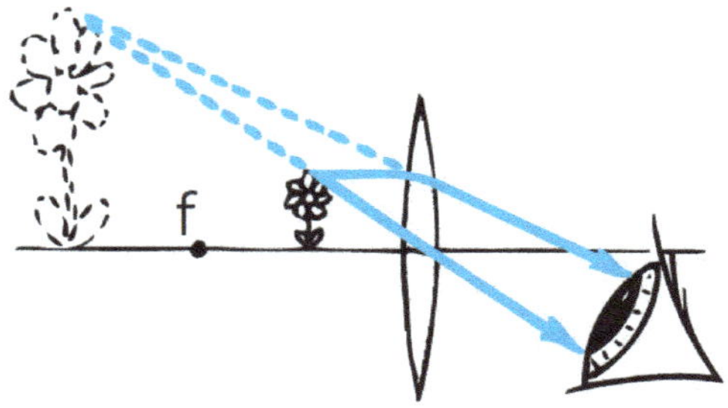

Figure 13.17 >
When an object is near a converging lens (inside its focal point *f*), the lens acts as a magnifying glass to produce a virtual image. The image appears larger and farther from the lens than the object.

When you use a magnifying glass, you hold it close to the object you wish to examine (**Figure 13.17**). This is because a converging lens provides an enlarged, right-side-up image only when the object is inside the focal point. If a screen is placed at the image distance, no image appears on it because no light is directed to the image position. The rays that reach your eye, however, behave virtually as if they originated at the image position. This is called a **virtual image**.

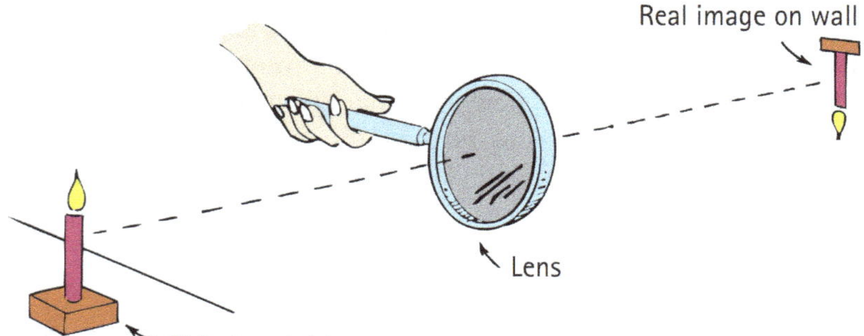

Figure 13.18 >
When an object is far from a converging lens (beyond its focal point), a real upside-down image is formed. Fun fact: The image formed on your retina is also upside down. Through the magic of biology, your brain works to interpret the image as upside right.

READING CHECK
Are real images formed with a single lens always, or sometimes, upside down?

When the object is distant enough to be outside the focal point of a converging lens, a **real image** is formed instead of a virtual image, **Figure 13.18** shows this case. A real image is upside-down. Likewise for projecting motion pictures on a screen. To see a right-side image the slide or film is upside-down. The same is true for the image in a camera. Real images formed with a single lens are always upside-down.

A diverging lens used alone produces a reduced virtual image. It makes no difference how far or how near the object is. The image is always virtual, right-side up, and smaller than the object. That's why a diverging lens is often used as a "finder" on a camera. When you look at the object to be photographed through such a lens, you see a virtual image that approximates the same proportions as the photograph.

Figure 13.19 >
A diverging lens forms a virtual, right-side-up image of Jamie and his cat.

CONCEPT **CHECK**

Why is the greater part of the photograph in Figure 13.19 out of focus?

CHECK YOUR ANSWER

Both Jamie and his cat and the virtual image of Jamie and his cat are "objects" for the lens of the camera that took this photograph. Since the objects are at different distances from the lens, images are at different distances relative to the focusing screen (film) in the camera. So only one object can be brought into focus. The same is true of your eyes. You cannot focus on near and far objects at the same time.

Lens Defects

No lens forms a perfect image. A distortion in an image is called an *aberration*. Aberrations can be minimized by combining lenses in particular ways. For this reason, most optical instruments use compound lenses instead of using single lenses. Each compound lens consists of several simple lenses.

Spherical aberration results when light passing through the edges of a lens focus at a slightly different place from where light passing through the center of the lens focuses (**Figure 13.20**). As a result, the image you see is blurred. This can be remedied by covering the edges of the lens. A camera does this with the use of a diaphragm. In good optical instruments, spherical aberration is corrected by a combination of lenses.

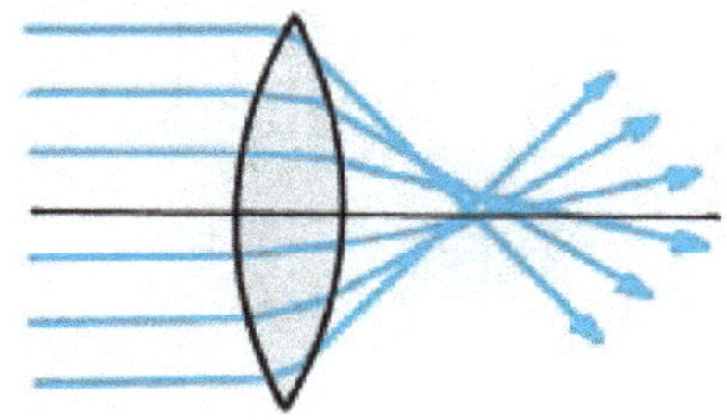

Figure 13.20
Spherical aberration.

Chromatic aberration is the result of different colors having different speeds. This means different refractions in the lens. Different colors of light focus in different places (**Figure 13.21**). As a result, some colors in the image you see may be in focus, while other colors are out of focus. This is corrected with achromatic lenses, which combine simple lenses of assorted glass.

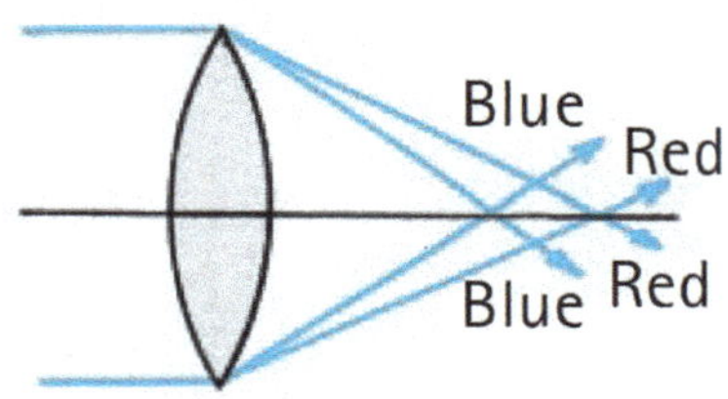

Figure 13.21
Chromatic aberration.

The pupil of your eye changes in size to regulate the amount of light that enters. Vision is sharpest when your pupil is smallest because light passes through only the center of your eye's lens. Through the center, spherical and chromatic aberrations are minimal. Also, the eye then functions like a pinhole camera, needing minimum focusing for a sharp image. Your vision is better in bright light because your pupils are smaller.

Astigmatism of the eye is a defect caused by an irregularity in the curvature of the cornea. It curves more in one direction than another—somewhat like the side of a barrel. Due to this defect, the eye doesn't form sharp images. The remedy is eyeglasses with cylindrical lenses that have more curvature in one direction than in another.

READING CHECK

Distinguish between spherical and chromatic aberration.

CONCEPT **CHECK**

1. If light traveled at the same speed in glass and in air, would glass lenses change the direction of light rays?
2. Why is chromatic aberration associated with a lens but not with a mirror?

CHECK YOUR ANSWERS

1. No.
2. Light of different frequencies travel at different speeds in a transparent medium and therefore refract at different angles. This is the cause of chromatic aberration. The angles of reflected light, however, have no relation to frequency. One color reflects the same as any other color. In telescopes, therefore, mirrors are preferred over lenses because of the absence of chromatic aberration for reflection.

One option for those with poor sight today is wearing eyeglasses. The advent of eyeglasses probably occurred in Italy in the late 1200s. (Curiously, the telescope wasn't invented until some 300 years later. If, in the meantime, anybody viewed objects through a pair of lenses separated along their axes, such as fixed at the ends of a tube, there is no record of it.) Another present-day option to wearing eyeglasses is contact lenses. And yet another is cornea laser surgery, where the cornea is shaved to a proper shape for normal vision. Improvements are ongoing.

Quick Activity

Seeing the Image of the Sun Cast by a Pinhole

Poke a small hole with the tip of a pencil or ball-point pen in the middle of a small piece of cardboard or index card. Hold the card in bright sunlight about a meter or so above the floor. The card casts a shadow on the floor. In the middle of the card's shadow is a circular spot of light—an image of the Sun. If the Sun is low in the sky, the circular image is lengthened to an ellipse. Hold the card higher and the image is bigger. Repeat the procedure, but instead of a round hole poked in the card, carefully cut a diamond shape. Notice that the solar image would not be diamond shaped, but the same as if cast through openings between leaves in a tree—circular. On the other hand, what will be the shape of these images at the time of a partial solar eclipse?

13.4 Light Diffraction

When you touch your finger to the surface of still water, circular ripples are produced. When you touch the surface with a straight edge, such as a horizontally held meter stick, you produce a plane wave. You can produce a series of plane waves by successively dipping a meter stick into the surface (**Figure 13.22**).

Figure 13.22 >
The vibrating meter stick makes plane waves in the tank of water. These waves diffract through the opening.

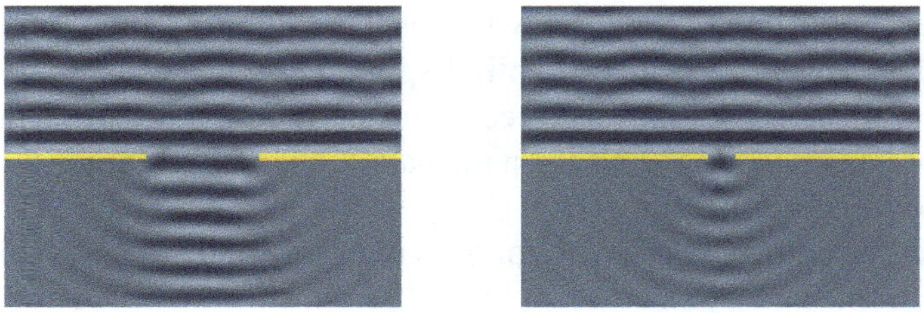

< Figure 13.23
Plane waves passing through openings of various sizes. The edges become more predominant with a smaller the opening, which makes the diffraction more apparent.

The photographs in **Figure 13.23** are top views of water ripples in a shallow glass tank (called a ripple tank). A barrier with an adjustable opening is in the tank. When plane waves meet the barrier, they continue through with some distortion. In the left image, where the opening is wide, the waves continue through the opening almost without change. At the two ends of the opening, however, the waves bend. This bending around an edge is called **diffraction**. Any bending of light by means other than reflection and refraction is diffraction. As the width of the opening is narrowed, as in the center image in Figure 13.23, the waves spread more. When the opening is small relative to the wavelength of the incident wave, they spread even more. Diffraction is the spreading of light around obstacles, a property of all kinds of waves that includes sound and light waves.

READING CHECK
What kinds of waves diffract?

Diffraction is not confined to narrow slits or to openings in general but can be seen for all shadows (**Figure 13.24**). On close examination, even the sharpest shadow is blurred slightly at the edge. When the light is of a single color (monochromatic), diffraction can produce fringes at the edge of the shadow, as in **Figure 13.25**. In white light, the fringes tend to merge together to create a fuzzy blur at the edge of a shadow.

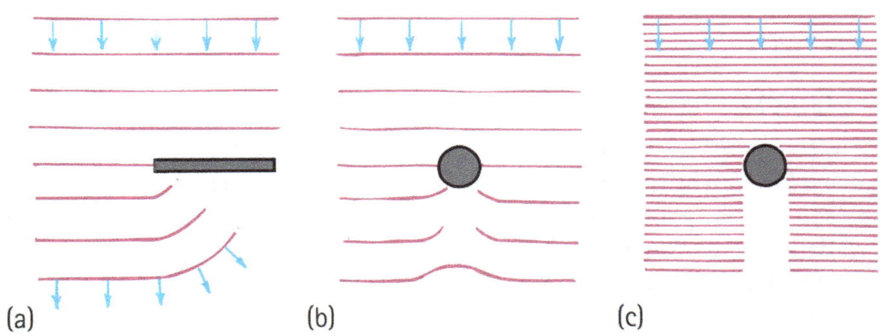

< Figure 13.24
(a) Waves tend to spread into the shadow region. (b) When the wavelength is about the size of the object creating the barrier, the shadow is soon filled in. (c) When the wavelength is short relative to the object's size, a sharp shadow is cast.

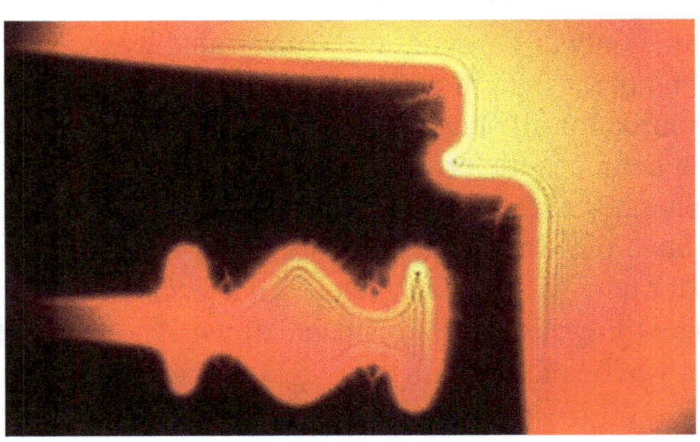

< Figure 13.25
Diffraction fringes are evident in the shadows of monochromatic (single-frequency) laser light.

The amount of diffraction depends on the wavelength of the wave compared with the size of the obstruction that casts the shadow. Long waves are better at filling in shadows, which is why foghorns emit low-frequency sound waves—to fill in any "blind spots." Likewise for light and other electromagnetic waves. When receiving signals on our electronic devices, we don't wish to "see" objects in the path of incoming signals. In this case, diffraction is a good thing.

Diffraction is not so nice for viewing very small objects with a microscope. If the size of an object is about the same as the wavelength of light, diffraction blurs the image. If the object is smaller than the wavelength of light, no structure can be seen. Instead, the entire image is lost due to diffraction. No amount of magnification or perfection of microscope design can defeat this fundamental diffraction limit.

To minimize this problem, microscopists illuminate tiny objects with electron beams rather than light. Compared with light waves, electron beams have extremely short wavelengths. Electron microscopes take advantage of the fact that all matter has wave properties. A beam of electrons has a wavelength smaller than those of visible light. In an electron microscope, electric and magnetic fields, rather than optical lenses, are used to focus and magnify images.

CONCEPT CHECK

Why does a microscopist use blue light rather than white light to illuminate objects being viewed?

CHECK YOUR ANSWER

There is less diffraction with blue light. This allows the microscopist to see more detail (just as a dolphin beautifully investigates fine detail in its environment by the echoes of ultra-short wavelengths of sound).

Huygens' Principle

The work of the 17th century Dutch physicist, Christian Huygens, provides insight as to why diffraction occurs. Toss a pebble into a still pond. It's easy to see that the pebble is the point source for a resulting wave that spreads radially away from where the pebble entered the water. But there's another way to generate that very same wave. Smash the pebble into a million tiny grains. From just above the water, magically toss those grains outward in all directions so they hit the pond all at once in a perfect circle. Where each grain hits, a tiny wavelet will be formed. All those wavelets combined will make the same outward moving wave generated by the single pebble.

Huygens great insight was that every point along a wave behaves as a point source for all subsequent waves. As shown in **Figure 13.26**, a central point source is the origin for the wave AA′. Each point along AA′, however, generates an outward moving wavelet. Combine all those wavelets and you'll have the wavefront BB′. So, every point along a wave serves as a source for new waves. This makes sense because each point along the wave is where things are jumping up and down. And where things jump up and down, a wave is formed.

We can relate this to diffraction by considering that any wave generated from a point source always travels radially. When a linear wavefront as shown in Figures 13.22 and 13.23 encounters an edge, the points near that edge will naturally generate waves in a radial fashion. We say that the wave has *diffracted* around the edge.

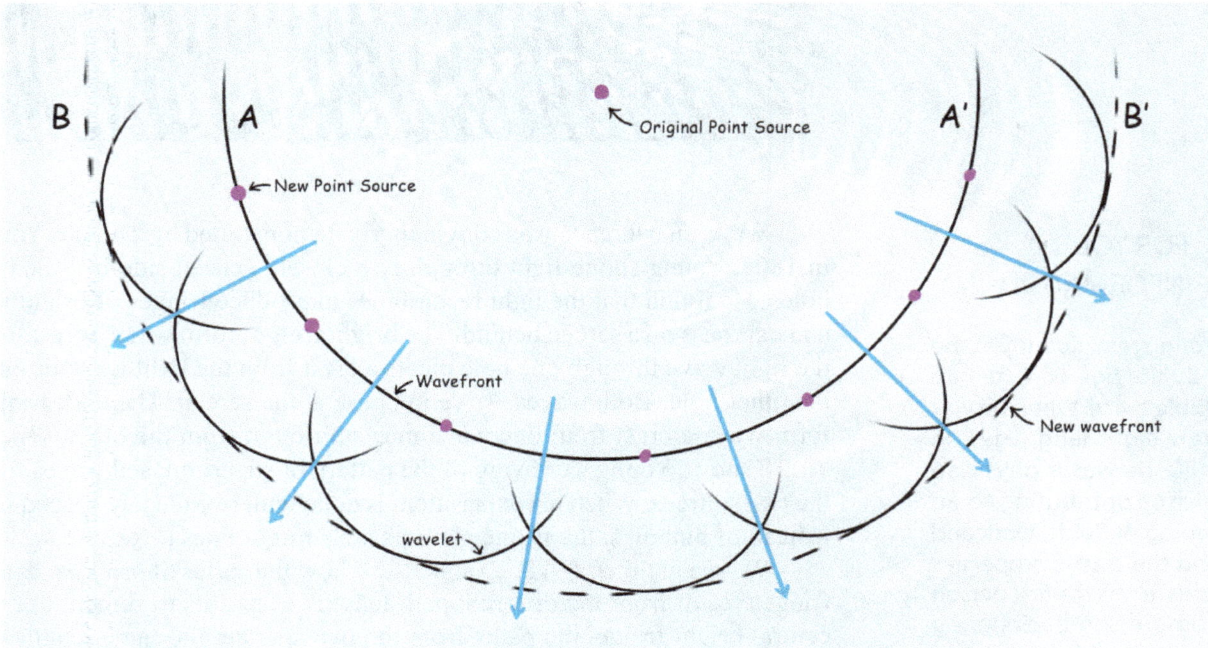

Λ Figure 13.26

Huygens' principle applied to a circular wavefront. Points all along wave front A to A' serve to generate wavefront B to B'. Likewise, points all along B to B' will act to generate subsequent wavefronts.

13.5 Light Interference

When two waves interact, the amplitude of the resulting wave is the sum of the amplitudes of the two individual waves. The overlapping waves produce a feature of waves that we call **interference**. As we discussed in our study of sound, waves can can combine constructively or destructively. Interference is a property of all waves, including light. Constructive and destructive interference is reviewed in **Figure 13.27**. We see that the adding, or superposition, of a pair of identical waves in phase produces a wave of the same frequency but twice the amplitude. If the waves are exactly one-half wavelength out of phase, their superposition results in complete cancellation. If they are out of phase by other amounts, partial cancellation occurs.

READING CHECK
What kinds of waves can undergo interference?

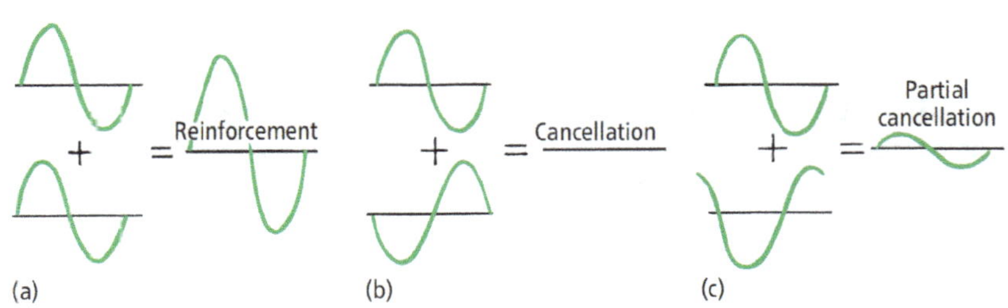

< Figure 13.27
Wave interference.

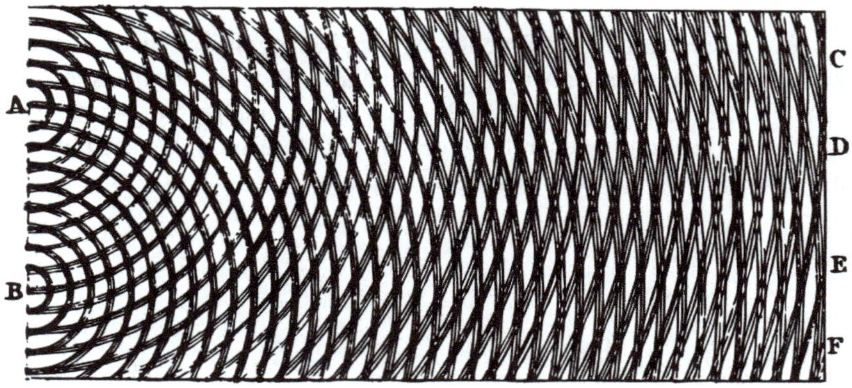

Figure 13.28 >
Thomas Young's original drawing of a two-source interference pattern.

FOR YOUR INFORMATION

Thomas Young read fluently at the age of 2. By the age of 4, he had read the Bible twice. When he was 14 he knew eight languages. In his adult life he was a physician and physicist, contributing to an understanding of fluids, work and energy, and the elastic properties of materials. He was the first person to make progress in deciphering Egyptian hieroglyphics. No doubt about it—Thomas Young was a bright guy!

Wave interference was convincingly demonstrated by Thomas Young in 1801. Young shone light through two closely spaced side-by-side pinholes. He found that the light recombines to produce fringes of brightness and darkness on a screen behind. The bright fringes form when a crest from the light wave through one hole meets a crest from the light wave through the other hole. Both waves arrive in phase at the screen. The dark fringes form when a crest from one wave meets a trough from the other. **Figure 13.28** shows Young's drawing of the pattern of superimposed waves from the two sources. When his experiment is done with two closely spaced slits instead of pinholes, the fringe patterns are straight lines (**Figure 13.29**).

We see in **Figures 13.30** and **13.31** how the series of bright and dark fringes result from the different path lengths from slits to screen. For the central bright fringe, the paths from the two slits are the same length and so the waves arrive in phase and reinforce each other. The dark fringes on either side of the central fringe result from one path being longer (or shorter) by one-half wavelength, so waves there arrive half a wavelength out of phase. The other sets of dark fringes occur where the paths differ by odd multiples of one-half wavelength: 3/2, 5/2, and so on.

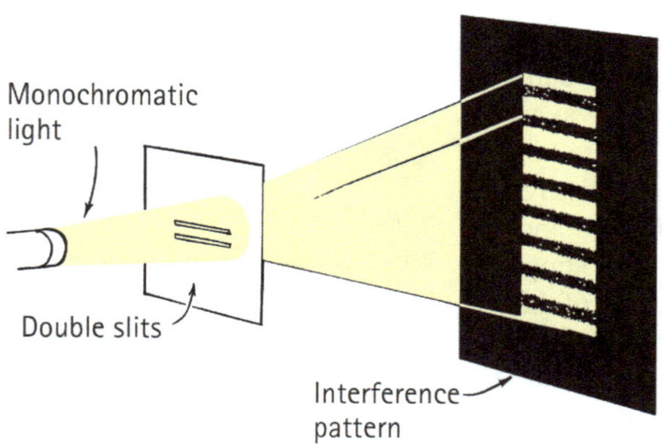

Λ Figure 13.29
When monochromatic light passes through two closely spaced slits, a interference pattern of fringes is produced.

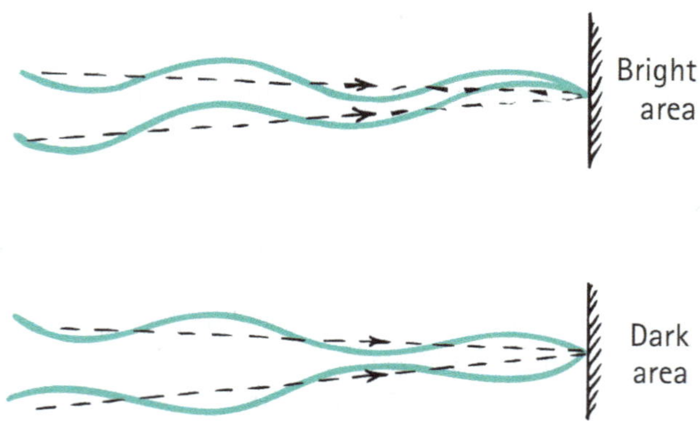

Λ Figure 13.30
A bright area occurs when waves from both slits arrive in phase; a dark area results from the overlapping of waves that are out of phase.

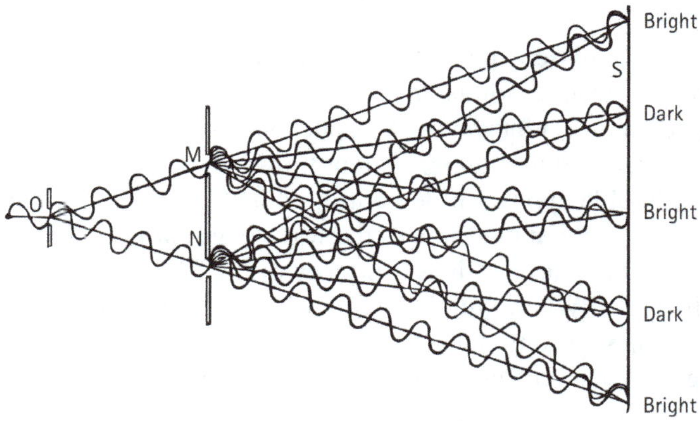

< Figure 13.31
Light from O passes through slits M and N and produces an interference pattern on the screen S.

CONCEPT **CHECK**

1. Which fringes would be wider apart, those of red light or blue light, assuming the red and blue are monochromatic (single frequency)?
2. Why is it important that monochromatic light be used?

CHECK YOUR ANSWERS

1. Red fringes would be more widely spaced. Can you see that longer waves in **Figure 13.31** would produce wider-apart fringes?
2. Light of various wavelengths would result in dark fringes for one wavelength filling in bright fringes for another. The result would be no distinct fringe pattern.

Interference patterns are not limited to one or two slits. A multitude of closely spaced slits makes up a *diffraction grating*, as shown in **Figure 13.32**.These devices, like prisms, disperse white light into colors. This happens because different colors of light have different frequencies, which means they diffract to different extents—higher frequency blues less so than lower frequency reds. Diffraction gratings are used in devices called *spectrometers*, which can be used to study the light coming from glowing sources, such as atoms or stars. The feathers of some birds get their color by acting as diffraction gratings. Likewise for the lined-up microscopic pits on the reflective surface of compact discs, which explains why they reflect a rainbow of color.

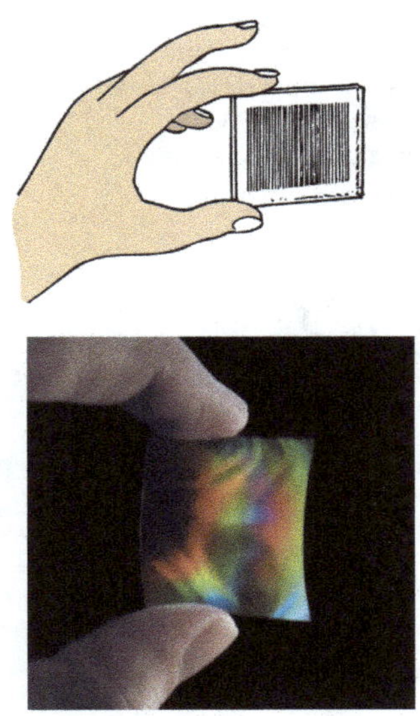

∧ Figure 13.32
Diffraction gratings employ wave interference. The one above is often used in place of a prism in a spectrometer.

Seeing Star-Shaped Stars

Have you ever wondered why stars are represented with spikes? All through the ages, stars have been drawn with spikes. Stars, however, do not have spikes. Each star is as round as our Sun. Because those stars are so far away, they appear as point sources of light in the night sky. The reason for the spikes is poor eyesight.

The surface of your eye, the cornea, becomes scratched by a variety of causes, such as sand particles flying in the wind. These scratches make up a diffraction grating. A scratched cornea, however, is not a very good diffraction grating, but its effects are evident. Look at a bright point source against a dark background—like a star in the night sky. Instead of seeing a point of light, you may see a spiky shape. The spikes even shimmer and twinkle if there are some temperature differences in the atmosphere to produce some refraction. And if you live in a windy desert region where sandstorms are frequent, your cornea will be even more scratched and you'll see more vivid star spikes.

Interference Colors from Thin Films

Perhaps you have noticed the beautiful spectrum of colors reflected from a soap bubble or from gasoline on a wet street. These colors are produced by the interference of light waves. This phenomenon is often called *iridescence* and is observed in thin transparent films.

A soap bubble appears iridescent in white light when the thickness of the soap film is about the same as the wavelength of light. Light waves reflected from the outer and inner surfaces of the film to your eye travel different distances. When illuminated by white light, the film may be just the right thickness at one place to cause the destructive interference of, say, red light. When red light is subtracted from white light, the mixture left appears as the complementary color—cyan. At another place, where the film is thinner, perhaps blue is cancelled. Then the light seen is the complement of blue—yellow. Whatever color is canceled by interference, the light seen is its complementary color.

READING CHECK
When white light illuminates a thin film and a particular color is cancelled by interference, what color remains?

Follow the paths of the two identical parallel waves in **Figure 13.33**. Although both waves undergo some transmission and reflection, we focus on their reflections. For clarity, one wave is drawn a bit darker than the other. The left-most (darker) incident wave is transmitted into the thin soapy film and reflects from the bottom surface to join the path of the (lighter) wave that reflects from the top surface. Aha! When the waves rejoin, the waves that reach the eye are out of step (out of phase) by a half wavelength. They are canceled by destructive interference! If green were the only color illuminating the soap film, no reflected light would reach the observer's eye. But if white light shines on the film, an observer sees the complementary color of green, which is magenta. So the color magenta is seen on the part of the soap film where green is canceled. On other parts where red is canceled, the film will appear cyan. Or where blue is canceled, yellow is seen. A variety of colors is produced.

Although we focus on destructive interference in Figure 13.33, constructive interference similarly contributes to the color mix. Constructive interference brightens incident colors instead of cancelling them. Partial interference also adds to soap-bubble colors. Waves of sunlight that undergo interference account for the colorful display of bubbles in a neighborhood park (**Figure 13.34**). Perhaps more study of wave interference awaits your further study.

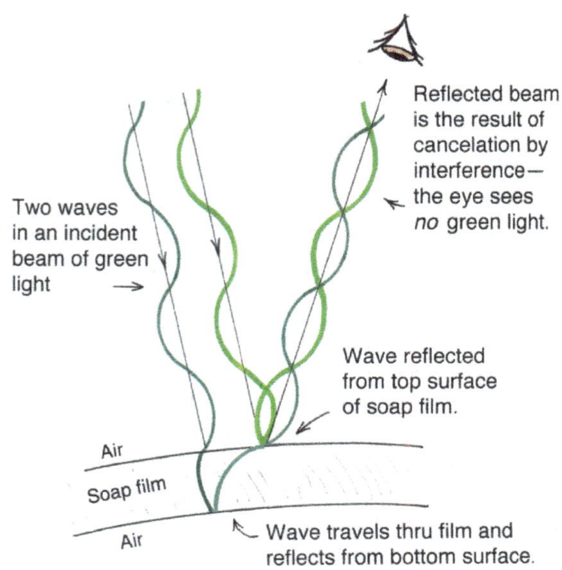

Figure 13.33 >
No green light reaches the eye when the pair of green waves cancel

< **Figure 13.34**.
Cheers to the colorful display of soap-bubble colors in the park!

Quick Activity

Swirling Colors

Dip the top of a dark-colored cup in dishwashing detergent so that a soap film forms across the mouth. Note: A dark cup makes the best background for viewing interference colors. Then hold the cup sideways and look at the reflected light from the soap film. Swirling colors appear as the soap runs down to form a wedge that grows thicker at the bottom. The top becomes thinner—so thin that it appears black. This occurs when the film is thinner than 1/4 the wavelength of the shortest waves of visible light. The film soon becomes so thin it pops.

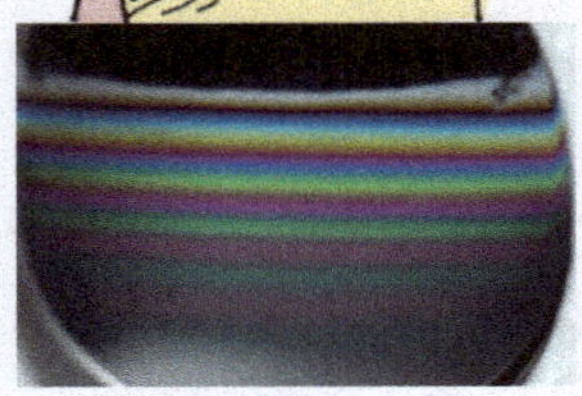

CONCEPT **CHECK**

1. What color appears to be reflected from a soap bubble in sunlight when its thickness is such that green light is canceled?

2. In the left column are the colors of certain objects. In the right column are various ways in which colors are produced. Match the right column to the left.

(a)yellow daffodil	(1)interference
(b)blue sky	(2)diffraction
(c)rainbow	(3)selective reflection
(d)peacock feathers	(4)refraction
(e)soap bubble	(5)scattering

CHECK YOUR ANSWERS

1. The composite of all the visible wavelengths except green is the complementary color, magenta. (Go back and review Section 12.6.)

2. a-3; b-5; c-4; d-2; e-1.

13.6 Light Polarization

Interference and diffraction provide the best evidence that light is wave-like. As we learned in Chapter 11, waves can be either longitudinal or transverse. Sound waves are longitudinal, which means the vibratory motion of the medium is along the direction of wave travel.

If you shake a rope either up and down or from side to side as shown in **Figure 13.35**, you'll produce a transverse wave along the rope. The plane of vibration is the same as the plane of the wave. If we shake it up and down, the wave vibrates in a vertical plane. If we shake it back and forth, the wave vibrates in a horizontal plane. We say that such a wave is *plane-polarized*—that the waves traveling along the rope are confined to a single plane. Polarization is a property of transverse waves. (Polarization does not occur among longitudinal waves—there is no such thing as polarized sound.) The fact that light waves exhibit **polarization** demonstrates that they are transverse.

Figure 13.35
A vertically plane-polarized plane wave and a horizontally plane-polarized plane wave.

A single vibrating electron can emit an electromagnetic wave that is plane-polarized. The plane of polarization matches the vibrational direction of the electron. That means a vertically accelerating electron emits light that is vertically polarized. A horizontally accelerating electron emits light that is horizontally polarized (**Figure 13.36**).

READING CHECK
What kind of waves can exhibit polarization?

A common light source, such as a fluorescent lamp or a candle flame, emits light that is unpolarized. This is because electrons that emit the light are vibrating in many random directions. There are as many planes of vibration as the vibrating electrons producing them. A few planes are represented in **Figure 13.37a**. We can represent all these planes by radial lines, shown in **Figure 13.37b**. Or, more simply, the planes can be represented by vectors in two mutually perpendicular directions, **Figure 13.37c**. Vertical vector represents all the components of vibration in the vertical direction. The horizontal vector represents all the components of vibration horizontally. The simple model of Figure 13.37c represents unpolarized light.

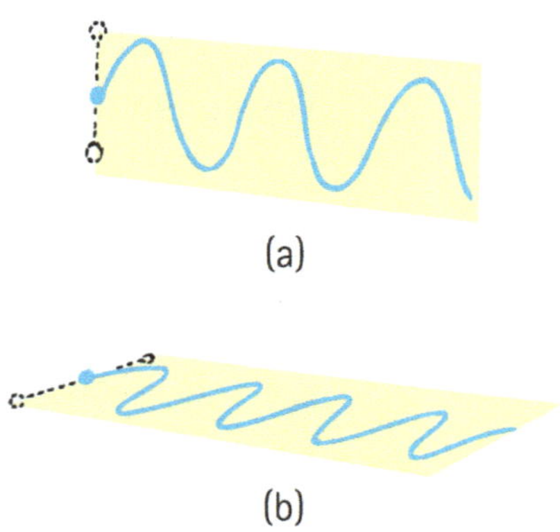

Figure 13.36
(a) A vertically plane-polarized wave from a charge vibrating vertically. (b) A horizontally plane-polarized wave from a charge vibrating horizontally.

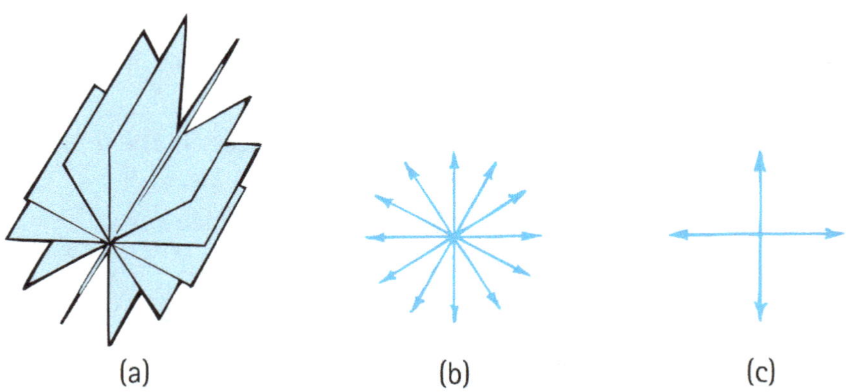

Figure 13.37
Representations of planes of waves. The (b) and (c) configurations show electric vectors that make up the electric part of electromagnetic waves.

Many crystals have the property of absorbing one polar orientation of light but not the other. The light that transmits through the crystal results in only one polar orientation and is said to be *polarized* (**Figure 13.38**). Herapathite is such a crystal. Microscopic herapathite crystals can be aligned and embedded between plastic sheets to produce polarizing filters, popular in sunglasses. More commonly, the plastic itself can be processed to serve as a polarizing filter.

If you look at unpolarized light through a polarizing filter, you can rotate the filter in any direction and the light appears unchanged. But if the light source itself is polarized, then as you rotate the filter, you progressively reduce more and more of the light until it is blocked out. An ideal filter transmits 50 percent of incident unpolarized light. That 50 percent is polarized. When two polarizing filters are arranged so that their polarization axes are aligned, light can pass through both (**Figure 13.39a**). If their axes are at right angles to each other, (in this case we say the filters are crossed) almost no light penetrates the pair (**Figure 13.39b**). (A small amount of shorter wavelengths do get through.) When polarizing filters are used in pairs like this, the first one is called the *polarizer* and the second one the *analyzer*.

Much of the light reflected from nonmetallic surfaces is polarized. The glare from glass or water is a good example. Except for light that hits vertically, the reflected ray has more vibrations parallel to the reflecting surface. The part of the ray that penetrates the surface has more vibrations at right angles to the surface (**Figure 13.40**). Skipping flat rocks off the surface of a pond is analogous. When the rocks hit parallel to the surface, they easily reflect. But when they hit with their faces at right angles to the surface, they "refract" into the water. The glare from reflecting surfaces can be dimmed a lot with the use of polarized sunglasses. The polarization axes of the lenses are vertical because most glare reflects from horizontal surfaces.

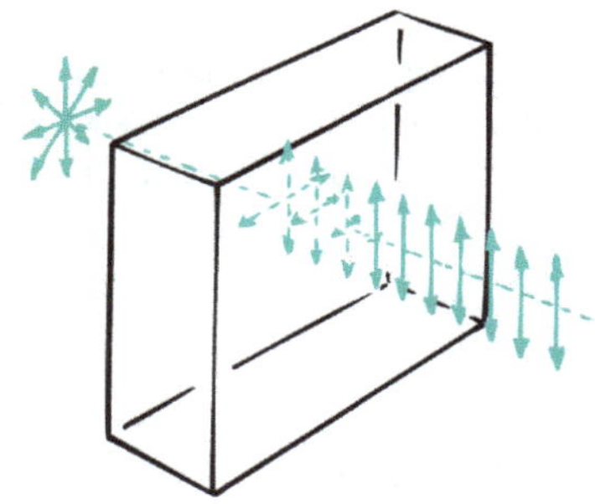

Λ Figure 13.38
One component of the incident unpolarized light is absorbed, resulting in emerging polarized light.

V Figure 13.39
(a) Both light and the vibrations of the rope pass through aligned filters. (b) Neither light nor vibrations of the rope pass through crossed filters. The blue dashed line shows the direction light would travel if the filters weren't crossed.

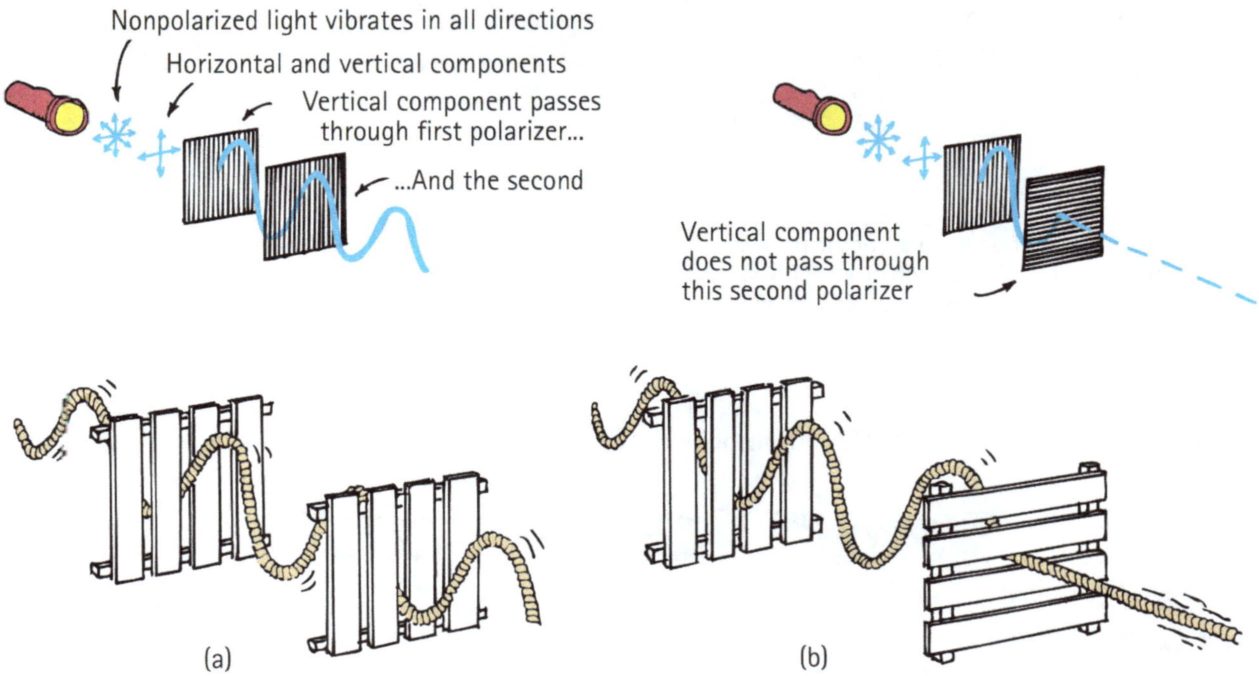

Figure 13.40 >
Are your sunglasses polarized? Here's the test: When polarized lenses overlap at right angles, no light gets through.

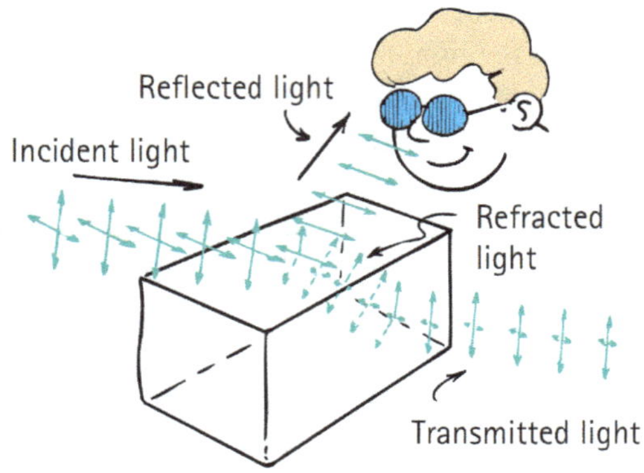

Figure 13.41 >
Most glare from nonmetallic surfaces is polarized. Note which components of incident light reflect and which transmit. Because most of the glare we encounter is from horizontal surfaces, the polarization axes of Polaroid sunglasses are vertical. Polarized glasses are particularly useful for driving on a sunny day and also for seeing clearly through the surface of water. Want to see the fish? Wear polarized glasses!

(a) (b) (c)

Λ Figure 13.42
Light (a) passes through when the axes of the polarized filters are aligned but (b) absorbed when Ludmila rotates one filter so that the axes are at right angles to each other. (c) When she inserts a third filter at an angle between the crossed ones, light again passes through. The reason for this involves vector components, which you will hopefully study in a follow-up physics course.

CONCEPT CHECK

Which pair of glasses is best suited for automobile drivers? (The polarization axes are shown by straight lines.)

CHECK YOUR ANSWER
Pair A is best suited because the vertical axis blocks horizontally polarized light, which makes up much of the glare from horizontal surfaces. Pair C is suited for viewing 3-D movies.

13.7 Wave-Particle

In ancient times, Plato and other Greek philosophers hypothesized that light was made up of tiny particles. Isaac Newton in the 1700s agreed that light had a particle nature. Then a hundred years later, the wave nature of light was demonstrated by Thomas Young in the double-slit experiment. This wave view was validated in 1862 by James C. Maxwell's finding that light is energy carried in the vibrating electric and magnetic fields of electromagnetic waves. The wave view of light was confirmed experimentally by Heinrich Hertz 25 years later. The wave nature of light seemed to be established.

Then in 1905, Albert Einstein published a Nobel Prize-winning paper that challenged the wave theory of light. Einstein stated that, in its interactions with matter, light strikes not as a wave, but as a tiny particle of energy called a **photon**.

Science had come full-circle in its view of light—particle to wave and back to particle. As we shall see, both views are correct! First let's look at Einstein's particle model of light, which explained a mystery to scientists in the early 1900s—*the photoelectric effect.*

When light shines on certain metal surfaces, electrons are ejected from the surfaces. This is the **photoelectric effect**, used in electric eyes, light meters, and some pre-digital motion-picture sound tracks. What puzzled investigators in the early 1900s was that only ultraviolet and violet light knock electrons from these surfaces. Lower-frequency light, like red, won't do it, even if the red light is very bright with much energy. They discovered that electrons are ejected by high frequency light, not by bright light. Very dim violet light ejects electrons, but a bright red or green light doesn't. If the brightness of violet light is increased, more electrons are ejected—but not at greater energies.

Einstein's explanation was that the electrons in the metal are bombarded by "particles of light"—photons. Einstein stated that the energy of each photon is proportional to its frequency:

$$E \sim f$$

So, Einstein viewed a beam of light as a hail of photons, each carrying energy proportional to its frequency. One photon is completely absorbed by each electron ejected from the metal.

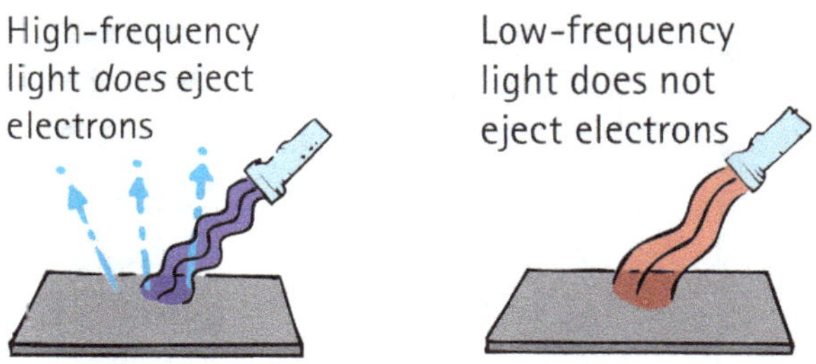

Figure 13.43
The photoelectric effect depends upon photon frequency.

All attempts by investigators at the time to explain the photoelectric effect by waves failed. However, Einstein's idea was experimentally verified 11 years after he announced it. The photoelectric effect proves convincingly that light has particle properties.

Recall Thomas Young's double-slit interference experiment, which we earlier discussed in terms of waves. When monochromatic light passes through a pair of closely spaced thin slits, an interference pattern is produced on photographic film (**Figure 13.44**). Now let's consider the experiment in terms of photons. Suppose we dim our light source so that in effect only one photon at a time reaches the thin slits. If the film behind the slits is exposed to the light for a very short time, the film becomes exposed as simulated in **Figure 13.45a**. Each spot shows where the film has been exposed to a photon. If the light is allowed to expose the film for a longer time, a pattern of fringes begins to emerge as in **Figure 13.45b** and **c**. This is quite amazing. We see spots on the film progressing photon by photon to form the same interference pattern characterized by waves!

Evidently, light has both a wave nature and a particle nature—a **wave-particle duality**. This duality is evident in the formation of optical images. Let's look carefully at the way a photographic image is formed, say on film, photon by photon (**Figure 13.46**). The photographic film consists of an emulsion that contains grains of silver halide crystal, each grain containing about 10^{10} silver atoms. Each photon that hits the film gives up its energy to a single grain in the emulsion. This energy activates surrounding crystals in the grain and develops the film. Many photons activating many grains produce the usual photographic exposure. When a photograph is taken with very feeble light, the image is built up by individual photons that arrive independently and at random. **Figure 13.46** beautifully simulates how an exposure progresses photon by photon.

What all this means is that light has both wave and particle properties. Simply put: Light behaves as a stream of photons when it interacts with the photographic film or other detectors. But light behaves as a wave in traveling from a source to the place where it is detected. In interference experiments, photons strike the film at places where we would expect to see constructive interference of waves.

READING CHECK

When does light behave as a stream of photons?

The fact that light shows both wave and particle behavior is one of the most interesting surprises that scientists discovered in the previous century. The finding that light comes in tiny bunches, tiny *quanta* as they are called,

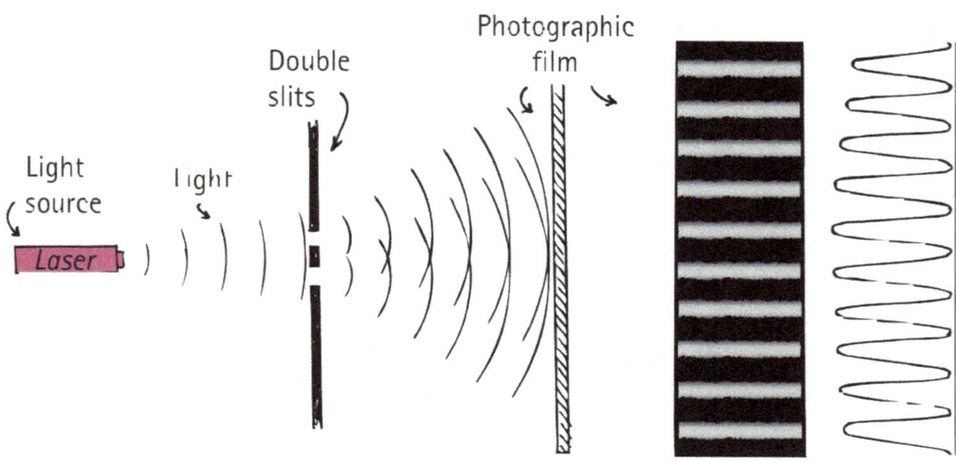

Figure 13.44 >

Arrangement for double-slit experiment. The black and white striped tall rectangle is a photograph of the interference pattern. To the far right is a graphic of the pattern.

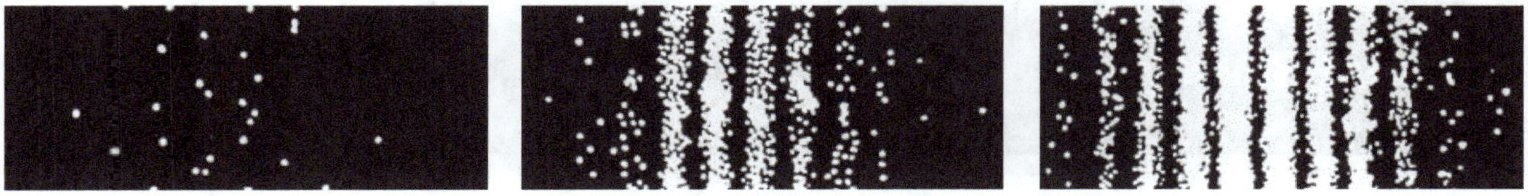

(a) (b) (c)

∧ Figure 13.45
Stages of a two-slit interference pattern. The pattern of individual exposed grains progresses from (a) 28 photons to (b) 1000 photons to (c) 10,000 photons. As more photons hit the screen, a pattern of interference fringes appears.

led to a whole new way of looking at nature—wave mechanics, or quantum mechanics. First electrons were found to have wave properties; a beam of electrons passing through slits exhibits the same type of diffraction pattern as light. Then other particles—even baseballs and orbiting planets—could be described by the new mechanics of waves. Quantum mechanics and Newtonian physics overlap in the macroworld, and both are seen as "correct." But only quantum mechanics, with its emphasis on waves, is wholly accurate in the microworld of the atom.

< Figure 13.46
Stages of film exposure reveal the photon-by-photon production of a photograph.

Concept Review

Chapter 13

Summary of Terms

Converging lens A lens that is thicker in the middle than at the edges and refracts parallel rays passing through it to a focus.

Diverging lens A lens that is thinner in the middle than at the edges, causing parallel rays passing through it to diverge.

Diffraction The bending of light as it passes around an obstacle or through a narrow slit, causing the light to spread and to produce light and dark fringes.

Dispersion Separation of light into colors according to their frequency.

Interference The result of superposing two or more waves of the same wavelength.

Photon A particle of light, or the basic packet of electromagnetic radiation.

Photoelectric Effect The emission of electrons from a metal surface when light shines on it.

Polarization The alignment of the transverse electric vectors that make up electromagnetic radiation.

Real image An image formed by light rays that converge at the location of the image—which can be displayed on a screen.

Virtual image An image formed by light rays that do not converge at the location of the image—which can be seen by an observer but cannot be projected onto a screen.

Review Questions

1. Which travels more slowly in glass, red light or violet light?
2. What color light bends most in a prism?
3. Where must the Sun be to view a rainbow?
4. Distinguish between a converging lens and a diverging lens.
5. What is the focal length of a lens?
6. Distinguish between a virtual image and a real image.
7. What kind of lens can be used to produce a real image? A virtual image?
8. Is diffraction more pronounced through a small opening or through a large opening?
9. For an opening of a given size, is diffraction more pronounced for a longer wavelength or a shorter wavelength?
10. Is interference restricted to only some types of waves or does it occur for all types of waves?
11. What is monochromatic light?
12. What causes the variety of colors seen in gasoline splotches on a wet street?
13. What accounts for the variety of colors in a soap bubble?
14. What phenomenon distinguishes between longitudinal and transverse waves?
15. How does the direction of polarization of light compare with the direction of vibration of the electrons that produced it?
16. How much unpolarized light does an ideal Polaroid filter transmit?
17. What evidence can you cite for the wave nature of light? For the particle nature of light?
18. Which are more successful in dislodging electrons from a metal surface, photons of violet light or photons of red light? Why?

19. Why won't a very bright beam of red light impart more energy to an electron than a feeble beam of violet light?

20. When does light behave as a wave? When does it behave as a particle?

Answers to Odd-Numbered Questions

1. Violet light travels more slowly.

3. Directly in back of your head.

5. Focal length is distance that parallel rays will converge.

7. A converging lens produces a real image; diverging lens produces a virtual image.

9. Longer wavelength.

11 Light of a single frequency (and wavelength).

13 The variety of path lengths between inner and outer reflecting surfaces.

15. Same.

17. Interference. Photoelectric effect.

19. The brightness of a beam involves the number of photons in it, but only one photon in the beam normally hits a single electron at one time. If a red one hits, low energy is given to it than if a higher-energy violet photon hits it.Think energy per photon, not overall energy.

Chapter 14

Atoms

THE MAIN IDEA

Atoms are building blocks of matter

Imagine you fall off your chair in slow motion while also shrinking in size. By the time you reach the floor, you have shrunk to one billionth of a meter, small enough to have fallen into a single atom. What do you see from inside this atom? Occasionally, you may encounter tiny bits of matter, such as fast-flying electrons or the central atomic nucleus that are tiny compared to the entire volume of the atom, which you come to realize is mostly empty space! Evidence for the existence of these miniscule, but most spacious atoms is overwhelming. Interestingly, there are only 118 known different kinds of atoms. Just as a multitude of words can be formed from only 26 letters, the multitude of materials around us arise from the various combinations of these fundamental particles. In this chapter, we explore the bizarre world of atoms and the amazing chart that tells their story—the periodic table.

14.1 Discovering Atoms

How small are atoms? Atoms are so small that the number of them in a baseball is about equal to the number of ping-pong balls that could fit inside a hollow sphere as big as the Earth. This is illustrated in **Figure 14.1.** Interestingly, humans stand midway in size between the atom and Planet Earth. Comprehending these size differences is challenging. In this chapter we study not only the small, but the very very small—atoms.

Atoms are so small, we could stack microscope on top of microscope and never "see" an atom. This is because atoms are smaller than the wavelengths of visible

Figure 14.1
If Earth were filled with nothing but ping-pong balls, the number would be huge and roughly equal to the number of atoms in a baseball. Said differently, if a baseball were the size of Earth, one of its atoms would be about the size of a ping-pong ball.

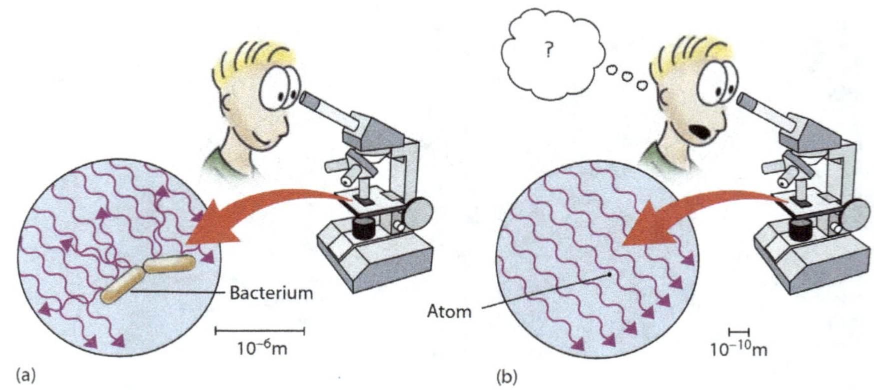

Figure 14.2 >
(a) A bacterium is visible because it is much larger than the wavelengths of visible light. As seen through a microscope, the bacterium reflects visible light back towards the eye. (b) An atom is invisible because it is smaller than the wavelengths of visible light, which pass by the atom with no reflection.

light. As illustrated in **Figure 14.2**, the diameter of an object visible under the highest magnification must be larger than the wavelengths of visible light.

Where do atoms come from? The origin of most atoms goes back to the birth of the universe. Hydrogen, H, the lightest atom, was probably the original atom, and hydrogen atoms make up more than 90 percent of the atoms in the known universe. Heavier atoms are produced in stars. There, enormous temperatures and pressures force hydrogen atoms to join, which makes heavier atoms. All that you see on Earth today is made of atoms that have been around since time zero, with heavier ones manufactured in stars that exploded billions of years before our solar system came into being. You are made of stardust, as is everything that surrounds you.

So, atoms in your body have literally existed since the beginning of time. They have been recycling throughout the universe for eons. In this way, you don't "own" the atoms that make up your body—you're simply the present caretaker. There will be many caretakers to follow.

There are about as many atoms of air in your lungs at any moment as there are breaths of air in the Earth's atmosphere. That's because about 10^{22} atoms make up a liter of air and there are about 10^{22} liters of air in the atmosphere. Here's what that means. Exhale a deep breath; the number of atoms exhaled approximately equals the number of breathfuls of air in the Earth's atmosphere. It will take several years for your breath to become uniformly mixed in the atmosphere. Then anyone, anywhere on Earth, who inhales a breath of air takes in, on average, one of the atoms you exhaled! Furthermore, you exhale many, many breaths, and so other people breathe in many, many atoms that were once in your lungs—that were once a part of you. And of course, vice versa: With each breath you take in, you recycle atoms that were once a part of everyone who ever lived. Some of the atoms breathed became atoms are part of our bodies, so it can be truly said that we are literally breathing one another. Science tells us, we're all one family.

READING CHECK

What is special about the atoms you inhale with each breathe?

Quick Activity

How Fast do Atoms Migrate?

Put a few drops of food coloring into a small container of cold water that is sitting still. Do not stir the water. Note the approximate time for the colored drops to spread to all parts of the water. Meanwhile, repeat this procedure using warm water. Explain any difference in the spreading in cold and warm water.

14.2 Evidence for Atoms

Because atoms are so tiny, scientists didn't fully recognize their existence until the early 1800s. They realized that if atoms were real, then it would provide insight into how some materials transform into other materials. Iron, for example, transforms into rust when iron atoms combine with oxygen atoms. Thinking in terms of atoms, scientists were soon able to create new materials, such as dyes, not found in nature. These successes led early scientists to accept atoms as real even though these tiny bits of matter could not be directly seen. Importantly, the idea of atoms led to modern chemistry, which itself has shaped the nature of modern living from agriculture, medicines, fuels, photovoltaics, batteries, computers, textiles, plastics, and much more.

> All things are made of atoms—little particles that move around in perpetual motion, attracting one another when they are a little distance apart, but repelling upon being squeezed into one another.
> — Richard Feynman

The first direct evidence for atoms was discovered in 1827 by a Scottish botanist, Robert Brown. He was studying grains of pollen in a drop of water under a microscope. He noticed that the grains were continually moving about. At first, he thought the grains were moving life forms. But later he found that dust particles and grains of soot moved the same way. This perpetual jiggling of particles is called *Brownian motion* and it wasn't until the early 1900s that Einstein was able to provide the explanation. Einstein correctly noted that Brownian motion results from collisions between visible particles of soot and invisible particles—atoms. Brown's grains of pollen were being bombarded by groups of atoms.

READING CHECK

What scientist discovered the first direct evidence for the existence of atoms?

Today, we still cannot see atoms directly. However, we can now generate images of them indirectly. In the mid 1980s, researchers developed the scanning probe microscope, which produces images by dragging an ultrathin needle back and forth over the surface of a sample. Interactions with the surface atoms are detected and translated by a computer into an image that corresponds to the positions of atoms on the surface as shown in **Figure 14.3**. A scanning probe microscope can also move individual atoms into desired positions.

During the early 1900s, scientists came to realize through their experiments that atoms are not the smallest particles of matter. Instead, atoms themselves are made of even smaller particles, which we call *subatomic particles*. Three examples include *protons*, *neutrons*, and *electrons*. Protons and neutrons are bound together at the atom's center to form the **atomic**

FOR YOUR INFORMATION

Atoms can stick together strongly to form a molecule, which is a collection of atoms. Water, for example, is made of water molecules where each water molecule is a combination of two hydrogen atoms with an oxygen atom, H_2O. The water molecule is a fundamental unit of water. Break the water molecule down into its constituent atoms and you'll no longer have water. The building of molecules from atoms is a focus of chemistry.

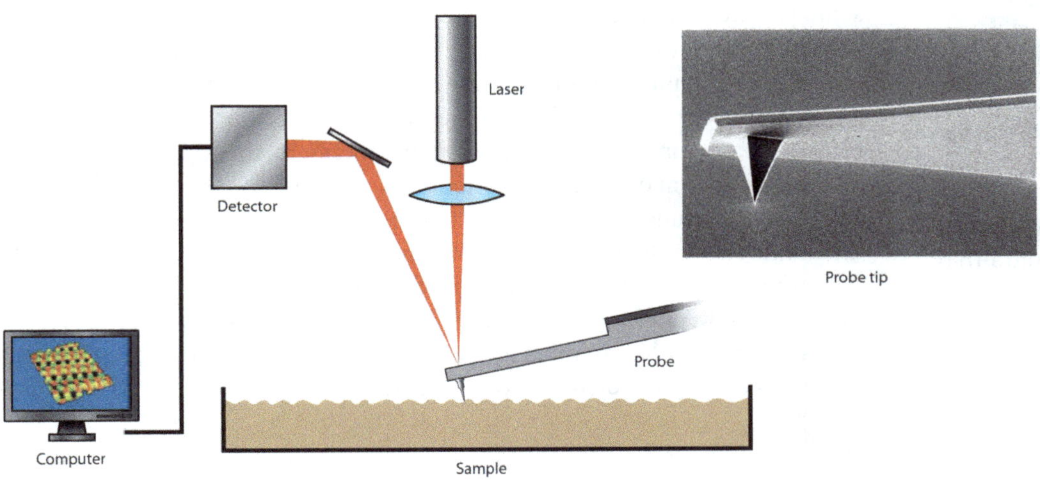

< Figure 14.3
A schematic of a scanning probe microscope that detects and characterizes the surface atoms of a material by way of an ultrathin prob tip attached to a miniature cantilever.

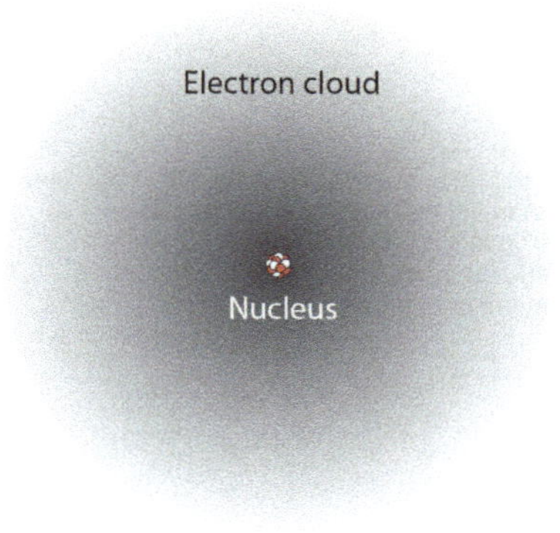

^ **Figure 14.4**
Electrons whiz around the atomic nucleus, forming what can be best described as a cloud. The cloud is most dense where the electrons tend to spend most of their time. Electrons, however, are invisible to us. Hence, such a cloud can only be imagined. Furthermore, if this illustration were drawn to scale, the atomic nucleus would be too small to be seen. Realistically, atoms are not well-suited to graphical illustrations.

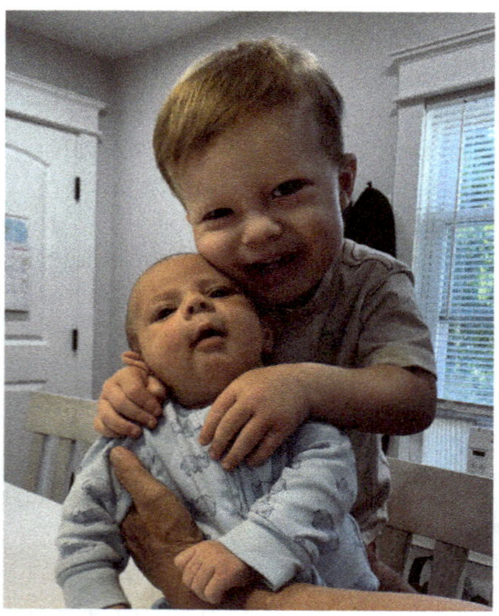

^ **Figure 14.5**
The Dooley brothers are as close as can be. But at the atomic level an ultra-thin gap exists between them.

nucleus. The nucleus is a relatively heavy core that makes up most of an atom's mass. Surrounding the nucleus are the tiny **electrons**, as indicated in **Figure 14.4**.

All atoms and all things made of atoms, including ourselves, are mostly empty space. How can this be? The answer has to do with the fact that the bulk of an atom's mass is concentrated within its nucleus. Surrounding that nucleus is empty space through which electrons are buzzing around. The size of an atom, therefore, is determined by how far away electrons move around the nucleus. Because electrons are even smaller than the nucleus, and because they are widely separated from each other (as well as from the nucleus), atoms are indeed mostly empty space—just as our solar system is mostly empty space.

So, if atoms are mostly empty space, why don't they simply pass through one another? They don't because electrons repel the electrons of neighboring atoms. Therefore, two atoms can get only so close to each other before they start repelling. This explains why gravity doesn't pull you through the floor as you stand. While the force of gravity pulls you down, the electric force of repulsion between the atoms of the floor and your feet pushes you up. As you stand on the floor, these two forces are balanced and you find yourself neither falling nor rising—there's more to standing than most people realize!

When the atoms of your hand push against the atoms of a wall, repulsions between electrons in your hand and electrons in the wall prevent your hand from passing through the wall. You sense this repulsion as a pressure that pushes back. Also, our sense of touch comes from these electrical repulsions. Interestingly, when you touch someone, your atoms and those of the other person do not meet. Instead, atoms from the two of you electrically repel. However close when touching, an imperceptible gap exists between the toucher and the touched (**Figure 14.5**).

Nanotechnology

Nanotechnology is the technology of tiny things. It began with microtechnology, which was ushered in during the 1980s. Engineers learned to build electronic circuits within the size of a micron (10^{-6} meters), thus the term *micro*technology. Such circuits have had a major impact on society—from personal computers, to smart phones, to the internet.

Today we are past the realm of microns to the realm of the nanometer (10^{-9} meters), which is the realm of individual atoms. Technology at this scale is called *nanotechnology*. Nanotechnology deals with incredibly tiny objects from 1 to 100 nanometers. For perspective, a water molecule is only about 0.2 nm. Nanotechnology is interdisciplinary, which means it requires the cooperative efforts of chemists, engineers, physicists, molecular biologists and many others. Interestingly, there are already many products on the market that contain components developed through nanotechnology. This includes sunscreens, dental bonding agents, stain-free clothing, and much more. Nanotechnology is still young and it may be decades before its potential is fully realized. Consider, for example, that personal computers didn't blossom until the 1990s, and in 2020 half of all people on Earth have access to them. Most experts agree, however, that continued benefits will occure in computer science and medicine.

The ultimate expert on nanotechnology is nature. Living organisms, for example, are complex systems of interacting biomolecules all functioning within the realm of nanometers. In this sense, the living organism is nature's nanomachine. We need look no further than our own bodies to find evidence of the power of nanotechnology. With nature as our teacher, we have much to learn. On medicine, for example, nanotechnology can help us discover the exact causes of nearly any disease or disorder (aging included) and their cures.

What are the limits of nanotechnology? As a society, how will we deal with the impending changes nanotechnology may bring? Consider the possibilities. Wall paint that can change color or be used to display video. Solar cells that capture sunlight so efficiently that they render fossil fuels obsolete. Robots with so much processing power that we begin to wonder whether they experience consciousness. Medicines that more than double the average human life span. And what of artificial intelligence potentially developing into artificial general intelligence? Stay tuned for a daring new revolution in human capabilities.

FOR YOUR INFORMATION

In the 1950s, physicist Richard Feynman foresaw the development of nanotechnology in his talk "There's Plenty of Room at the Bottom" where he noted the great potential of being able to manipulate matter by its individual atoms.

Richard Feynman

CONCEPT CHECK

Why are atoms invisible?

CHECK YOUR ANSWER

Atoms are invisible because visible light passes right by them. An individual atom is smaller than the wavelengths of visible light and so is unable to reflect that light. The atomic images generated by scanning probe microscopes are computer-generated images produced by the movements of an ultra-thin needle.

14.3 The Periodic Table

Atoms make up the matter around us, from stars to steel to chocolate ice cream. Given all these different types of material, your grandparents might have thought that there must be many different kinds of atoms. Not true. The number of different kinds of atoms is surprisingly small. The great variety of substances results from the many ways a few kinds of atoms can be combined. Just as the three colors red, green, and blue can be combined to form any color on a computer screen or the 26 letters of the alphabet make up all the words in a dictionary, only a few kinds of atoms combine in different ways to produce all substances. To date, we know of 118 distinct atoms. Of these, about 90 are found in nature. The remaining atoms have been created in the laboratory.

Any material made of only one kind of atom is described as an **element**. A few examples are shown in **Figure 14.6**. All of the elements are listed in a chart called the **periodic table**, shown in **Figure 14.7**.

As you can see from the periodic table, each element is designated by its **atomic symbol**, which comes from the letters of the element's name. For example, the atomic symbol for carbon is C, and that for chlorine is

READING CHECK

When is a material properly called an element?

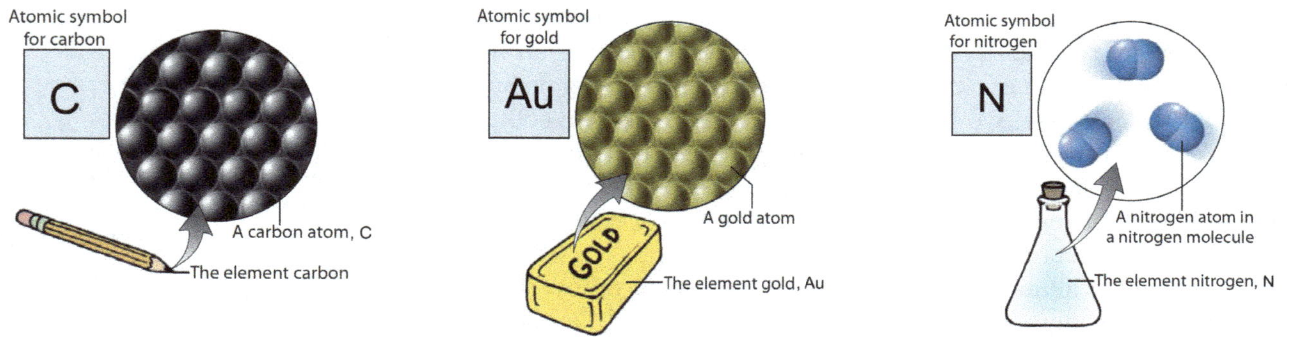

Λ Figure 14.6

Any element consists of only one kind of atom. The carbon of a graphite pencil consists of only carbon atoms. Gold consists of only gold atoms, and a flask of gaseous nitrogen consists of only nitrogen atoms.

Cl. In many cases, the atomic symbol is derived from the element's Latin name. Gold has the atomic symbol Au after its Latin name, *aurum*. Lead has the atomic symbol Pb after its Latin name, *plumbum* (**Figure 14.8**). Elements having symbols derived from Latin names are usually those discovered earliest.

Note that only the first letter of an atomic symbol is capitalized. The symbol for the element cobalt, for instance, is Co, while CO is a combination of two elements: carbon, C, and oxygen, O.

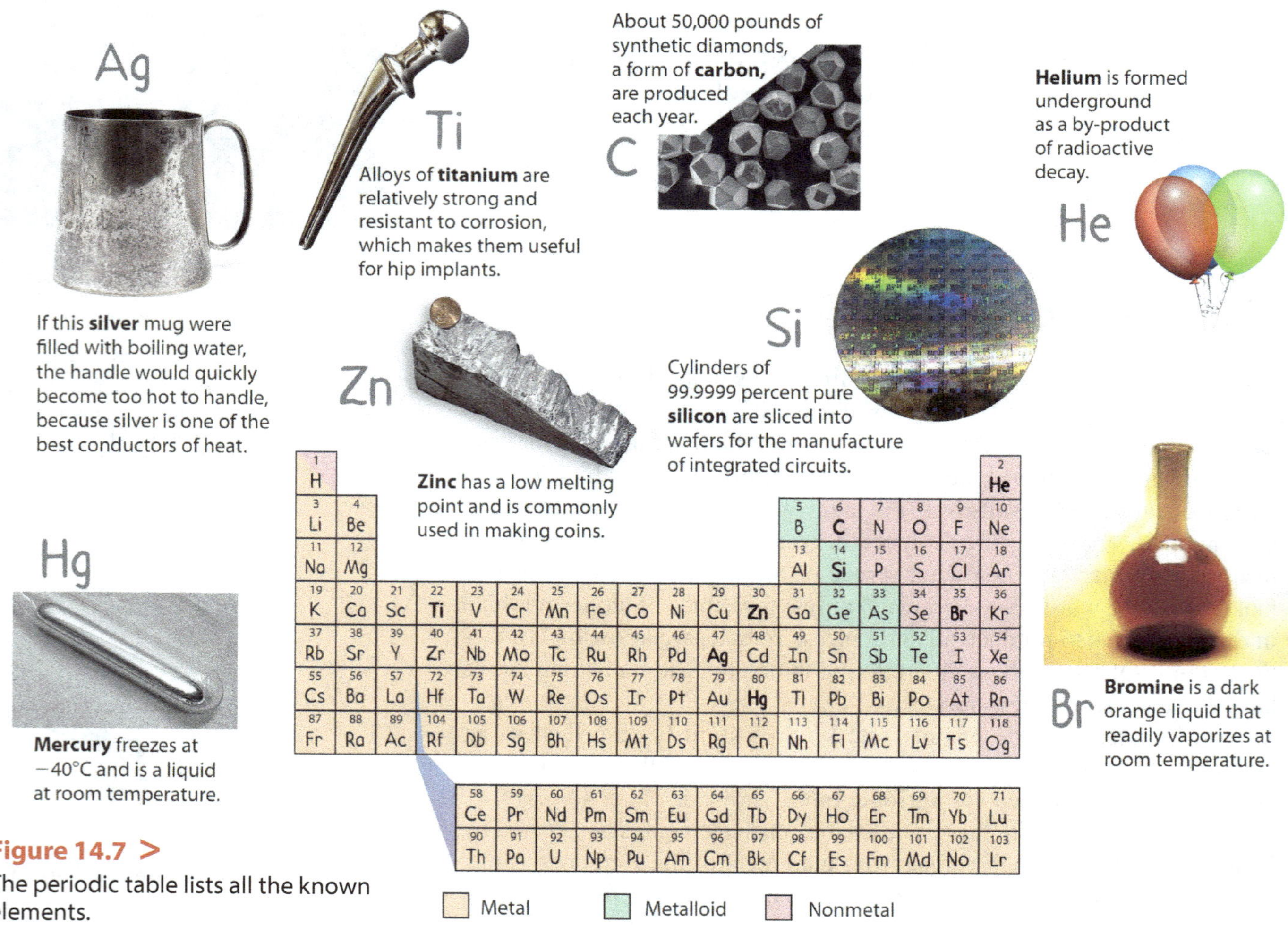

1 H																	2 He
3 Li	4 Be											5 B	6 C	7 N	8 O	9 F	10 Ne
11 Na	12 Mg											13 Al	14 Si	15 P	16 S	17 Cl	18 Ar
19 K	20 Ca	21 Sc	22 Ti	23 V	24 Cr	25 Mn	26 Fe	27 Co	28 Ni	29 Cu	30 Zn	31 Ga	32 Ge	33 As	34 Se	35 Br	36 Kr
37 Rb	38 Sr	39 Y	40 Zr	41 Nb	42 Mo	43 Tc	44 Ru	45 Rh	46 Pd	47 Ag	48 Cd	49 In	50 Sn	51 Sb	52 Te	53 I	54 Xe
55 Cs	56 Ba	57 La	72 Hf	73 Ta	74 W	75 Re	76 Os	77 Ir	78 Pt	79 Au	80 Hg	81 Tl	82 Pb	83 Bi	84 Po	85 At	86 Rn
87 Fr	88 Ra	89 Ac	104 Rf	105 Db	106 Sg	107 Bh	108 Hs	109 Mt	110 Ds	111 Rg	112 Cn	113 Nh	114 Fl	115 Mc	116 Lv	117 Ts	118 Og

58 Ce	59 Pr	60 Nd	61 Pm	62 Sm	63 Eu	64 Gd	65 Tb	66 Dy	67 Ho	68 Er	69 Tm	70 Yb	71 Lu
90 Th	91 Pa	92 U	93 Np	94 Pu	95 Am	96 Cm	97 Bk	98 Cf	99 Es	100 Fm	101 Md	102 No	103 Lr

Figure 14.7 >

The periodic table lists all the known elements.

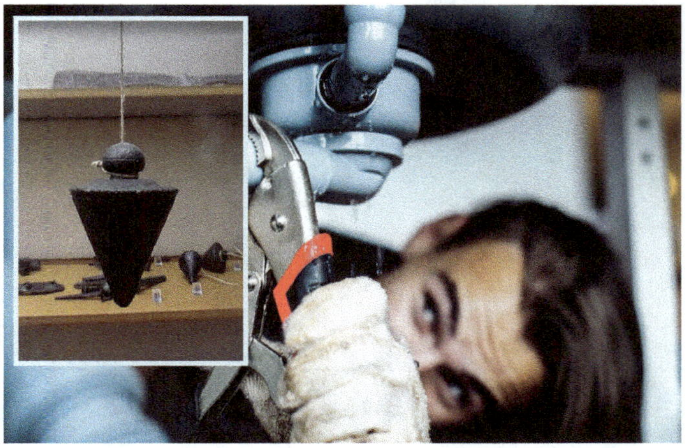

< **Figure 14.8**
A plumb bob, a heavy weight attached to a string and used by carpenters and surveyors to establish a straight vertical line, gets it name from the lead (plumbum, Pb) that is still sometimes used as the weight. Plumbers got their name because they once worked with lead pipes.

The elements are organized within the periodic table in a remarkable fashion. You will learn about this organization in much more detail in chemistry class. For now, you should know that each vertical column of the periodic table is called an **atomic group** of elements (or sometimes a *family* of elements), as shown in **Figure 14.9**. Elements within the same group (vertical column) have similar properties. It is no coincidence, for example, that copper, Cu, silver, Ag, and gold, Au, are all within the same group (number 11). Note how these elements are all "precious metals" used for both jewelry and money.

Another important characteristic of the periodic table is that each horizontal row is called an **atomic period** of elements. Across any period (horizontal row), the properties of elements gradually change. This gradual change is called a *periodic trend*. As **Figure 14.10** shows, one periodic trend is that the size of atoms becomes smaller as you move from left to right across any period. Note that the trend repeats from one horizontal row to the next.

FOR YOUR INFORMATION

Most of the elements of the periodic table have been detected in stars and interstellar gases.

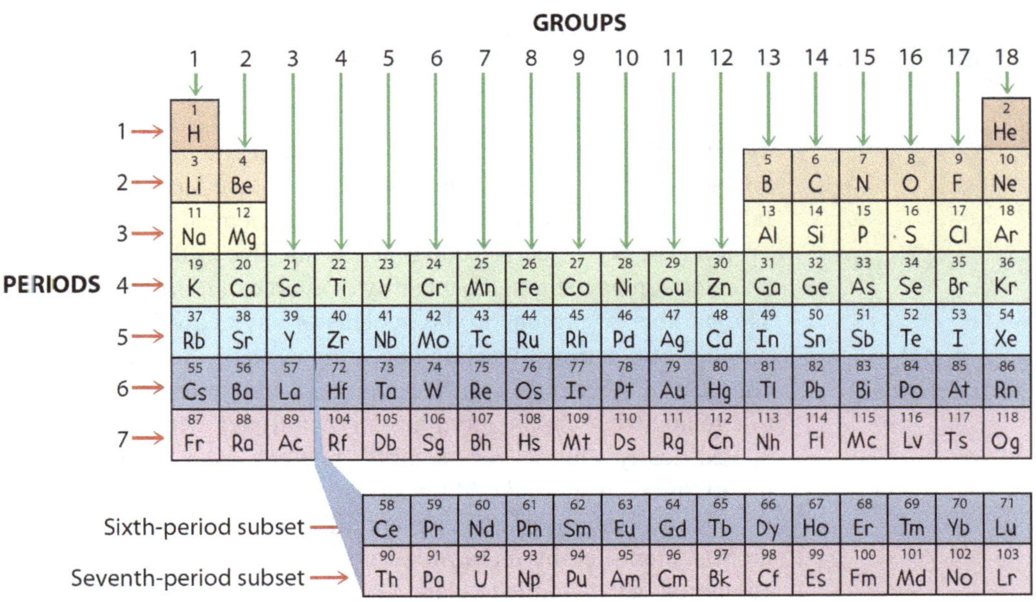

^ **Figure 14.9**
The 18 groups (vertical columns) and 7 periods (horizontal rows) of the periodic table. Note that not all periods contain the same number of elements. Also note that the sixth and seventh periods each include a subset of elements, which are listed apart from the main body.

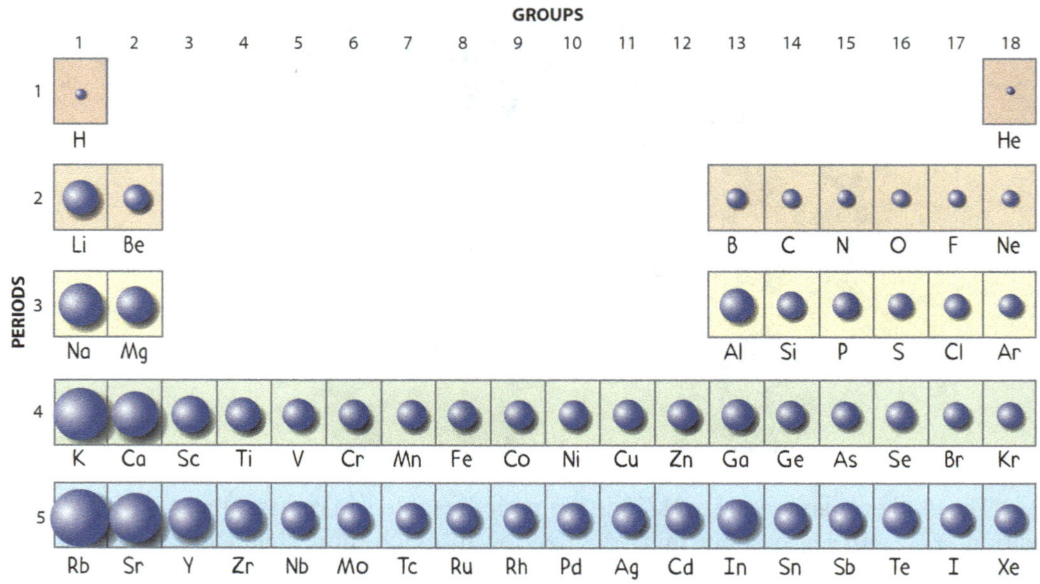

Figure 14.10 >
The size of atoms gradually decreases from left to right across any period. Atomic size is a periodic (repeating) property.

An important aspect of a periodic trend is that it allows us to predict the property of an element based upon its relative location within the periodic table. For example, we find that elements toward the lower left tend to have the largest atoms. Conversely, elements toward the upper right tend to have the smallest atoms. The relative size of atoms, therefore, can be predicted merely by looking at where the elements are positioned. Which are smaller: The atoms of copper, Cu, or zirconium, Zr? Because copper is closer to the upper right, we can assess that copper atoms are the smaller. So, the periodic table is something we learn to read. Much like a dictionary, it need not be memorized, but it should be readily available. This is why you'll see the periodic table often posted within any science class or laboratory.

14.4 Subatomic Particles

To get a closer look at the atom, let's investigate the particles found in its core—the *atomic nucleus*. First, consider protons. A **proton** is the fundamental particle in the atomic nucleus that carries a positive electric charge. The proton is heavy—nearly 2000 times as massive as the electron. The proton and electron have the same quantity of charge, but the charges are opposite. The number of protons in the nucleus of any neutral atom is equal to the number of electrons whirling about the nucleus. So, the proton's positive charge and electron's negative charges balance each other. Can you see this guarantees that the atom has an overall net electric charge of zero? It is electrically neutral. For example, an electrically balanced oxygen atom has eight electrons and eight protons.

READING CHECK
How do protons and neutrons compare in terms of mass and electric charge?

How do atoms of one element differ from another element? Each element is identified by its **atomic number**, which is the number of protons contained in the atomic nucleus. Hydrogen, with one proton in the nucleus, has atomic number 1. Helium, with two protons in its nucleus, has atomic number 2; and so on. Look at the periodic table in Figure 14.7. You will see that it lists the elements in order of increasing atomic number.

In addition to the proton, the atomic nucleus also includes another fundamental particle—the neutron. The **neutron** is a nuclear particle with about the same mass as the proton, but with no electric charge. Any object with no net electric charge is said to be electrically neutral (which is how the neutron got its name). Protons and neutrons are both examples of **nucleons**, which is the general term given to the protons and neutrons within the nucleus.

CONCEPT CHECK

How many protons are there in an iron atom? (Iron has the chemical symbol Fe and the atomic number 26.)

CHECK YOUR ANSWER

The atomic number of an atom and its number of protons are the same. Thus, there are 26 protons in an iron atom. Another way to say this is that all atoms that contain 26 protons are, by definition, iron atoms.

14.5 Isotopes

An element has a definite number of protons, but its number of neutrons may vary. For example, most hydrogen atoms (atomic number 1) have no neutrons. A small percentage, however, have one neutron, and a smaller percentage have two neutrons. Similarly, most iron atoms (atomic number 26) have 30 neutrons, but a small percentage have 29 neutrons. Atoms of the same element that contain different numbers of neutrons are **isotopes** of one another.

How can we tell isotopes apart? We identify isotopes by their mass number, which is the total number of protons and neutrons they contain. In other words, mass number is the number of *nucleons*. As **Figure 14.11** shows, a hydrogen isotope with only one proton is called hydrogen-1, where 1 is the mass number. A hydrogen isotope with one proton and one neutron is therefore hydrogen-2, and a hydrogen isotope with one proton and two neutrons is hydrogen-3. Similarly, an iron isotope with 26 protons and 30 neutrons is called iron-56, and one with only 29 neutrons is iron-55.

READING CHECK

How does the mass number of an atom compare with the number of nucleons it contains?

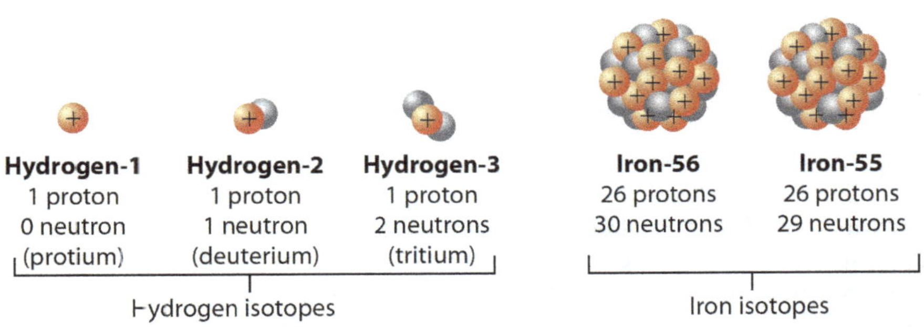

< **Figure 14.11**
Isotopes of hydrogen and iron.

Another way to express isotopes is by **mass number**—the number of nucleons within the nucleus of an isotope. We write the mass number as a superscript and the atomic number as a subscript to the left of the atomic symbol. For example, an iron isotope with a mass number of 56 and atomic number of 26 is written

Mass number ↘ $^{56}_{26}\mathbf{Fe}$—Atomic symbol
Atomic number ↗

> Life is not measured by the number of breaths we take, but by the moments that take our breath away.
> — *George Carlin*

The total number of neutrons in an isotope can be calculated by subtracting its atomic number from its mass number:

mass number
− atomic number
number of neutrons

For example, uranium-238 has 238 nucleons. The atomic number of uranium is 92, which tells us that 92 of these 238 nucleons are protons. The remaining 146 nucleons must be neutrons:

238 protons and neutrons
− 92 protons
146 neutrons

Isotopes of an element differ only by mass, not by electric charge. That means they have all the same chemical properties, which makes them chemically indistinguishable. For example, about 1% of the carbon we eat is the carbon-13 isotope with seven neutrons per nucleus. The remaining 99% of the carbon in our diet is common carbon-12 isotope, with six neutrons per nucleus. Our bodies can't tell the difference.

CONCEPT CHECK

How many neutrons are in the lithium-7 isotope? (Li, atomic number 3)

CHECK YOUR ANSWER

Subtract the total number of protons from the total number of nucleons to obtain the total number of neutrons. For lithium-7, there are 7 − 3 = 4 neutrons.

FOR YOUR INFORMATION

The value of isotopes in scientific research is quite significant. For example, how are we able to decipher past global temperatures? Water molecules containing a heavier oxygen isotope do not evaporate as readily as those with a lighter oxygen isotope. As the global climate becomes warmer, these heavier water molecules evaporate more readily. So, drill down to ancient ice or snow within a glacier. Measure the amount of denser water and you'll have a direct assessment of a past temperature.

CONCEPT **CHECK**

What is the mass number of iron-56? How about uranium-238?

CHECK YOUR ANSWERS

The mass number of any element is the total number of protons and neutrons, which is shown as a whole number after the name of the element. For iron-56, the mass number is 56. For uranium-238, the mass number is 238.

14.6 Atomic Mass

The total mass of an atom is called its **atomic mass**. This is the sum of the masses of all the atom's electrons, protons, and neutrons. Because the mass of electrons are so little compared with the mass of protons and neutrons, electrons contribute practically nothing to atomic mass. The periodic table in **Figure 14.12** lists sample atomic masses of elements.

The atomic mass of an element is defined as the average atomic mass of its various isotopes. This is why the atomic masses shown in the periodic table are not whole numbers. To understand why, consider the following analogy. Imagine 100 students take a test worth 100 points. If all students scored 80 points, then the average score would be 80 percent. If instead one of the students scored a perfect 100 points, what would happen to the class average? The answer is that the class average would rise a little bit because of that one higher-scoring student. Similarly, we find that most all carbon atoms are the carbon-12 isotope. About one out of every 100 carbon atoms, however, is the heavier carbon-13 isotope. This small amount of carbon-13 raises the average mass of carbon from exactly 12.0000 to the slightly higher value of 12.011, which is what you see listed in the periodic table. Fun fact: Atomic mass is the average mass of the atoms in an element measured in a special unit known as the *atomic mass unit*, or amu, which you will learn more about in a chemistry course.

Nucleon, neutron, atomic mass, mass number, atomic number and more! Whew! Why all these new terms? Remember, we're describing an invisible world of atoms. Atoms are not part of our common language. To describe this tiny and very real world of the atom means that we expand our vocabulary. The same thing is true when you're introduced to the world of a new online game—you need new words to describe the new objects and the new experiences. The more you learn, the more words you need to describe what you learn!

READING CHECK

How is atomic mass defined?

FOR YOUR INFORMATION

Don't confuse *atomic mass* with *mass number*. Mass number is an integer equal to the number of nucleons in a nucleus. The average atomic mass of carbon, for example, is 12.011 amu. Interestingly, 1 amu is an exceedingly small quantity equal to 1.661×10^{-24} grams.

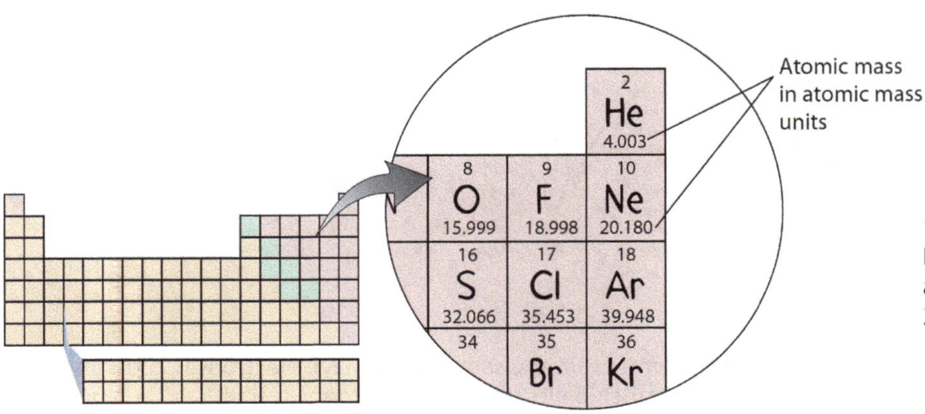

Figure 14.12
Helium, He, has an atomic mass of 4.003, and neon, Ne, has an atomic mass of 20.180

14.7 Electron Shells

The identity of a song is determined by the arrangement of its musical notes. In a similar way, the identity of an atom is determined by how its electrons are arranged. Within the atom, electrons behave as though they are arranged in **electron shells**. There are at least seven shells, and each shell can hold only a limited number of electrons. As shown in **Figure 14.13**, the innermost shell can hold 2 electrons, the second and third shells, 8 electrons each; the fourth and fifth shells, 18 electrons each; and the sixth and seventh shells, 32 electrons each.

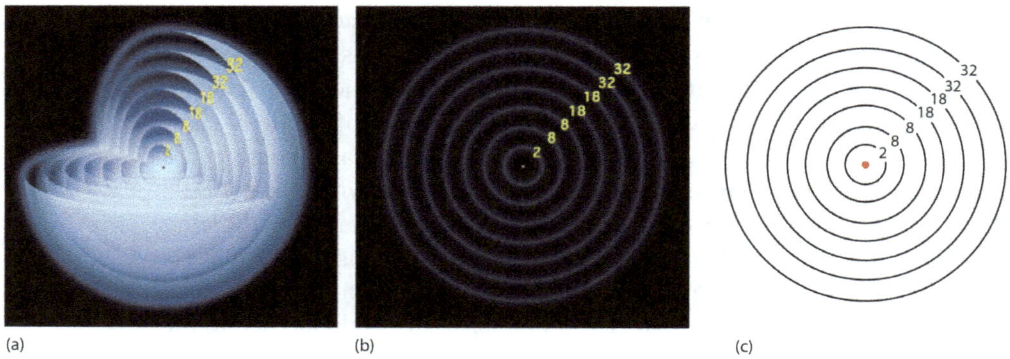

Figure 14.13 >
(a) A cutaway view of the seven shells, indicating the number of electrons each shell can hold. (b) A two-dimensional, cross-sectional view of the shells. (c) A simple cross-sectional view resembling a planetary model of the atom.

These seven shells correspond to the seven periods in the periodic table. Note that the number of elements in each period is equal to the shell's capacity for electrons: 2, 8, 8, 18, 18, 32, 32, which all adds up to 118—the number of known elements. The first shell, for example, can hold only two electrons. That's why we find only two elements, hydrogen and helium, in the first period **(Figure 14.14)**. The second and third shells each can hold eight electrons, and so eight elements are found in both the second and third periods.

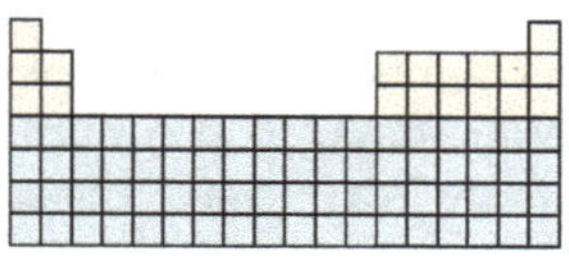

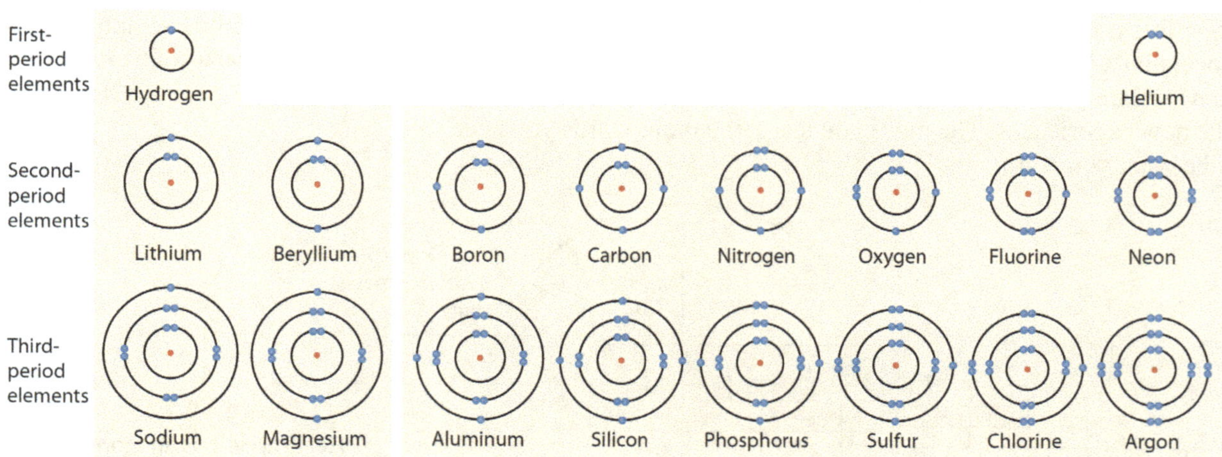

Λ Figure 14.14
Shown here are the first three periods of the periodic table. Elements in the same period have electrons in the same shells. Elements in the same period differ from one another by the number of electrons in the outermost shell.

The electrons of the outermost shell in any atom are the ones directly exposed to their external environment. These important outermost electrons are called **valence electrons**. They are the first to interact with other atoms, and the ones that participate in chemical bonding, which is discussed in a chemistry course. The term *valence* comes from the Latin valentia, "strength", and it refers to the "combining power" of an atom. Atoms combine to form **molecules**, which are tightly-held groups of atoms, by way of valence electrons.

READING CHECK

What are valence electrons and why are they important?

Look carefully at Figure 14.14. Can you see that the valence electrons of atoms above and below one another (within the same group) are similarly organized? For example, atoms of the first group, which include hydrogen, lithium, and sodium, each have a single valence electron. The atoms of the second group, including beryllium and magnesium, each have two valence electrons. Similarly, atoms of the last group, including helium, neon, and argon, each have their outmost shells filled to capacity with valence electrons—two for helium, and eight for both neon and argon. So, we see that the valence electrons of atoms in the same group of the periodic table are similarly organized. This explains why elements directly above or below one another in the periodic table have similar properties.

Most materials are made from more than one kind of atom. Water, H_2O, for example, is made from the combination of hydrogen and oxygen atoms. These materials are called *compounds*.

Electron shells are an example of a **conceptual model**, which is any model that helps us to understand and predict the behavior of a system. For example, a conceptual model of Earth's atmosphere enables understanding and prediction of the weather. Likewise, a conceptual model of the flow of money helps us to make decisions to keep the economy healthy. While a conceptual model helps us to grasp an atom's properties and behavior, this model does little to tell us what the atom "looks like." Therefore, the electron shells are not what atoms look like. Instead, these shells are merely pictorial representations providing us a handle for understanding the world of super, super small. But with this understanding we have accomplished much.

CONCEPT CHECK

Do atoms really look like the shells of Figure 14.13?

CHECK YOUR ANSWER

No! Remember, atoms are smaller than the wavelengths of visible light, and so they are not well-suited to graphical illustrations. The shells of Figure 14.13 are not actual images of atoms. Instead, these are just pictures to help us understand the behavior of atoms.

FOR YOUR INFORMATION

Two-time Nobel laureate Linus Pauling (1901–1994) was an early proponent of teaching beginning chemistry students a shell model, from which the organization of the periodic table could be described. In 1954 Pauling won the Nobel Prize in Chemistry for his research into the nature of the chemical bond in 1954. In 1962 he was awarded the Nobel Peace Prize for his campaign against the testing of nuclear bombs, which was introducing massive amounts of radioactivity into the environment.

14.8 Atomic Spectra

> **FOR YOUR INFORMATION**
>
> A star's age is revealed by its elemental makeup. The first and oldest stars were composed of hydrogen and helium, because those were the only elements that existed at that time. Heavier elements were produced after many of these early stars exploded in supernovae. Later stars incorporated these heavier elements in their formation. In general, the younger a star, the greater the amounts of these heavier elements it contains.

Atoms give off light as they are subjected to various forms of energy, such as heat or electricity. The atoms of any given element in the gaseous phase emit only certain frequencies of light, however. As a consequence, each element emits its own distinctive glow when energized. Sodium atoms emit bright yellow light, which makes them useful as the light source in street lamps because our eyes are very sensitive to yellow light. To name just one more example, neon atoms emit a brilliant red-orange light, which makes them useful as the light source in neon signs.

A **spectroscope**, shown in **Figure 14.15**, is an instrument used to observe the color components of any light source. A spectroscope allows us to analyze the frequencies of light emitted by a light source. A classic incandescent bulb shining white light reveals all the colors of the rainbow, as shown in Figure 14.15b.

A glowing element, as shown in **Figure 14.16**, reveals a spectrum of only particular frequencies. The pattern of those frequencies is referred to as that element's **atomic spectrum**, also known as an *emission spectrum*. The atomic spectrum is an element's fingerprint. You can identify the elements in a light source by analyzing the light through a spectroscope and looking for characteristic patterns.

The Quantum Hypothesis

An important step toward our present-day understanding of atoms and their spectra was taken by the German physicist Max Planck (1858–1947). In 1900, Planck hypothesized that light energy is *quantized* in much the same way matter is quantized. The mass of a gold brick, for example, equals some whole-number multiple of the mass of a single gold atom. Similarly, an electric charge is always some whole-number multiple of the charge on a single electron. Mass and electric charge are therefore said to be *quantized*, in that they consist of some number of fundamental units.

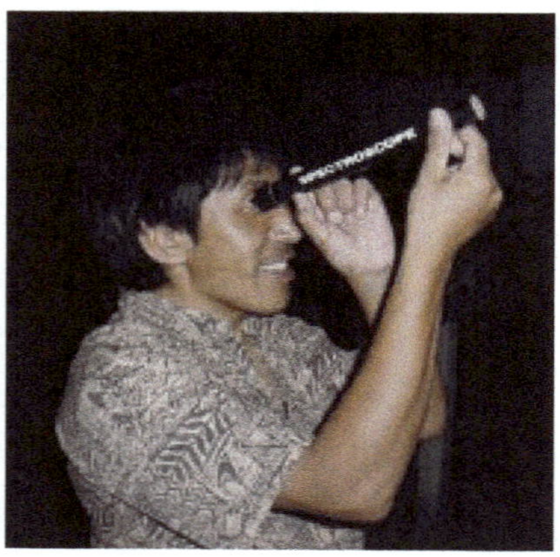

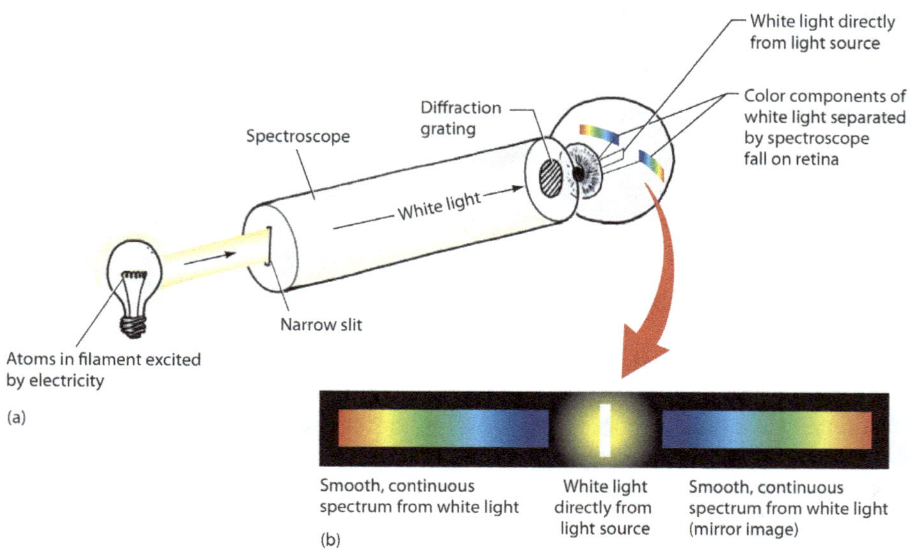

Figure 14.15 >
(a) In a spectroscope, light emitted by atoms passes through a narrow slit before being separated into particular frequencies by a prism or (as shown here) a diffraction grating. (b) This is what the eye sees when the slit of a diffraction-grating spectroscope is pointed toward a white-light source. Spectra of colors appear to the left and right of the slit.

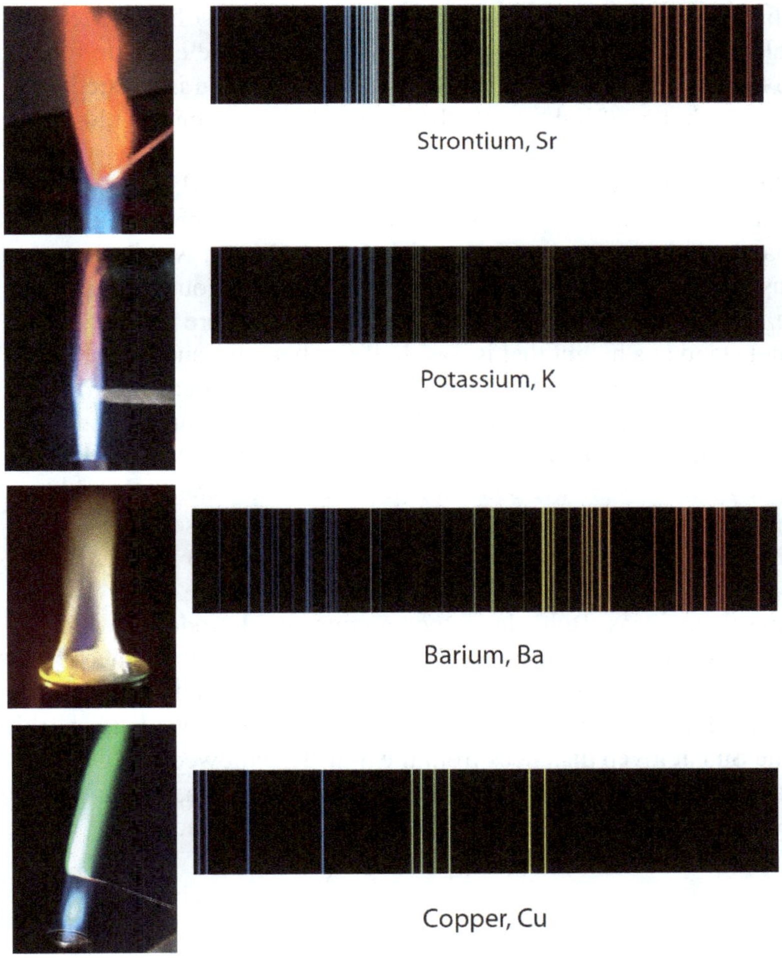

< Figure 14.16
Elements heated by a flame glow their characteristic color. This is commonly called a flame test and is used to test for the presence of an element in a sample. When viewed through a spectroscope, the color of each element is revealed to consist of a pattern of distinct frequencies known as an atomic spectrum.

Planck identified each discrete parcel of light energy as a **quantum**, represented in **Figure 14.17**. A few years later, Einstein recognized that a quantum of light energy behaves much like a tiny particle of matter. To emphasize its particulate nature, each quantum of light was called a *photon*, a name coined because of its similarity to the word electron.

Using Planck's quantum hypothesis, the Danish scientist Niels Bohr (1885–1962) explained the formation of atomic spectra as follows. First, an electron has more potential energy when farther from the nucleus. This is analogous to the greater potential energy an object has when held higher above the ground. Second, Bohr recognized that when an atom absorbs a photon of light, it is absorbing energy. This energy is acquired by one of the electrons. Because this electron has gained energy, it must move away from the nucleus.

READING CHECK
What does an electron have when farther from the nucleus?

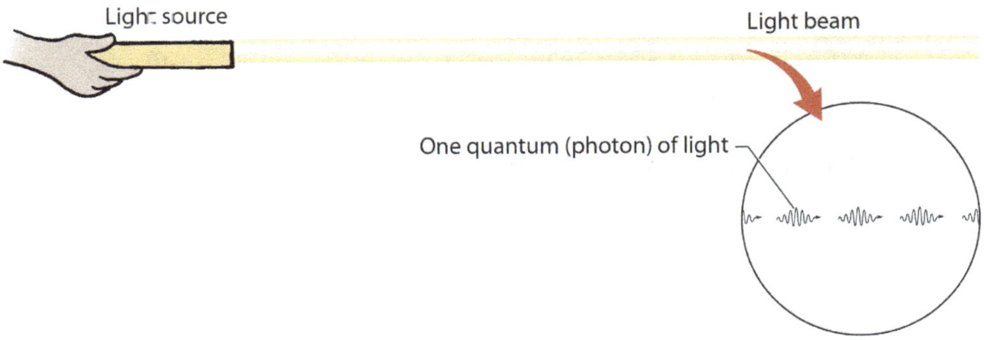

< Figure 14.17
Light is quantized, which means it consists of a stream of energy packets. Each packet is called a quantum, also known as a photon.

∧ Niels Bohr (1885–1962) and Albert Einstein (1879–1955) were good friends and colleagues, but they differed on their views of quantum theory and its philosophical implications. Einstein accepted quantum theory but was one of its strongest critics. His critiques were answered for the most part by Niels Bohr through a series of exchanges, called the Bohr–Einstein debates, spanning the latter parts of their lives.

Bohr realized that the opposite is also true: when a high-potential-energy electron in an atom loses some of its energy, the electron moves closer to the nucleus and the energy lost from the electron is emitted from the atom as a photon of light. Both absorption and emission are illustrated in **Figure 14.18.**

Bohr reasoned that because light energy is quantized, the energy of an electron in an atom must also be quantized. In other words, an electron cannot have just any amount of potential energy. Rather, within the atom there must be a number of distinct energy levels, analogous to steps on a staircase. Where you are on a staircase is restricted to where the steps are—you cannot stand at a height that is, say, halfway between any two adjacent steps. Similarly, there are only a limited number of permitted energy levels in an atom, and an electron can never have an amount of energy between these permitted energy levels. Bohr gave each energy level a **quantum number** n, where n is always some integer. The lowest energy level has a quantum number $n = 1$. An electron for which $n = 1$ is as close to the nucleus as possible, and an electron for which $n = 2$, $n = 3$, and so forth is farther away, in a step-wise fashion, from the nucleus.

Using these ideas, Bohr developed a conceptual model in which an electron moving around the nucleus is restricted to certain distances from the nucleus, with these distances determined by the amount of energy the electron has. Bohr saw this as similar to the way the planets are held in orbit around the Sun at given distances from the Sun. The allowed energy levels for any atom, therefore, could be graphically represented as orbits around the nucleus, as shown in **Figure 14.19**. Bohr's quantized model of the atom thus became known as the *planetary model.*

Figure 14.18 >
An electron is lifted away from the nucleus as the atom it is in absorbs a photon of light and drops closer to the nucleus as the atom releases a photon of light.

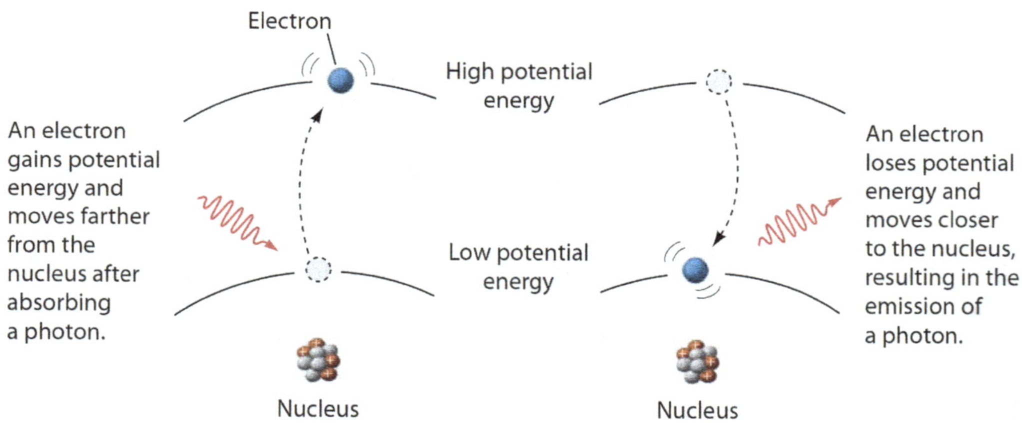

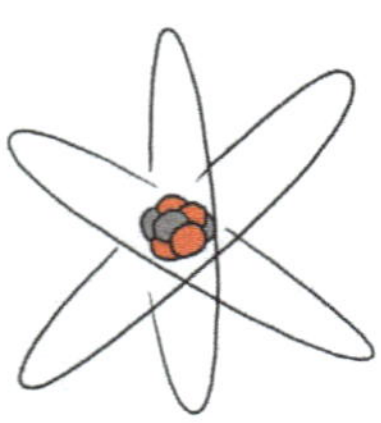

Not to be taken literally

CONCEPT **CHECK**
What is released as an electron transitions from a higher to a lower energy level?

CHECK YOUR ANSWER A photon of light.

Bohr used his planetary model to explain why atomic spectra contain only a limited number of light frequencies, as shown in **Figure 14.20**. According to the model, photons are emitted by atoms as electrons move from higher-energy outer orbits to lower-energy inner orbits. The energy of an emitted photon is equal to the difference in energy between the two orbits. Because an electron is restricted to discrete orbits, only particular light frequencies are emitted, as atomic spectra show.

Interestingly, any transition between two orbits is always instantaneous. In other words, the electron doesn't "jump" from a higher to a lower orbit the way a squirrel jumps from a higher branch in a tree to a lower one. Rather, it takes no time for an electron to move between two orbits. Bohr was serious when he stated that electrons could never exist between permitted energy levels!

Bohr's planetary atomic model proved to be a tremendous success. By utilizing Planck's quantum hypothesis, Bohr's model solved the mystery of atomic spectra. Despite its successes, though, Bohr's model was limited, because it did not explain why energy levels in an atom are quantized. Bohr himself was quick to point out that his model was to be interpreted only as a crude beginning and that the picture of electrons whirling about the nucleus like planets about the Sun was not to be taken literally (a warning to which popularizers of science paid no heed).

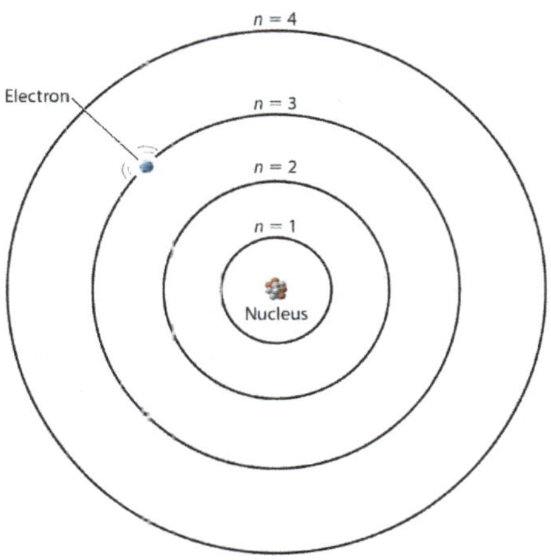

∧ Figure 14.19
Bohr's planetary model of the atom, in which electrons orbit the nucleus much as planets orbit the Sun, is a graphical representation that helps us understand how electrons can possess only certain quantities of energy.

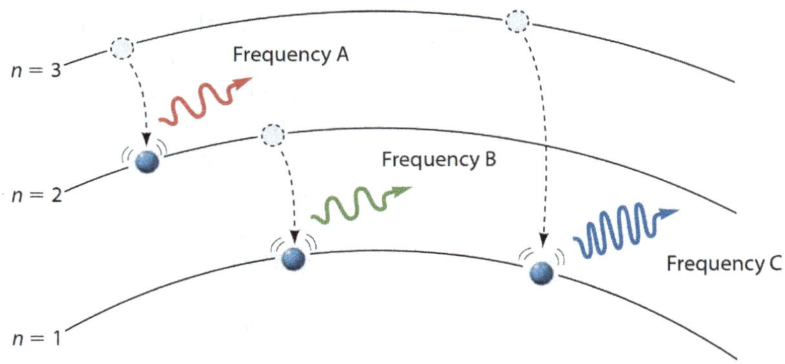

∧ Figure 14.20
The frequency of light emitted (or absorbed) by an atom is proportional to the energy difference between electron orbits. Because the energy differences between orbits are discrete, the frequencies of light emitted (or absorbed) are also discrete. The electron here can emit only three discrete frequencies of light—A, B, and C. The greater the transition, the higher the frequency of the photon emitted.

CONCEPT **CHECK**
Is the Bohr model of the atom a physical model or a conceptual model?

CHECK YOUR ANSWER The Bohr model is a conceptual model. It is not a scaled-up version of an atom but instead is a representation that accounts for the atom's behavior.

14.9 Electron Waves

If light has both wave properties and particle properties, why can't a material particle, such as an electron, also have both? This question was posed by the French physicist Louis de Broglie (1892–1987) while he was still a graduate student in 1924. His revolutionary answer was that every moving particle of matter—by virtue of its energy of motion—is endowed with the characteristics of a wave.

We now speak of waves as an essential feature of any bit of matter. An electron, or any particle, can show itself as a wave or as a particle, depending on how we examine it. This is called the wave–particle duality. Just a few years after de Broglie's suggestion, researchers in Great Britain and the United States confirmed the wave nature of electrons by observing diffraction and interference effects when electrons bounced from crystals. A practical application of the wave properties of electrons is the electron microscope, which focuses not visible-light waves but rather electron waves. Electron microscopes are able to show far greater detail than optical microscopes, as **Figure 14.21** shows.

READING CHECK

What does the wave nature of an electron help explain?

An electron's wave nature can be used to explain why electrons in an atom are restricted to particular energy levels. Permitted energy levels are a natural consequence of the need for electron waves to form stable patterns around the atomic nucleus.

As an analogy, consider the wire loop shown in **Figure 14.22a**. This loop is affixed to a mechanical vibrator that can be adjusted to create a series of waves traveling through the wire. If the lengths of the waves are equal to the length of the wire (or some fraction thereof), then these waves form what is called a *standing wave*, as shown in **Figure 14.22b**. Within a standing wave, identically sized waves overlap one another, and they do so in a manner that creates regions of maximum intensity and other regions of minimum intensity. A region of maximum intensity occurs where crests are overlapping with crests and troughs are overlapping with troughs, as shown in **Figure 14.23**. (Note: a "crest" is the high point of a wave while a "trough" is the low point of a wave.) A region of minimum intensity occurs where crests and troughs overlap so as to cancel each other out. At the center of minimum intensity is a point of zero intensity, called a node.

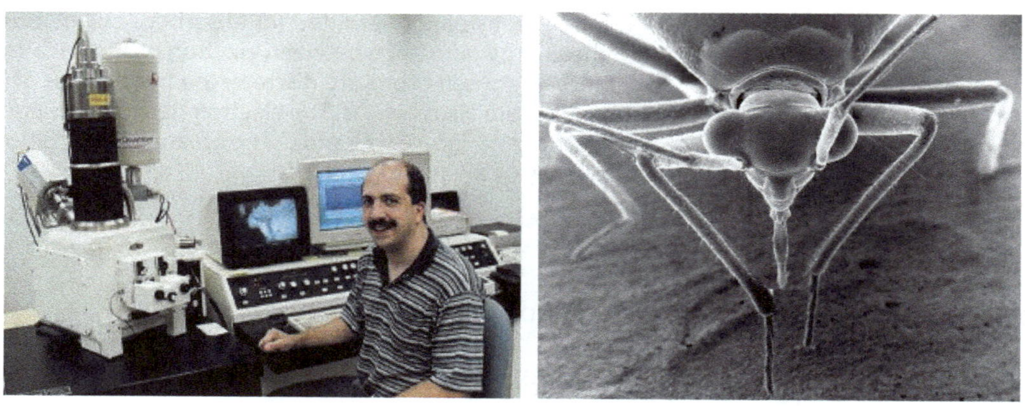

∧ Figure 14.21

An electron microscope makes practical use of the wave nature of electrons. The wavelengths of electron beams are typically thousands of times shorter than the wavelengths of visible light, so the electron microscope is able to distinguish detail not visible with optical microscopes. (b) Detail of an insect's head as seen with an electron microscope at a "low" magnification of 200 times. Note the remarkable resolution.

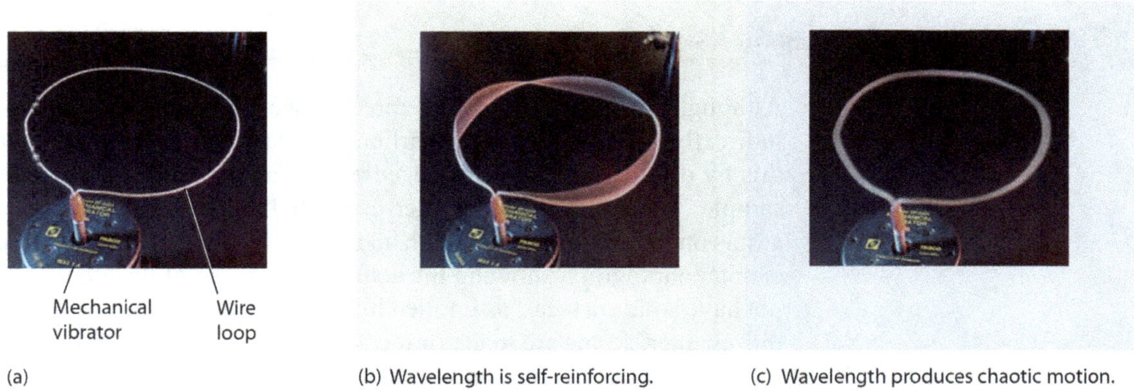

Figure 14.22
A wire loop affixed to the post of a mechanical vibrator at rest. Waves are sent through the wire when the post vibrates. (b) Only waves of certain wavelengths are able to form standing waves. The one shown here is 2/3 the length of the loop. (c) Most other wavelengths result in chaotic motion with no specific regions of maximum and minimum intensities.

For a given size of the wire loop, only certain wavelengths will give rise to a stable standing wave. Waves of any other wavelength are unable to align properly and the result is chaotic motion, as suggested in **Figure 14.22c**. These same ideas apply to electron waves within an atom. We find that only certain electron wavelengths form standing waves. Because of their stability, these particular electron wavelengths are favored.

But why does an electron have wavelike characteristics? Recall that according to de Broglie, the wavelength of an electron is a consequence of its energy. Significantly, if only certain wavelengths are favored, then it follows that only certain amounts of energy are favored. In other words, the energy of an electron confined to an atom is *quantized*. Consider what happens when whistling down a long tube. The whistler soon discovers that only certain frequencies can be generated. Interestingly, when a sound wave is confined to a tube, the consequence is a quantization of its frequencies. This is an important principle behind any wind instrument. Likewise, when an electron is confined to an atom, the consequence is a quantization of the electron's energy.

Each standing wave of an electron corresponds to one of the permitted energy levels seen in Figures 14.19 and 14.20. Only the frequencies of light that match the difference between any two of these permitted energy levels can be absorbed or emitted by an atom. The wave nature of electrons also explains why they do not spiral closer and closer to the positive nucleus that attracts them. By viewing each electron as a standing wave, we see that the circumference of the smallest orbit can be no smaller than a single wavelength.

In summary, the properties of electrons—and all atoms made of those electrons—arise from their quantum nature. In this sense, the bizarre quantum nature of the super-small is what makes our natural world possible.

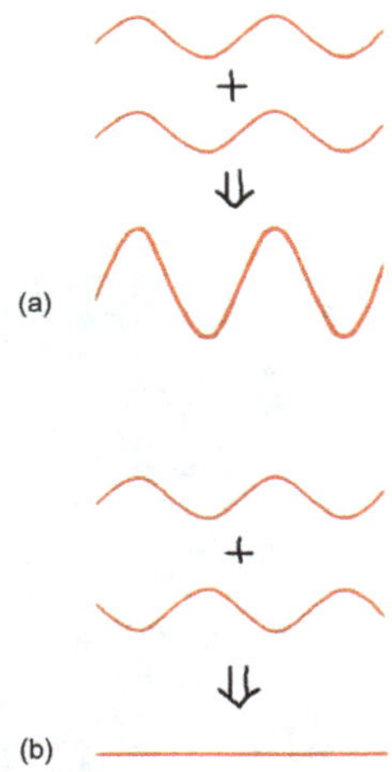

Figure 14.23
(a) Crests overlapping crests and troughs overlapping troughs results in a wave of maximum intensity. (b) Crests overlapping troughs results in a wave of zero intensity.

CONCEPT **CHECK**
What must an electron be doing in order to have wave properties?

CHECK YOUR ANSWER
According to de Broglie, particles of matter behave like waves by virtue of their motion, which is a form of energy. An electron must therefore be moving in order to have wave properties.

14.10 Physical and Conceptual Models

Although we cannot see atoms directly, we can generate images of them indirectly. As was discussed earlier, the scanning probe microscope does this by dragging an ultra-thin needle back and forth over the surface of a sample. The result is a computer-generated image of the positions of the atoms on the surface. While such an image is both remarkable and useful, it is not a photograph showing the actual appearance of the atoms. Atoms do not have solid surfaces, as implied by these images. Rather, as we discuss in this chapter, atoms are made of mostly empty space. Envision what empty space looks like. Then you will have a more accurate picture of what it's really like down there at the level of atoms and molecules.

A very small or very large visible object can be represented with a physical model, which is a model that replicates the object at a more convenient scale. **Figure 14.24a**, for instance, shows a large-scale physical model of a microorganism that a biology student might use to study the structure. Because atoms are invisible, however, we cannot use a physical model to represent them. In other words, we cannot simply scale up the atom to a larger size, as we might with a microorganism. So, rather than describing the atom with a physical model, scientists use a conceptual model, which is a representation of a system that helps us predict how the system behaves. The more accurate a conceptual model, the more accurately it predicts the behavior of the system. So while a main advantage of a physical model is allowing us to visual on a different scale, a main advantage of a conceptual model is allowing us to predict the behavior of a system.

READING CHECK
What is a conceptual model?

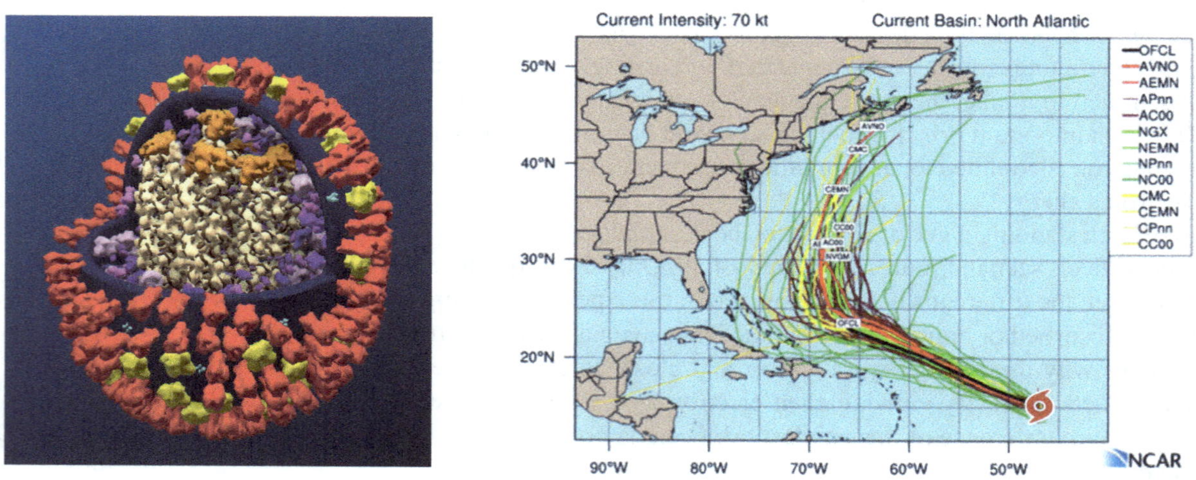

(a) (b)

Figure 14.24

(a) This large-scale model of an influenza virus is a physical model rendered by a computer. To play with this model in 3D, visit: https://3d.nih.gov/entries/3DPX-013373. (b) Weather forecasters rely on conceptual models to predict the behavior of weather systems, such as the potential paths of a hurricane.

For example, the weather is best described using a conceptual model like the one shown in **Figure 14.24b**. Such a model shows how the various components of the system—humidity, atmospheric pressure, temperature, electric charge, the motion of large masses of air—interact with one another. Other systems that can be described by conceptual models are the economy, population growth, the spread of diseases, and team sports.

Like the weather, the atom is a complex system of interacting ultra-tiny components, notably electrons, protons, and neutrons. The atom, therefore, is best described with a conceptual model. Thus, you should be careful not to interpret any visual representation of an atomic conceptual model as a recreation of an actual atom. In Section 14.8, for example, you were introduced to the planetary model of the atom, wherein electrons are shown orbiting a dense center of protons and neutrons (the atomic nucleus) much as planets microorganism's orbit the Sun. **This planetary model is limited, however, in that it fails to explain many properties of atoms.** Therefore, newer and more accurate (and more complicated) conceptual models of the atom have since been introduced, such as the electron-shell model. But even more advanced models have their limitations. But despite their limitations, such models are excellent guides to understanding.

CONCEPT **CHECK**

A basketball coach describes a playing strategy to her team by way of sketches on a game card. Do the illustrations represent a physical model or a conceptual model?

CHECK YOUR ANSWER

The sketches are a conceptual model the coach uses to describe a system (the players on the court), with the hope of achieving an outcome (winning the game).

14.11 Quantum Phenomena

You can measure the speed of a pitched baseball by having it fly through a pair of photogates that are a known distance apart, as shown in **Figure 14.25**. The ball is timed as it interrupts beams of light in the gates. The accuracy of the ball's measured speed has to do with uncertainties in the measured distance between the gates and in the timing mechanisms. Interactions between the macroscopic ball and the photons it encounters are insignificant.

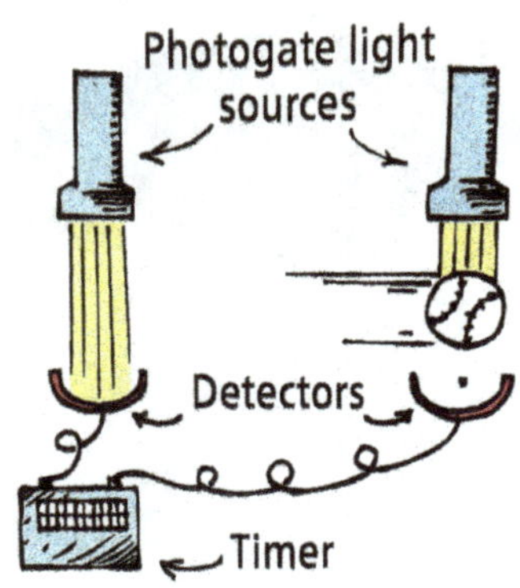

∧ Figure 14.25
The ball's speed is measured by dividing the distance between the photogates by the time difference between crossing the two light paths. Photons hitting the ball alter its motion much less than the motion of an oil supertanker being altered by a few fleas bumping into it.

In the case of measuring submicroscopic things like electrons we have a much different situation. Even a single photon bouncing off of an electron appreciably alters the motion of the electron, and in an unpredictable way. If you wish to observe an electron and determine its whereabouts with light, the wavelength of the light would have to bee very short. You fall into a dilemma. Light of a short wavelength, which can "see" the tiny electron better, corresponds to a large quantum of energy, which ,in turn, greatly alters the electron's state of motion. If, on the other hand, you use a long wavelength that corresponds to a smaller quantum of energy, the change you induce will be smaller, but the determination of its position by means of the coarser wave will be less accurate. The act of probing something as tiny as an electron produces a considerable uncertainty in either its position or its motion.

The uncertainty of measurement in the atomic domain, which was first stated mathematically by the German physicist Werner Heisenberg, is called the *uncertainly principle*. It is a fundamental principle in quantum mechanics. Heisenberg found that when the uncertainties in the measurement of the momentum and position of a particle are multiplied, the product must be equal to our greater than Planck's constant, h, divided by 2π:

Werner Heisenberg (1901 - 1976)

$$(\Delta p)(\Delta x) \geq h/2\pi$$

Here, the Δ symbol means "uncertainty of": Δp is the uncertainty of momentum (the symbol for momentum is conventionally p), and Δx is the uncertainty of position. The product of these two uncertainties must be equal to or greater than ($\geq$) the size of $h/2\pi$.

The significance of the uncertainly principle is that, even in the best of conditions, the lower limit of uncertainly is $h/2\pi$. This means that if we wish to know the momentum of an electron with great accuracy (small Δp), the corresponding uncertainly in position will be large. Or, if we wish to know the position with great accuracy (small Δx), the corresponding uncertainly in momentum will be large. The shaper one of these quantities is, the less sharp is the other.

The uncertainly principle is relevant only to quantum phenomena. The inaccuracies in measure the position and momentum of a baseball due to quantum effects are completely negligible. But the inaccuracies in measuring the position and momentum of an electron are far from negligible. This is because the uncertainties in the measurements of these subatomic quantities are comparable to the magnitude of the quantities themselves.

READING CHECK

Is Heisenberg's uncertainty principle applicable to the practical case of using a thermometer to measure the temperature of a glass of water?

Quantum Computing

A computer is a system that allows for computations. The abacus is an early form of a computer. It allows for the storage of information based upon the position of the beads. Also, the combination of various positions allows for computations to be completed in an efficient manner.

Abacus

For the modern computer, storage of information and the processing of that information is achieved by using an electronic device called the *transistor*. Simply put, a transistor can hold information as either a 1 or a 0. A single transistor by itself is no big deal. But billions of them working together squeezed into a tiny integrated circuit is a very big deal. They form the basis of modern computers and many other modern electronic devices. They can hold massive amounts of information, and this includes programs that allow us to do amazing things with that information—drawing programs, writing programs, virtual reality programs, or any app you might be using on your smart phone.

There have been many advances in transistor technology over the past several decades. Yet transistors face a fundamental limit when it comes to the speed at which they can operate. Most notably, because they are electronic circuits, they run on electricity, which necessarily means they get hot. The higher the speed, the hotter (and less efficient) they get.

Enter quantum computers. While a traditional computer stores bits of information within miniature electronic gadgets, a quantum computer stores bits of information within what are known as "quantum bits", or qubits. Here's the key: while a transistor can be set to a "0" or a "1", a quantum bit can be set to both. This is a purely quantum phenomenon known as coherent superposition where two complementary states can exist simultaneously. What this means is that the qubit can hold twice the information. While a system of 10 transistors can be described by 10 bits of information, a system of 10 qubits would be described by $2^{10} = 10^{24}$ bits of information, which is explosively larger. The memory capacity of a quantum computers would thus also be explosively larger.

Also impressive is the speed at which calculations can be performed. While the transistors of a traditional computer are connected electronically, the qubits of a quantum computer are connected by a process known as entanglement, which is also a purely quantum phenomenon. Two entangled qubits respond to each other instantaneously, regardless of the distance between them. Einstein called entanglement "spooky action at a distance.". Because entangled qubits act together instantaneously, calculations made using groups of entangled quibits are exceedingly fast.

Cybersecurity—An Application of Quantum Phenomenon

Unfortunately, quantum computers are still at a very basic research level. There are many technical hurdles to a viable quantum computer. But it's something to keep an eye upon over the upcoming decades. That said, aspects of quantum computing are already here with us today when it comes to protecting highly sensitive data being sent from one place to another. Photons within fiber optic networks can be entangled. This allows for cryptographic keys that make it impossible for a third party to "peek in" on a transfer of private data. The moment they look, is the moment that the quantum entanglement is destroyed, which cuts off access to the data.

FOR YOUR INFORMATION

Planck's constant, *h*, is a fundamental constant of nature that serves to set a lower limit on the smallness of things. It ranks with the velocity of light and Newton's gravitational constant as a basic constant of nature. The equation $E = hf$ gives the smallest amount of energy that can be converted to light with frequency *f*. The radiation of light is not emitted by an atom continuously but is emitted as a sequence of photons, with each photon throbbing at a frequency *f* and carrying an energy *hf*.

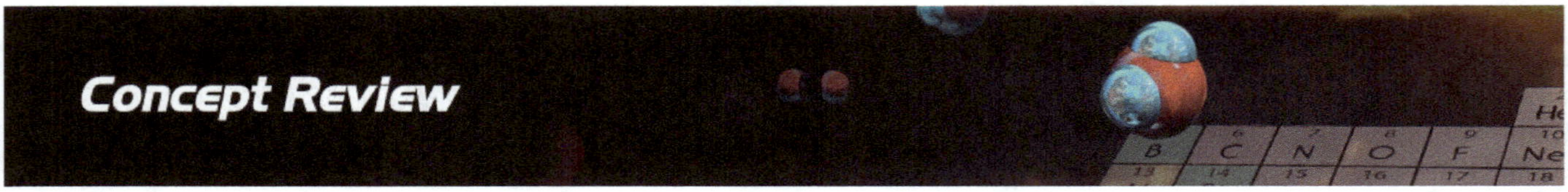

Chapter 14

Summary of Terms

Atomic Group A vertical column in the periodic table, also known as a family of elements.

Atomic Mass The mass of an element's atoms listed in the periodic table as an average value based on the relative abundance of the element's isotopes.

Atomic Nucleus The dense, positively charged center of every atom.

Atomic Number A count of the number of protons in the atomic nucleus.

Atomic Period A horizontal row in the periodic table.

Atomic Spectrum The pattern of frequencies of electromagnetic radiation emitted by the atoms of an element, considered to be an element's "fingerprint."

Atomic Symbol An abbreviation for an element or atom.

Conceptual Model A model that helps us to understand and predict the behavior of a system.

Electron An extremely small, negatively charged subatomic particle found outside the atomic nucleus.

Electron Shell A region of space around the atomic nucleus in which electrons may reside.

Element Any material that is made up of only one type of atom.

Atomic Group A vertical column in the periodic table, also known as a family of elements.

Isotopes Any member of a set of atoms of the same element whose nuclei contain the same number of protons but different numbers of neutrons.

Mass Number The total number of protons and neutrons within an isotope.

Molecule A group of tightly held atoms and also the fundamental unit of many materials, such as water.

Nucleon Any subatomic particle found in the atomic nucleus. Another name for either proton or neutron.

Neutron An electrically neutral subatomic particle of the atomic nucleus.

Periodic Table A chart in which all known elements are listed in order of atomic number.

Physical Model A representation of an object on some convenient scale.

Proton A positively charged subatomic particle of the atomic nucleus.

Quantum A small, discrete packet of energy.

Quantum Number An integer that specifies the quantized energy level within an atom.

Spectroscope A device that uses a prism or diffraction grating to separate light into its color components and measure their frequencies.

Valence Electron An electron located in the outermost occupied shell in an atom and can participate in chemical bonding.

Review Questions

1. Which atom is most plentiful in the universe?
2. What is the source of Brownian motion?
3. Is it possible to see an atom using visible light? Why or why not?
4. What type of microscope allows us to visualize the positions of individual atoms?
5. What exists at the center of every atom?
6. Why is pure gold classified as an element?
7. What is the atomic symbol for the element cobalt?
8. How many periods are in the periodic table?
9. How many groups are in the periodic table?
10. How does the size of atoms vary across any period of the periodic table?
11. What is the role of atomic number in the periodic table?
12. What is a nucleon?
13. Distinguish between a proton and a neutron in terms of electric charge.
14. What occurs for two atoms to be isotopes of each other?
15. Distinguish between atomic number and mass number.
16. Distinguish between mass number and atomic mass.
17. How many nucleons are there in Hydrogen-3?
18. How many shells make up the periodic table?
19. Which electrons in an atom are most responsible for its properties?
20. Why do the group 18 elements neon, Ne, and argon, Ar, have such similar properties?
21. What is the difference between a conceptual model and a physical model?
22. What is uncertain in the uncertainty principle?
23. Why do we say atomic spectra are like fingerprints of the elements?
24. What causes an atom to emit light?
25. Did Bohr think of his planetary model as an accurate representation of what an atom looks like?
26. What was Planck's quantum hypothesis?
27. About how fast does an electron travel around the atomic nucleus?
28. Who first proposed that electrons exhibit the properties of a wave?
29. What is a fundamental limit of the transistor?
30. How does the speed of an electron change its fundamental nature?

Answers to Odd-Numbered Questions

1. Hydrogen is the most plentiful.

3. No. Atoms are smaller than the wavelengths of visible light and so cannot be seen using any visible light microscope.

5. The atomic nucleus is at the center.

7. As can be found in the periodic table, the atomic symbol for the element cobalt is Co in the periodic table.

9. The periodic table has 18 groups.

11. The periodic table lists elements in order of increasing atomic number.

13. Proton has one unit of charge and the neutron has none.

15. Atomic number is the number of protons in the nucleus. Mass number is the total number of protons and neutrons in the nucleus.

17. Three nucleons in hydrogen-3.

19. The electrons in the outer most shell of an atom, the valence electrons, are the ones most responsible for the properties of an atom.

21. A conceptual model describes the behavior of a system while a physical model depicts an object on a different scale.

23. The atoms of each element emit only select frequencies of light. The pattern of these frequencies is unique to that element.

25. No, Bohr's model merely illustrated the different energy levels of an electron in an atom.

27. An electron moves around the nucleus at around 2 million meters per second.

29. A limiting factor for a transistor is the heat it generates.

"Curiosity is a delicate little plant that, aside from stimulation, stands mainly in need of freedom."

—Albert Einstein
(1879 - 1955)

Chapter 15

Nuclear Energy

THE MAIN IDEA

The atomic nucleus is a source of tremendous energy.

The atomic nucleus is the source of nuclear energy, which comes in different forms. Radioactivity is nuclear energy that occurs naturally in the air we breathe, in the rocks around us, and even in the food we eat. Some nuclear energy is human-controlled and is an important part of medical diagnostics and radiation therapy. Another source of nuclear energy is nuclear power plants, which transform nuclear energy into electricity. Of course, a nuclear bomb is another form of nuclear energy. Our most significant source of nuclear energy, however, is the Sun. We'll see in this chapter how solar energy comes from thermonuclear fusion and how it relates to Einstein's famous equation, $E = mc^2$. In this nuclear age it makes good sense to have a basic understanding of the atomic nucleus and the nuclear energy that touches every aspect of everyday life. Let's begin.

15.1 Unstable Nuclei

As we learned in the previous chapter, atoms are made of electrons, neutrons, and protons. The neutrons and protons are located at the highly dense center of each atom—its nucleus. Most atoms have stable nuclei, which means they hold together well. This stability is due to an optimal balance of the number of protons with the number of neutrons. These stable nuclei remain unchanged.

Some atoms, however, have nuclei that are unstable because they contain "off-balance" numbers of protons and neutrons. They may either contain too many neutrons or not enough protons or vice versa. Sooner or later unbalanced nuclei undergo spontaneously transformations wherein neutrons become protons, or vice versa. During these transformations, the nucleus ejects highly energetic particles while also emitting highly energetic electromagnetic radiation.

A material containing unstable nuclei undergoing these sorts of transformations is said to be *radioactive*. The high-energy particles and radiation emitted by a radioactive substance is called **radioactivity**. Notably, a radioactive nucleus disintegrates as it spontaneously rearranges its protons and neutrons. A nucleus changes its identity. For example, a uranium nucleus ultimately transforms into a lead nucleus. The nucleus is no longer what it was. We say that the nucleus has "decayed," which is why the process of emitting radioactivity is often called *radioactive decay*.

Radioactive atoms emit three distinct types of radiation, named by the first three letters of the Greek alphabet, α, β, γ—alpha, beta, and gamma. Alpha rays carry a positive electric charge, beta rays carry a negative charge, and gamma rays carry no charge. The three rays can be separated by placing a magnetic field across their paths (**Figure 15.1**).

READING CHECK

What types of radiation are emitted by the nuclei of radioactive elements?

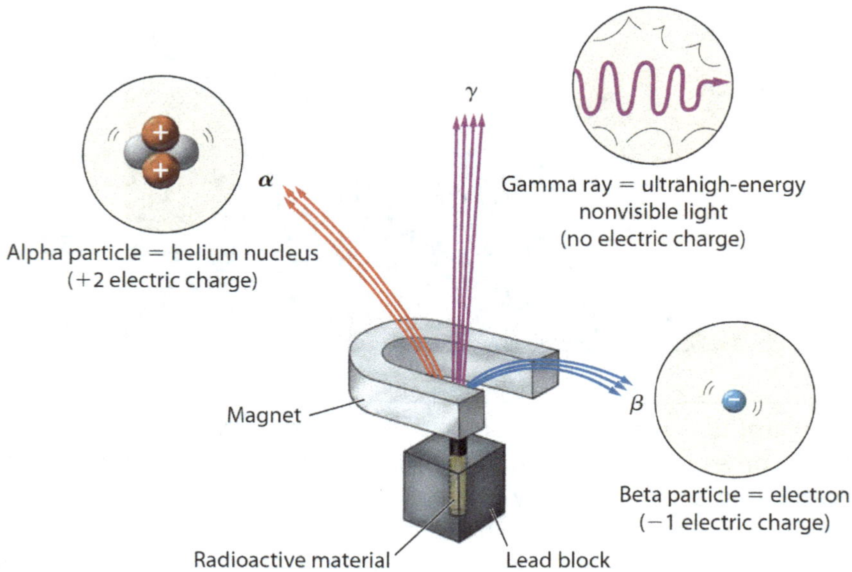

Figure 15.1 >
In a magnetic field, alpha rays bend one way, beta rays bend the other way, and gamma rays don't bend at all. Note that the alpha rays bend less than the beta rays. This occurs because alpha particles have more inertia (mass) than beta particles. The combined beam comes from a radioactive material placed at the bottom of a hole drilled in a block of lead.

Alpha radiation is a stream of alpha particles. An **alpha particle** consists of the combination of two protons and two neutrons (in other words, it is the nucleus of the helium atom, atomic number 2). Alpha particles are relatively easy to shield against due to their relatively large size and their double positive charge (+2). For example, they do not normally penetrate such lightweight materials as paper or clothing. Because of their great kinetic energies, however, alpha particles can cause significant damage to the surface of a material, especially living tissue. When traveling through only a few centimeters of air, alpha particles pick up electrons and become nothing more than harmless helium. Amazingly, that's the origin of the helium in a child's balloon—practically all Earth's helium atoms were at one time energetic alpha particles.

Beta radiation is a stream of beta particles. A **beta particle** is merely an electron ejected from a radioactive nucleus. We will soon discuss how an electron arises from the nucleus. For now, just keep in mind that a beta particle is a fast-flying electron. A beta particle is normally faster than an alpha particle, and it carries only a single negative charge (−1). Beta particles are not as easily stopped as alpha particles, and they are able to penetrate light materials such as paper or clothing. They can penetrate fairly deeply into skin, where they have the potential to harm or kill living cells, but they are unable to penetrate even thin sheets of

FOR YOUR INFORMATION

Most helium atoms produced within the Earth find their way to the surface and then upward to outer space. Some helium, however, collects within natural gas deposits, which can contain as much as 7 percent helium. Most helium used in the world is isolated from the natural gas reserves of the Great Plains of the United States. Prior to World War II, the United States stopped supplying helium to the Germans, who then needed to use combustible hydrogen gas to fill their large zeppelin airships. One such airship, the Hindenburg, famously exploded in 1937.

denser materials, such as aluminum. Beta particles, once stopped, simply become a part of the material they are in, like any other electron.

The **gamma ray** is the high-frequency electromagnetic radiation emitted by radioactive nuclei. Like photons of visible light, a gamma ray is pure energy. The amount of energy in a gamma ray, however, is much greater than that in visible light, ultraviolet light, or even X rays (**Figure 15.2**). Because they have no mass or electric charge, and because of their high energies, gamma rays are able to penetrate through most materials. However, they cannot easily penetrate very dense materials, such as lead. Lead is commonly used as a shielding material in laboratories or hospitals to combat against gamma radiation exposure. Delicate molecules inside cells throughout our bodies that are zapped by gamma rays suffer structural damage. Hence, gamma rays generally cause more damage to our cells than does alpha or beta radiation.

Figure 15.3 shows the relative penetrating power of the three types of radiation. **Figure 15.4** shows an interesting practical use for gamma radiation.

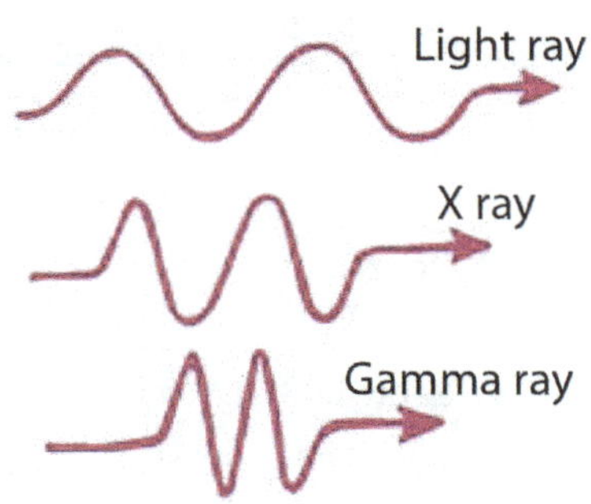

Figure 15.2
A gamma ray is simply electromagnetic radiation, much higher in frequency and energy than light and X rays.

CONCEPT**CHECK**

Pretend that you are given three radioactive rocks. One is an alpha emitter, one is a beta emitter, and one is a gamma emitter, and you know which is which. You can throw away one, but of the remaining two, you must hold one in your hand and place one in your pocket. What can you do to minimize your exposure to radiation?

CHECK YOUR ANSWER

Throw away the gamma emitter, because it would penetrate your body from any of these locations. Hold the alpha emitter in your hand, because the skin on your hand is enough to shield you. Put the beta emitter in your pocket, because beta particles will likely be stopped by the combined thickness of your clothing and skin. (Ideally, of course, you should distance yourself as much as possible from all of the rocks.)

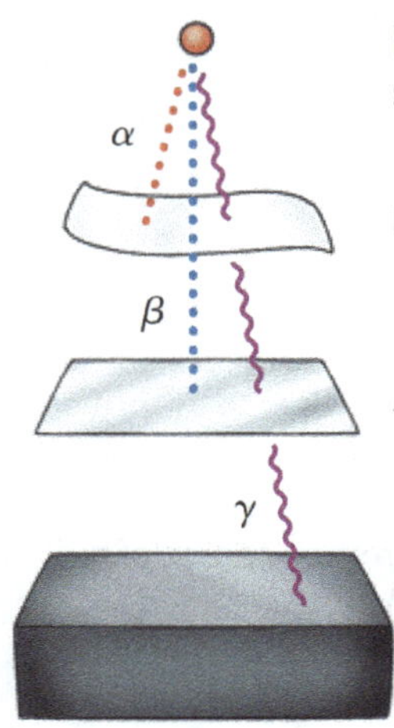

Figure 15.3
Alpha particles are the least penetrating and can be stopped by a few sheets of paper. Beta particles will readily pass through paper but not through a sheet of aluminum. Gamma rays penetrate several centimeters into solid lead.

Figure 15.4
The shelf life of fresh strawberries and other perishables is markedly increased when the food is subjected to gamma rays from a radioactive source, which kills the microorganisms that normally lead to spoilage. The strawberries are only a receiver of radiation and are not transformed into an emitter of radiation, as can be confirmed with a radiation detector.

15.2 Radioactivity Is Natural

It is a common misconception that radioactivity is something new in the environment. Actually, radioactivity has been around far longer than humans. It has always been in the soil we walk on and in the air that we breathe, and it warms Earth's interior. In fact, radioactive decay in Earth's interior heats the water that spurts from a geyser, wells up from a natural hot spring, or powers a volcano.

As **Figure 15.5** shows, most of the radiation we encounter is natural background radiation that originates from Earth, the Sun, and from other stars. At sea level, the protective blanket of the atmosphere reduces this radiation, but radiation is more intense at higher altitudes, where the air is thinner. In Denver, the "Mile-High City," a person receives more than twice as much cosmic radiation as occurs at sea level.

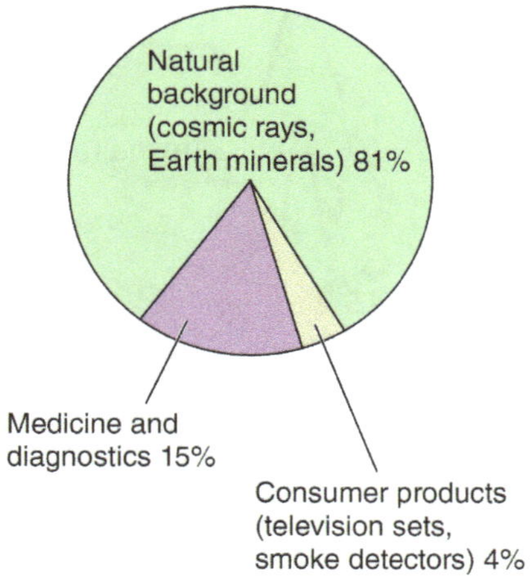

Figure 15.5
Origins of radiation exposure for an average individual in the United States.

A common source of radiation of an Earthly origin is radon-222, an inert gas arising from uranium, which is widely found at low levels within all rock, soil, and water. Radon is a heavy gas that tends to accumulate in basements after it seeps up through cracks in the floor. Levels of radon vary from region to region, depending upon local geology. You can check the radon level in your home with a radon detector kit (**Figure 15.6**). If levels are high, corrective measures, such as sealing the basement floor and walls and maintaining adequate ventilation, should be taken. Radon gas poses a serious health risk.

READING CHECK

What is a common source of radiation arising from Earth?

About 19 percent of our annual exposure to radiation comes from sources outside of nature, primarily medical procedures. Fallout from nuclear testing and the coal and nuclear power industries are also contributors. Interestingly, the coal industry outranks the nuclear power industry as a source of radiation. The global combustion of coal annually releases about 13,000 tons of radioactive thorium and uranium into the atmosphere (in addition to other environmentally damaging molecules, including greenhouse gases). Both of these elements are found naturally in coal deposits, so their release is a natural consequence of burning coal. Nuclear power plants also produce radioactive by-products. Worldwide, the nuclear power industries generate about 10,000 tons of radioactive waste each year. Most of this waste, however, is contained and not released into the environment.

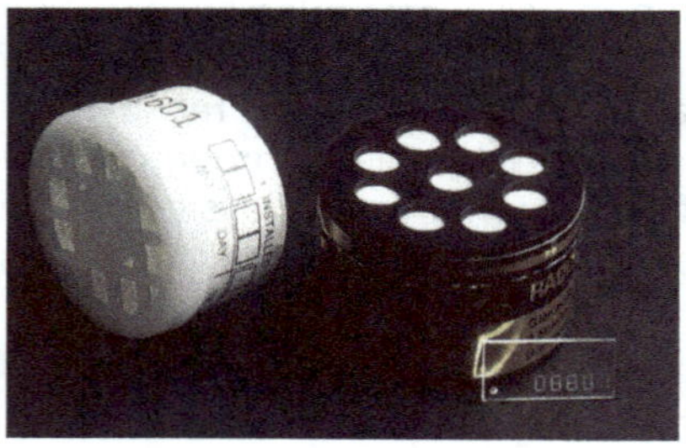

Figure 15.6
A commercially available radon test kit for the home. The canister is unsealed in the area to be sampled. Radon seeping into the canister is adsorbed by activated carbon within the canister. After several days, the canister is resealed and sent to a laboratory for determination of the radon level.

When radiation encounters our intricately structured cells, it can create chaos. Cells are able to repair most kinds of damage caused by radiation if the damage is not too severe. A cell can survive an otherwise lethal dose of radiation if the dose is spread over a long period of time to allow intervals for healing. When radiation is sufficient to kill cells, the dead cells can be replaced by new ones. Sometimes a radiated cell will survive with damaged DNA. This can alter the genetic information contained in a cell, producing one or more mutations. Although the effects of most mutations are inconsequential in terms of a person's health, some mutations affect the functioning of cells. Genetic mutations are the cause of most cancers, for example. In addition, mutations that occur in an individual's reproductive cells can be passed to the individual's offspring. In this case, the mutation will be present in every cell in the offspring organism's body—and may well have an effect on the functioning of the organism.

Rems Are Units of Radiation

The unit of measure for radiation dosage based on potential damage is measured using the **rem** (roentgen equivalent man). Lethal doses of radiation begin at 500 rems. A person has about a 50 percent chance of surviving a dose of this magnitude received over a short period of time. During radiation therapy, a patient may receive localized doses in excess of 200 rems each day for a period of weeks (**Figure 15.7**).

Throughout our lives all the radiation we receive from natural sources and medical procedures is only a fraction of 1 rem. For convenience, the smaller unit *millirem* is used; 1 millirem (mrem) is 1/1000 of a rem.

The average person in the United States is exposed to about 360 millirems a year, as **Table 15.1** indicates. About 80 percent of this radiation comes from natural sources, such as cosmic rays (radiation from our Sun as well as other stars) and the Earth. A typical diagnostic X ray exposes a person to between 5 and 30 millirems (0.005 and 0.030 rem), less than 1/10,000 of the lethal dose.

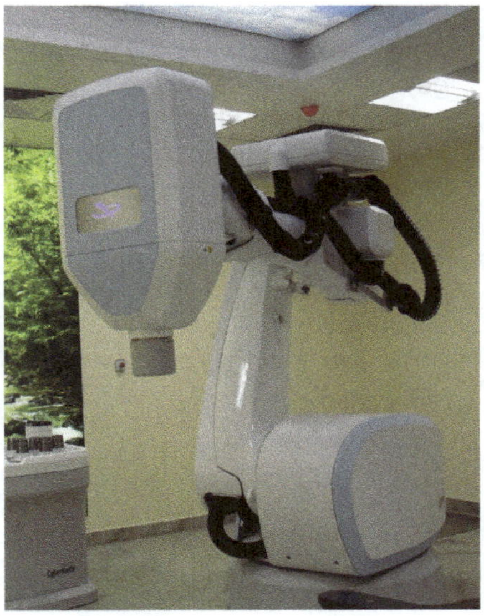

Figure 15.7
Nuclear radiation is focused on harmful tissue, such as a cancerous tumor, to selectively kill or shrink the tissue in a technique known as radiation therapy. This application of nuclear radiation has saved millions of lives—a clear-cut example of the benefits of nuclear technology. The CyberKnife System shown here uses robotics to deliver precise doses with extreme accuracy.

Table 15.1 Annual Radiation Exposure

SOURCE	TYPICAL AMOUNT RECEIVED IN 1 YEAR (MILLIREMS)
Natural Origin	
Cosmic radiation	26
Ground	33
Air (radon-222)	198
Human tissues (potassium-40; radium-226)	35
Human Origin	
Medical procedures	
Diagnostic X rays	40
Nuclear medicine	15
Television tubes, other consumer products	11
Weapons-test fallout	1

Source: U.S. Nuclear Regulatory Commission, http://www.nrc.gov/reading-rm/doc-collections/fact-sheets/bio-effects-radiation.html

Radioactive Tracers and Medical Imaging

Radioactive isotopes can be incorporated into molecules whose location can then be traced by the radiation they emit. When used in this way, radioactive isotopes are called *tracers*, and **Figure 15.8** shows one use for them. To check the action of a fertilizer, researchers incorporate radioactive isotopes into the molecules of the fertilizer and then apply the fertilizer to plants. The amount taken up by the plants can be measured with radiation detectors. From such measurements, scientists can tell farmers how much fertilizer to use, because fertilizer uptake is a physical and chemical process that is not affected by the radioactivity of the materials involved.

Fertilizer with radioactive isotope applied to crop

Radioactivity detected in plant

Figure 15.8
Tracking fertilizer uptake with a radioactive isotope.

FOR YOUR INFORMATION

The deeper you go below the Earth's surface, the hotter it gets. At a depth of merely 30 kilometers the temperature is over 500°C. At greater depths it is so hot that rock melts into magma, which can rise to the Earth's surface to escape as lava. Superheated subterranean water can escape violently to form geysers or more gently to form a soothing natural hot spring. The main reason for Earth's hot interior is radiation emitted by radioactive isotopes. So, volcanoes, geysers, and hot springs are all powered by radioactivity. Even the drifting of continents (plate tectonics) is a consequence of the Earth's internal radioactivity.

Tracers are also used in industry. Motor oil manufacturers can quantify the lubricating qualities of their products by running oil in engines containing small but measurable amounts of radioactive isotopes. As the engine runs and the pistons rub against the inner chambers, some of the metal from the engine invariably makes its way into the motor oil, and this metal carries with it the embedded radioactive isotopes. The better the lubricating qualities of a motor oil, the fewer radioactive isotopes it will contain after running in the engine for a given length of time.

In a technique known as *medical imaging*, isotopes are used for the diagnosis of internal disorders. Small amounts of a radioactive material such as sodium iodide, NaI, which contains the radioactive isotope iodine-131, are administered to a patient and traced through the body with a radiation detector. The result, shown in **Figure 15.9**, is an image that shows how the material is distributed in the patient's body. This technique works because the path the tracer material takes is influenced only by its physical and chemical properties, not by its radioactivity. The tracer may be introduced alone or along with some other chemical, known as a *carrier compound*, that helps target the isotope to a particular type of tissue in the body.

Table 15.2 Uses for Various Radioactive Isotopes

ISOTOPE	USAGE
Fluorine-18	Medical PET scans
Calcium-47	The study of bone formation in mammals
Californium-252	Inspection of airline luggage for explosives
Hydrogen-3 (tritium)	Drug-metabolism studies
Iodine-131	Treatment of thyroid disorders
Iridium-192	Testing the integrity of metal parts
Thallium-201	Cardiology and for tumor detection
Xenon-133	Lung-ventilation and blood-flow studies

Source: The Regulation and Use of Radioisotopes in Today's World (NUREG/BR-0217, Revision 1), U.S. Nuclear Regulatory Commission

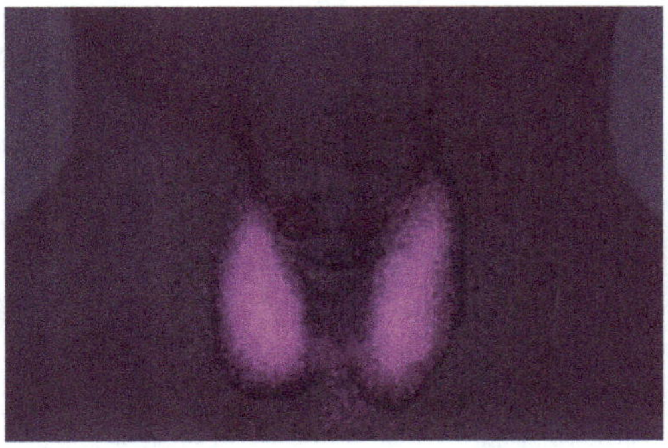

Figure 15.9
The thyroid gland, located in the neck, absorbs much of the iodine that enters the body through food and drink. Its image is obtained by giving a patient the radioactive isotope iodine-131. These images are useful in diagnosing metabolic disorders.

FOR YOUR INFORMATION

The only element beyond uranium to find a commercial application is americium, Am, which is a key component of nearly all household smoke detectors. This element completes an electric circuit by ionizing air within a chamber. Smoke particles interfere with this ionization, thus breaking the circuit and triggering the alarm.

15.3 Imbalanced Forces

We know that electric charges of the same charge repel one another. So how is it possible that positively charged protons in the nucleus stay clumped together? This question led to the discovery of an attraction called the **strong nuclear force**, which acts between all nucleons. This force is very strong but only over extremely short distances. Repulsive electric forces, on the other hand, are relatively long-ranged. **Figure 15.10** suggests a comparison of the strengths of these two forces over distance.

Because the strong nuclear force decreases over distance, a large nucleus is not as stable as a small one, as shown in **Figure 15.11**. For protons that are close together, as in small nuclei, the attractive strong nuclear force easily overcomes the repulsive electric force. But for protons that are far apart, like those on opposite edges of a large nucleus, the attractive strong nuclear force may be weaker than the repulsive electric force.

A large atomic nucleus is more susceptible to the repulsive forces among protons. This means there is a limit to the size of the atomic nucleus. As evidence of this, we find that all nuclei having more than 83 protons are radioactive. Furthermore, the superheavy elements, such as

READING CHECK
Are protons on opposite sides of a large nucleus attracted to each other?

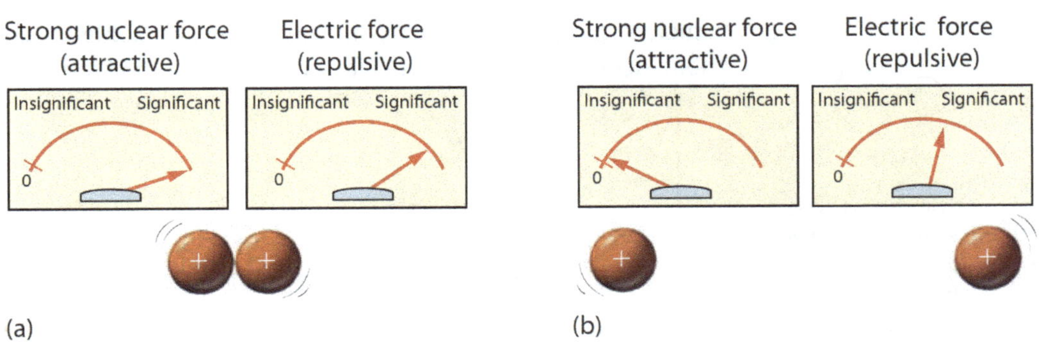

Figure 15.10
(a) Two protons near each other experience both an attractive strong nuclear force and a repulsive electric force. At this tiny separation distance, the strong nuclear force overcomes the electric force, resulting in their staying together. (b) When the two protons are relatively far from each other, the electric force is more significant. The protons repel each other. This proton–proton repulsion in large atomic nuclei reduces nuclear stability.

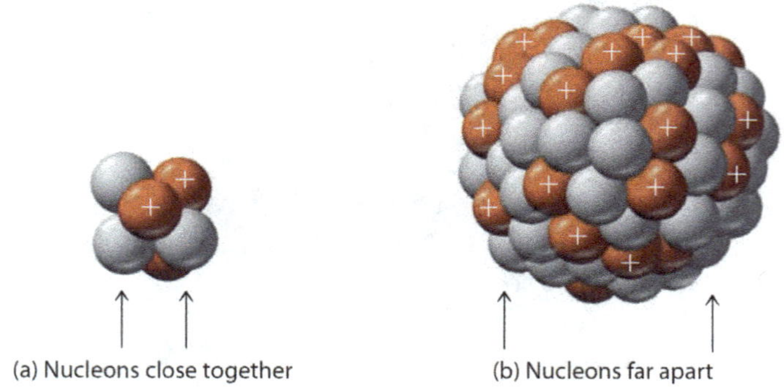

Figure 15.11 >
(a) All nucleons in a small atomic nucleus are close to one another; hence, they experience an attractive strong nuclear force. (b) Nucleons on opposite sides of a larger nucleus are not as close to one another, so the attractive strong nuclear forces holding them together are much weaker. The result is that the large nucleus is less stable.

those above uranium, atomic number 92, are not found in nature. These superheavy elements are difficult to make in the laboratory. When they are produced, they exist for only fractions of a second.

Neutrons serve as the "nuclear cement" holding an atomic nucleus together. Protons attract both protons and neutrons by the strong nuclear force. Protons also repel other protons by the electric force. Neutrons, on the other hand, have no electric charge and so attract protons and other neutrons only by the strong nuclear force. Therefore, the presence of neutrons adds to the attraction among nucleons and helps hold the nucleus together (**Figure 15.12**).

Nuclei with larger numbers of protons require larger numbers of neutrons to help balance the repulsive electric forces. For light elements, it is sufficient to have about as many neutrons as protons. For instance, the most common isotope of carbon, C-12, has equal numbers of each—six protons and six neutrons. For large nuclei, more neutrons than protons are needed. Because the strong nuclear force diminishes rapidly over distance, nucleons must be practically touching in order for the strong nuclear force to be effective. Nucleons on opposite sides of a large atomic nucleus are not so attracted to one another. The electric force, however, does not diminish by much across the diameter of a large nucleus and so begins to win out over the strong nuclear force. To compensate for the weakening of the strong nuclear force across the diameter of the nucleus, large nuclei have more neutrons than protons. Lead nuclei, for example, have about one-and-a-half times as many neutrons as protons.

So, we see that neutrons are stabilizing and large nuclei require an abundance of them. But neutrons are not always successful in keeping a nucleus intact. Interestingly, neutrons are not stable without protons. A lone neutron is radioactive and spontaneously transforms to a proton and

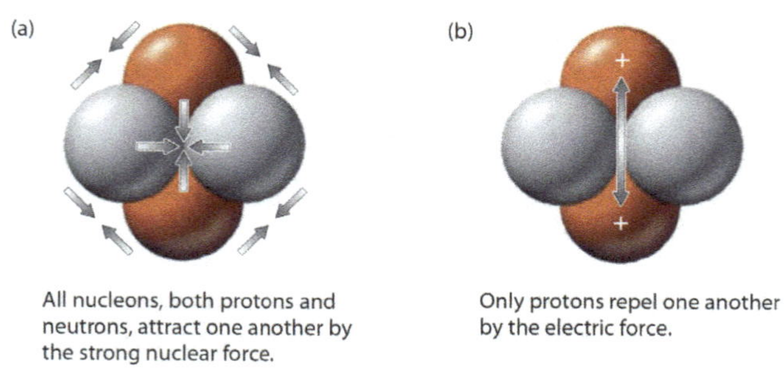

Figure 15.12 >
(a) The presence of neutrons helps hold the nucleus together by increasing the effect of the strong nuclear force, represented by the single-headed arrows. (b) The total strength of the attractive nuclear force, therefore, exceeds that of the repulsive electric force occurring between the two protons, represented by the double-headed arrow.

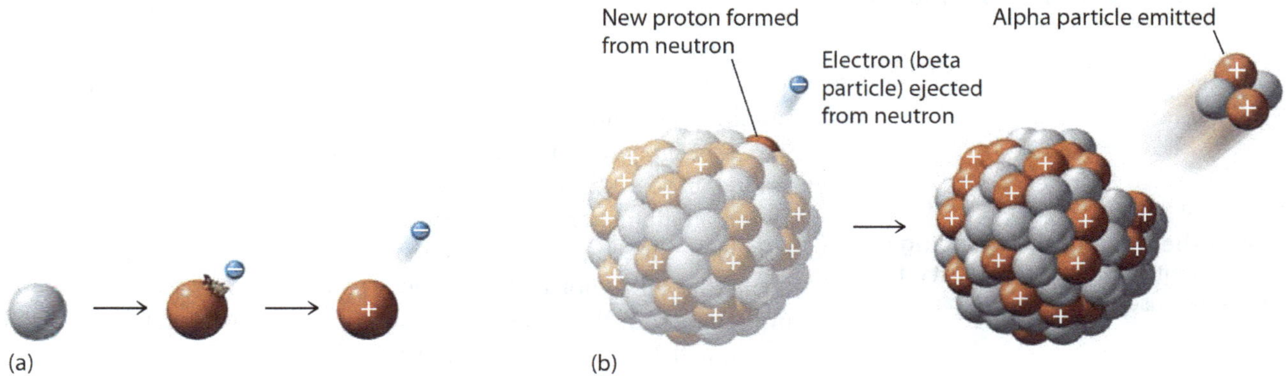

Figure 15.13
(a) A neutron without an adjacent proton is unstable and decays to a proton by emitting an electron. (b) Large nuclei have more neutrons than protons, which means that some of these neutrons don't have a sufficient number of adjacent protons. One of these extra neutrons may transform into a proton. Destabilized by an increase in the number of protons, the nucleus begins to shed fragments, such as alpha particles.

an electron (**Figure 15.13a**). A neutron seems to need protons around to keep this from happening. After the size of a nucleus reaches a certain point, the neutrons outnumber the protons to the point that there are not sufficient protons in the mix to prevent the neutrons from turning into protons. As neutrons in a nucleus change into protons, the stability of the nucleus decreases because the repulsive electric force becomes increasingly significant. The result is that pieces of the nucleus fragment away in the form of radiation, as indicated in **Figure 15.13b**. This sets an upper limit on the mass and size of the atomic nucleus, which explains why the periodic table is limited in its number of elements.

CONCEPT**CHECK**

1. What role do neutrons serve in the atomic nucleus? What is the fate of a neutron when alone or distant from one or more protons?

2. Two protons in the atomic nucleus repel each other, but they are also attracted to each other. Why?

CHECK YOUR ANSWERS

1. Neutrons serve as a nuclear cement in nuclei and add to nuclear stability. But when alone or away from protons, a neutron becomes radioactive and spontaneously transforms to a proton and an electron.

2. While two protons repel each other by the electric force, they also attract each other by the strong nuclear force. These forces act simultaneously. So long as the attractive strong nuclear force is stronger than the repulsive electric force, the protons will remain together. Under conditions in which the electric force overcomes the strong nuclear force, however, the protons fly apart from each other.

15.4 Transmutation

READING CHECK

What is the term for the changing of one element into another? Hint: It is the same word used by the alchemists.

When a radioactive nucleus emits an alpha or beta particle, the atomic number of the nucleus is changed. It becomes another element.

The changing of one element to another is called transmutation. Consider a uranium-238 nucleus, which contains 92 protons and 146 neutrons. When an alpha particle is ejected, the nucleus loses two protons and two neutrons. Since an element is defined by the number of protons in its nucleus, the 90 protons and 144 neutrons remaining no longer constitute a uranium atom. What we have now is the nucleus of a different element—thorium. This transmutation can be written as a nuclear equation:

$$^{238}_{92}\mathrm{U} \longrightarrow {}^{234}_{90}\mathrm{Th} + {}^{4}_{2}\mathrm{He}$$

We see that uranium-238 ($^{238}_{92}$U) transmutes to thorium-234 with the emission of an alpha particle, as written to the right of the arrow. When this transmutation occurs, energy is released, mostly in the form of the kinetic energy of the alpha particle, $^{4}_{2}$He, partly in the kinetic energy of the thorium atom, and partly in the form of gamma radiation. In this and all such equations, the mass numbers at the top balance ($238 = 234 + 4$) and the atomic numbers at the bottom also balance ($92 = 90 + 2$).

Thorium-234, the product of this reaction, is also radioactive. When it decays, it emits a beta particle. Because the formation of a beta particle results in an additional proton, the atomic number of the resulting nucleus is increased by 1. So, after beta emission by thorium with 90 protons, the resulting element has 91 protons. It is no longer thorium but the element protactinium. Although the atomic number has increased by 1 in this process, the mass number (protons + neutrons) remains the same. The nuclear equation is:

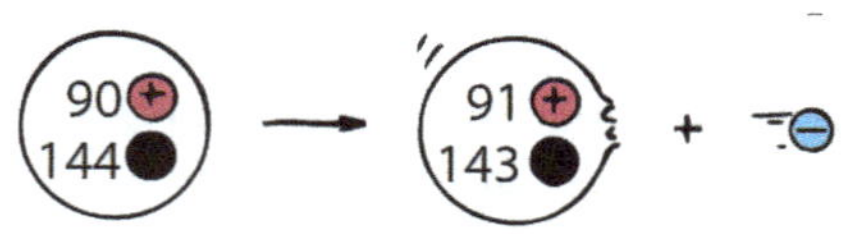

$$^{234}_{90}\mathrm{Th} \longrightarrow {}^{234}_{91}\mathrm{Pa} + {}^{0}_{-1}\mathrm{e}$$

We write an electron as $^{0}_{-1}$e. The superscript 0 indicates that the electron's mass is insignificant relative to that of protons and neutrons. The subscript -1 is the electric charge of the electron.

So, we see that when an element ejects an alpha particle from its nucleus, the mass number of the remaining atom is decreased by 4 and its atomic number is decreased by 2. The resulting atom is an atom of the element two spaces back in the periodic table, because this atom has two fewer protons. When an element ejects a beta particle from its nucleus, the mass of the atom is practically unaffected, meaning there is no change in its mass number but its atomic number increases by 1. The resulting atom becomes the element one space forward in the periodic table because it has one more proton.

FOR YOUR INFORMATION

Beta emission is also accompanied by the emission of a *neutrino*, which is a neutral particle with nearly zero mass that travels at about the speed of light. Neutrinos are hard to detect because they interact very weakly with matter—a piece of lead about eight light-years thick would be needed to stop half the neutrinos produced in typical nuclear decays. Thousands of neutrinos are flying through you every second of every day, because the universe is filled with them. Only occasionally, one or two times a year, does a neutrino interact with an atom of your body. Your body is mostly empty space!

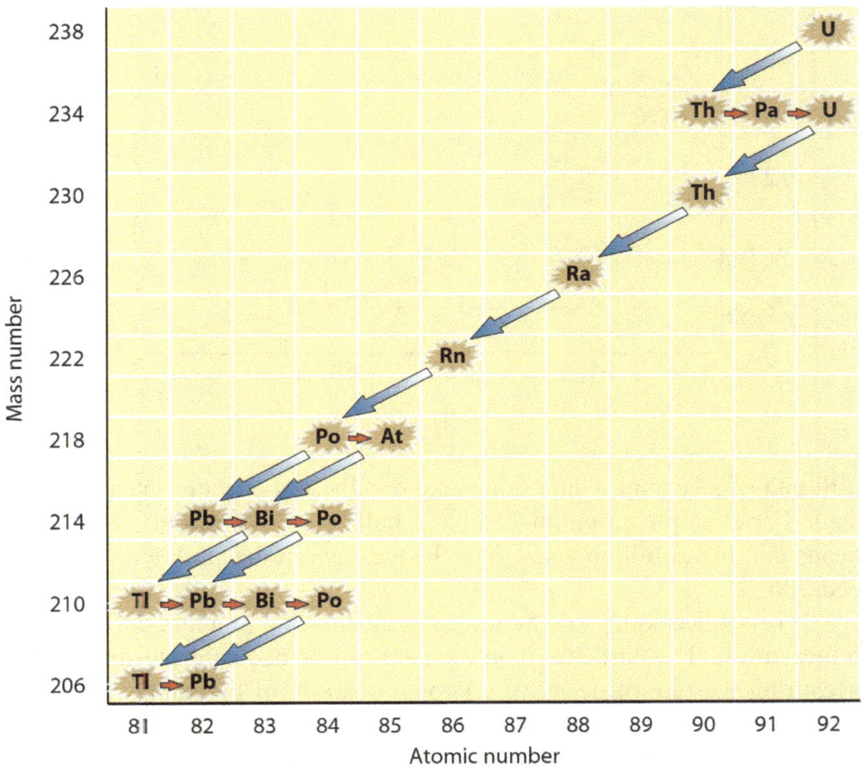

< Figure 15.14
U-238 decays to Pb-206 through a series of alpha and beta decays.

The decay of uranium-238 to lead-206 is shown in **Figure 15.14**. Each large blue arrow shows an alpha decay, and each small red arrow shows a beta decay.

CONCEPTCHECK
What finally becomes of all the uranium that undergoes radioactive decay?

CHECK YOUR ANSWER
All uranium ultimately becomes lead. Along the way, it exists as the elements indicated in Figure 15.14.

15.5 Radioactive Half-Life

Radioactive isotopes decay at different rates. The radioactive decay rate is measured in terms of a characteristic time, the **half-life**.

The half-life of a radioactive material is the time required for half of the radioactive atoms to decay. Radium-226, for example, has a half-life of 1620 years. This means that half of any given specimen of Ra-226 decays by the end of 1620 years. In the next 1620 years, half of the remaining radium decays, leaving only one-fourth the original number of radium atoms. The other three-fourths convert, by a succession of decays, to lead. After 20 half-lives, an initial quantity of radioactive atoms is diminished to about one-millionth of the original quantity (**Figure 15.15**).

READING CHECK
How is radioactive half-life defined?

Half-lives are remarkably constant and not affected by external conditions. Some radioactive isotopes have half-lives that are less than a

Figure 15.15 >
Radium-226 has a half-life of 1620 years, meaning that every 1620 years the amount of radium decreases by half as it transmutes to other elements.

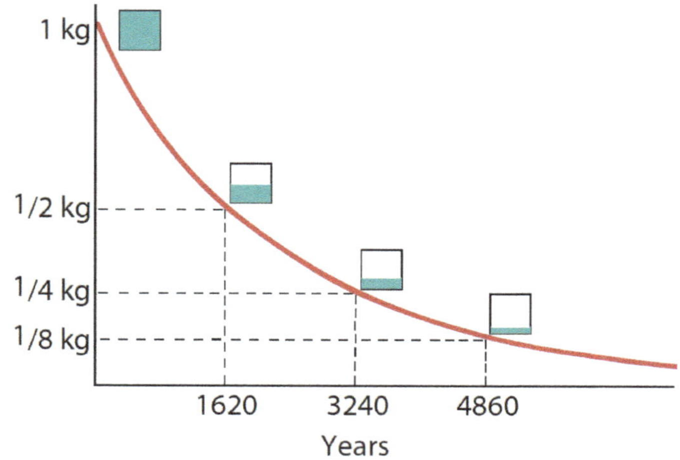

millionth of a second, while others have half-lives of more than a billion years. For example, uranium-238 has a half-life of 4.5 billion years. This means that in 4.5 billion years, half the uranium in the earth today will be lead-206.

It is not necessary to wait through the duration of a half-life in order to measure it. The half-life of an element can be accurately estimated by measuring the rate of decay of a known quantity of the element. This is easily done using a radiation detector. In general, the shorter the half-life of a substance, the faster it disintegrates and the more radioactivity per minute is detected. **Figures 15.16** and **15.17** show two ways of detecting radiation.

The half-life of radium-226 is 1620 years. Does this mean that an individual Ra-226 nucleus has to wait 1620 years before it decays? The answer is no. Some Ra-226 nuclei will decay within a few minutes. Others

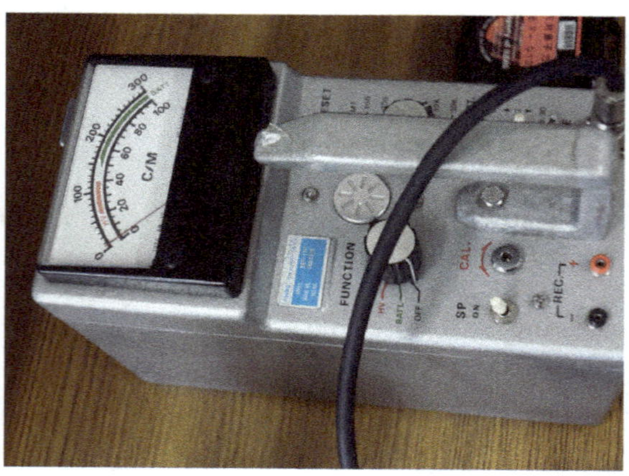

^ Figure 15.16
The handheld tube of a Geiger counter contains a gas that becomes ionized by incoming nuclear radiation. Audible clicks indicate radiation intensity.

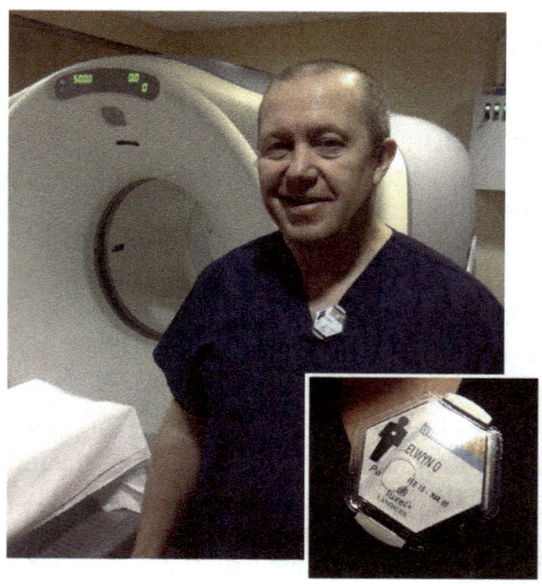

^ Figure 15.17
The film badge worn by Elwyn contains audible alerts for radiation surges and accumulated exposure. Information from the individualized badges is periodically downloaded to a database for analysis.

won't decay for tens of thousands of years. But if all the nuclei of a radioactive element decay at different times, then why is the half-life of a radioactive substance always the same? Half-life applies only to macroscopic quantities of a radioactive substance. In such quantities, there are billions upon billions of nuclei. Half-life is a statistical measure of what happens to all these nuclei *on average*. If, on average, the nuclei decay quickly, then the half-life will be short, which tells us that these nuclei are not very stable. Conversely, if, on average, it takes a long time for the nuclei to decay, then the half-life will be long, telling us that these nuclei are more stable.

CONCEPT**CHECK**

1. If a sample of radioactive isotopes has a half-life of 1 day, how much of the original sample will remain at the end of the second day? The third day?

2. Which will give a higher counting rate on a radiation detector—radioactive material that has a short half-life or a long half-life?

CHECK YOUR ANSWERS

1. In the second day, one-fourth of the original sample will be left. The three-fourths that underwent decay becomes a different element altogether. At the end of 3 days, one-eighth of the original sample will remain.

2. The material with the shorter half-life is more active and will show a higher counting rate on a radiation detector.

15.6 Isotopic Dating

Cosmic rays continually bombard Earth's atmosphere. This bombardment causes many atoms in the upper atmosphere to transmute. These transmutations result in a "spraying out" of protons and neutrons into the environment. Most of the protons are stopped as they collide with the atoms of the upper atmosphere. These protons strip electrons from the atoms they collide with and thus become hydrogen atoms. The neutrons, however, continue for longer distances because they have no electric charge and therefore do not interact electrically with matter. Eventually, many of them collide with atomic nuclei in the lower atmosphere. A nitrogen atom that captures a neutron, for instance, becomes an isotope of carbon by emitting a proton:

$$^{1}_{0}n + ^{14}_{7}N \longrightarrow ^{14}_{6}C + ^{1}_{1}H$$

This carbon-14 isotope, which makes up less than one-millionth of 1 percent of the carbon in the atmosphere, is radioactive and has eight neutrons. (The most common isotope, carbon-12, has six neutrons and is not radioactive.) Because both carbon-12 and carbon-14 are forms of carbon, they have the same chemical properties. Both of these isotopes,

READING CHECK

Plants contain a little carbon-14, but why do humans also contain carbon-14?

for example, chemically react with oxygen to form carbon dioxide, which is consumed by plants through the process of photosynthesis. This means that all plants contain a tiny quantity of radioactive carbon-14.

All animals eat either plants or plant-eating animals; therefore, all animals, including humans, have a little carbon-14 in them. So, we see why all living things on Earth contain some carbon-14. This radioactive isotope decays back into nitrogen as it emits a beta particle:

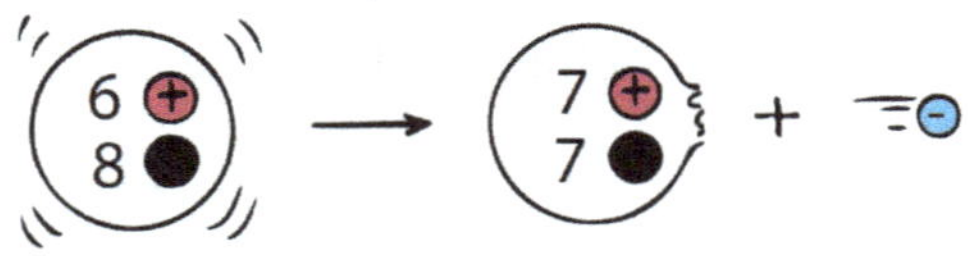

$$^{14}_{6}\text{C} \longrightarrow \, ^{14}_{7}\text{N} + \, ^{0}_{-1}\text{e}$$

Because plants absorb carbon dioxide while alive, any carbon-14 that decays is immediately replenished with fresh carbon-14 from the atmosphere. In this way, a radioactive equilibrium is reached in which there is a constant ratio of about 1 carbon-14 atom to every 100 billion carbon-12 atoms. When a plant dies, replenishment of carbon-14 ends. Then the percentage of carbon-14 decreases at a constant rate determined by its half-life. The amount of carbon-12, however, does not change, because this isotope does not undergo radioactive decay. The longer a plant or other organism is dead, the less carbon-14 it contains relative to the constant amount of carbon-12.

The half-life of carbon-14 is about 5730 years. This means that half of the carbon-14 atoms now present in a plant or animal that dies today will decay in the next 5730 years. Half of the remaining carbon-14 atoms will then decay in the following 5730 years, and so on.

With this knowledge, scientists are able to calculate the age of carbon containing artifacts, such as wooden tools or the skeleton shown in **Figure 15.18**. Their age is measured by their current level of radioactivity. This process, known as **carbon-14 dating**, enables investigators to probe as far as 50,000 years into the past. Beyond this time span, there is too little carbon-14 remaining for accurate analysis.

∧ Figure 15.18

The amount of radioactive carbon-14 in the skeleton is reduced by one-half every 5730 years. The result is that the same skeleton today contains only a trace amount of the original carbon-14. The arrows represent the relative amounts of carbon-14.

Because of fluctuations in cosmic ray bombardment rates over the centuries, carbon dating has an uncertainty of about 15 percent. This means, for example, that the straw of an old adobe brick dated to be 500 years old may really be only 425 years old on the low side or 575 years old on the high side. For many purposes, this is an acceptable level of uncertainty.

Artifacts derived from nonliving materials can also be dated based on the radioactive minerals they contain. The naturally occurring mineral isotopes uranium-238 and uranium-235, for example, decay very slowly and ultimately become lead—but not the common isotope lead-208. Instead, as was shown in **Figure 15.14**, uranium-238 decays to lead-206. Uranium-235, on the other hand, decays to lead-207. Thus, the lead-206 and lead-207 that now exist in a uranium-bearing rock were at one time uranium. Older rocks contain higher percentages of these trace isotopes.

If you know the half-lives of uranium isotopes and the percentage of lead isotopes in some uranium-bearing rock, you can calculate the date of the rock's formation. Rocks dated in this manner have been found to be as much as 3.7 *billion* years old. Samples from the Moon have been dated at 4.2 billion years, which is close to the estimated age of our solar system: 4.6 billion years.

∧ Figure 15.19
Carbon-14 dating was developed by the American chemist Willard F. Libby (1908–1980) at the University of Chicago in the 1950s. For this work, he received the Nobel Prize in Chemistry in 1960.

CONCEPT**CHECK**
Suppose that an archaeologist extracts 1 gram of carbon from an ancient ax handle and finds that it is one-fourth as radioactive as 1 gram of carbon extracted from a freshly cut tree branch. About how old is the ax handle?

CHECK YOUR ANSWER
Assuming the ratio of C-14/C-12 was the same when the ax was made, the ax handle is as old as two half-lives of C-14, or about 11,460 years old.

15.7 Nuclear Fission

Biology students know that living tissue grows by the division of cells. The splitting in half of living cells is called fission. In a similar way, the splitting of an atomic nucleus into two smaller halves is called **nuclear fission**.

Nuclear fission involves the delicate balance between two forces within the nucleus. One force is the *strong nuclear force*, which is a force that holds all the nucleons together. The second force is the repulsive *electric force* occurring among all the like-sign protons. In most nuclei the nuclear strong force dominates. In uranium, however, this domination is weak. If the uranium nucleus is stretched into an elongated shape (**Figure 15.20**), the nuclear force quickly weakens and the electric forces may push it into an even more elongated shape. If the elongation passes a critical point, the electric forces overwhelm the strong nuclear forces, and the nucleus splits. This splitting process is called nuclear fission.

The absorption of a neutron by a uranium nucleus supplies enough energy to cause such an elongation. The resulting fission process may produce many different combinations of smaller nuclei. More significantly,

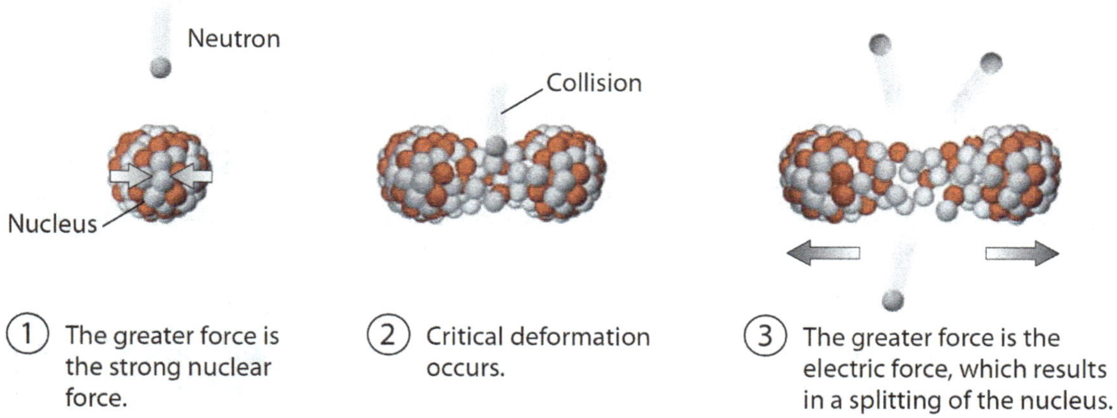

^ Figure 15.20
An elongation of the nucleus may result in the repulsive electric force's overcoming the attractive strong nuclear force, in which case fission occurs.

the energy released by fission is enormous—about 7 million times that of a TNT molecule explosion. This energy is mainly in the form of the kinetic energy of the fission fragments, which fly apart from one another. A much smaller amount of energy is released as gamma radiation.

Here is a pictorial with the equation for a typical uranium fission reaction:

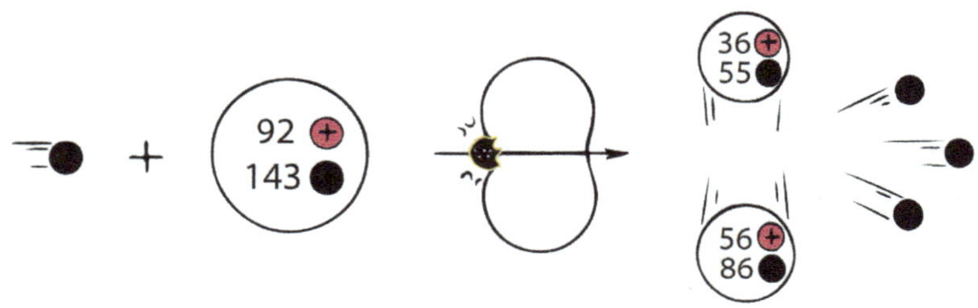

$$ {}^{1}_{0}n + {}^{235}_{92}\text{U} \longrightarrow {}^{91}_{36}\text{Kr} + {}^{142}_{56}\text{Ba} + 3({}^{1}_{0}n) $$

Note that in this reaction 1 neutron starts the fission of a single uranium nucleus, which produces nuclear fragments and 3 neutrons. These 3 neutrons can cause the fission of 3 more uranium atoms, releasing 9 more neutrons. If each of these 9 neutrons succeeds in splitting a uranium atom, the next step in the reaction produces 27 neutrons, and so on. Such a sequence, illustrated in **Figure 15.21**, is called a **chain reaction**. A chain reaction is a self-sustaining reaction in which the products of one reaction event initiate further reaction events.

READING CHECK
What exactly is a chain reaction?

Why do chain reactions not happen in naturally occurring uranium ore deposits? They would if all uranium atoms fissioned so easily. Fission occurs mainly in the rare isotope U-235, which makes up only 0.7 percent of the uranium in pure uranium metal, as shown in **Figure 15.22**. When the more abundant isotope U-238 absorbs neutrons created by fission of U-235, the U-238 typically does not undergo fission. So, any chain reaction is snuffed out by the neutron-absorbing U-238, as well as by the rock in which the ore is imbedded. The rock is slightly warmed.

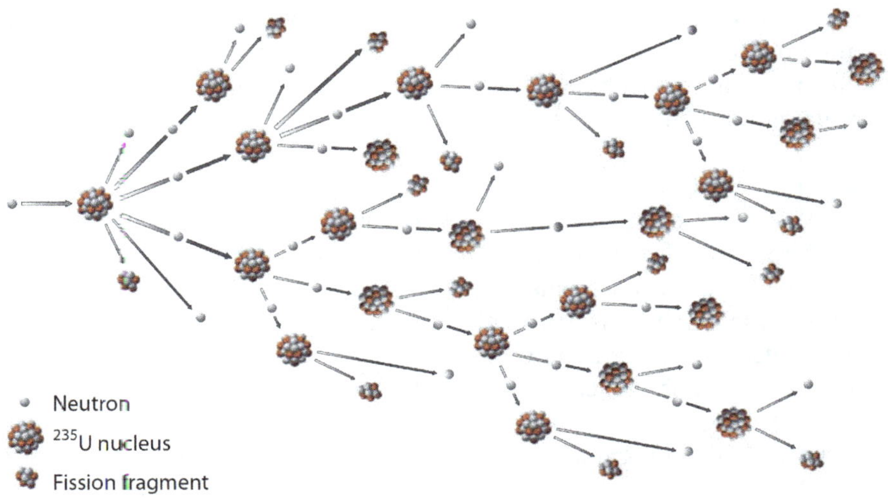

^ **Figure 15.21**
A chain reaction.

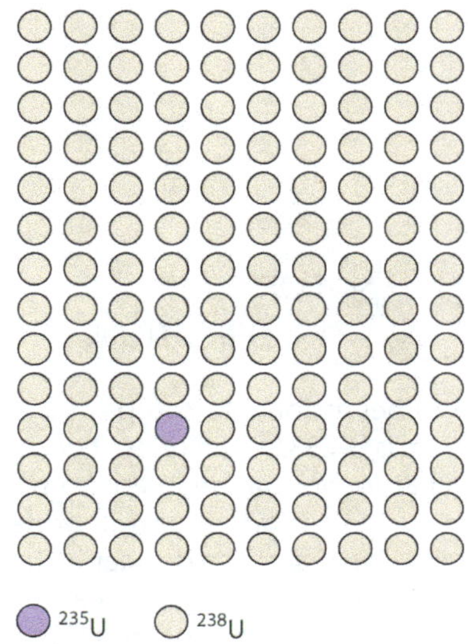

^ **Figure 15.22**
Only 1 part in 140 of naturally occurring uranium is U-235.

Chain reactions are more effective in large chunks of uranium than in smaller chunks. In smaller chunks, neutrons easily reach the opposite surface and escape, as shown in **Figure 15.23**. As the neutrons escape, the chain reaction no longer builds up.

If two small chunks of uranium are suddenly pushed together, they make a larger chunk. Within this larger chunk, neutrons are no longer able to escape as easily. Instead, they continue the chain reaction, which becomes sustainable. The minimum size of a chunk needed for a sustainable chain reaction is called the **critical mass**. Any chunk at or above the critical mass produces a steady release of energy. With the correct engineering, this release of energy can be controlled, which is what happens within a nuclear power plant. A different sort of engineering, as shown in **Figure 15.24**, produces an explosion of energy, which is the basis of a nuclear fission bomb.

Constructing a fission bomb is a formidable task. The difficulty is in separating enough uranium-235 from the more abundant uranium-238. Separation is difficult because, except for their slightly different masses, the two isotopes have the same physical and chemical properties. Scien-

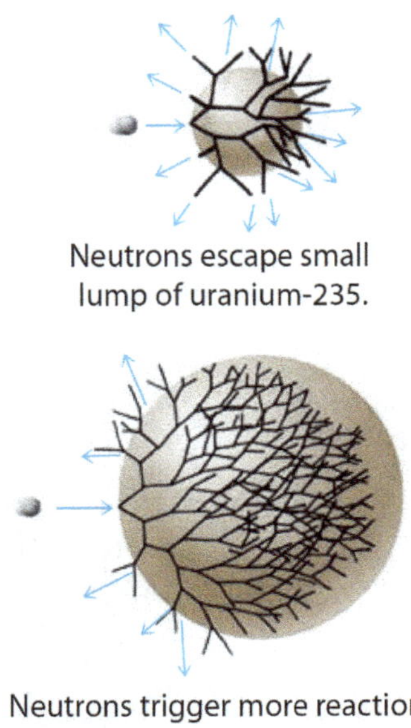

^ **Figure 15.23**
The buildup of chain reactions is greater in large chunks of U-235 than in small chunks.

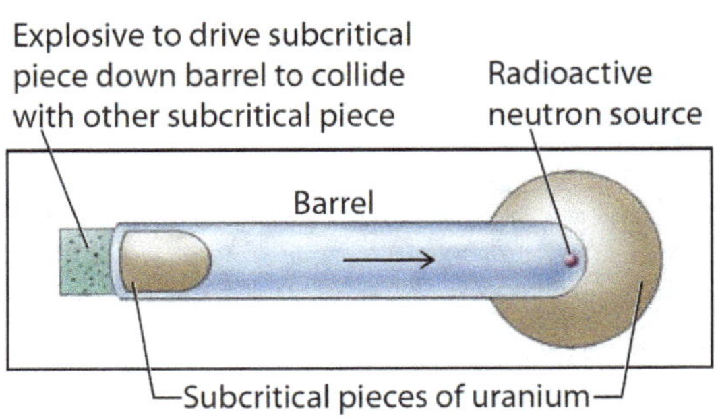

^ **Figure 15.24**
Simplified diagram of an idealized uranium "gun-type" fission bomb.

tists took more than 2 years to extract enough of the 235 isotope from uranium ore to make the bomb that was detonated at Hiroshima in 1945. To this day, uranium isotope separation, also known as uranium enrichment, remains a difficult process.

Nuclear Fission Reactors

The awesome energy of nuclear fission was introduced to the world in the form of nuclear bombs, and this violent image still colors our thinking about nuclear power, making it difficult for many people to recognize its potential usefulness. Currently, about 20 percent of electric energy in the United States is generated by *nuclear fission reactors* (whereas most electric power is nuclear in some other countries—about 75 percent in France). These reactors are simply nuclear furnaces. They, like fossil fuel furnaces, do nothing more elegant than boil water to produce steam for a turbine (**Figure 15.25**). The greatest practical difference is the amount of fuel involved: a mere 1 kilogram of uranium fuel, less than the size of a baseball, yields more energy than 30 freight-car loads of coal.

A fission reactor contains four components: nuclear fuel, control rods, moderator (to slow the velocity of the neutrons, making them effective for the fission process), and a liquid (usually water) to transfer heat from the reactor to the turbine and generator. The nuclear fuel is primarily U-238, plus about 3 percent U-235. Because the U-235 isotopes are so highly diluted with U-238, an explosion like that of a nuclear bomb is not possible. The reaction rate, which depends on the number of neutrons that initiate the fission of other U-235 nuclei, is controlled by rods inserted into the reactor. The control rods are made of a neutron-absorbing material, usually metals like cadmium or boron.

Heated water around the nuclear fuel is kept under high pressure to keep it at a high temperature without boiling. It transfers heat to a second, lower-pressure water system, which operates the turbine and electric

fyi FOR YOUR INFORMATION

Another fissionable isotope of uranium is U-233, which doesn't occur in nature but can be "bred" from a mixture of thorium and fluoride salts. The reactor to do that is the liquid fluoride thorium reactor, LFTR (often pronounced "lifter"). Watch for it as a contender for safe nuclear power.

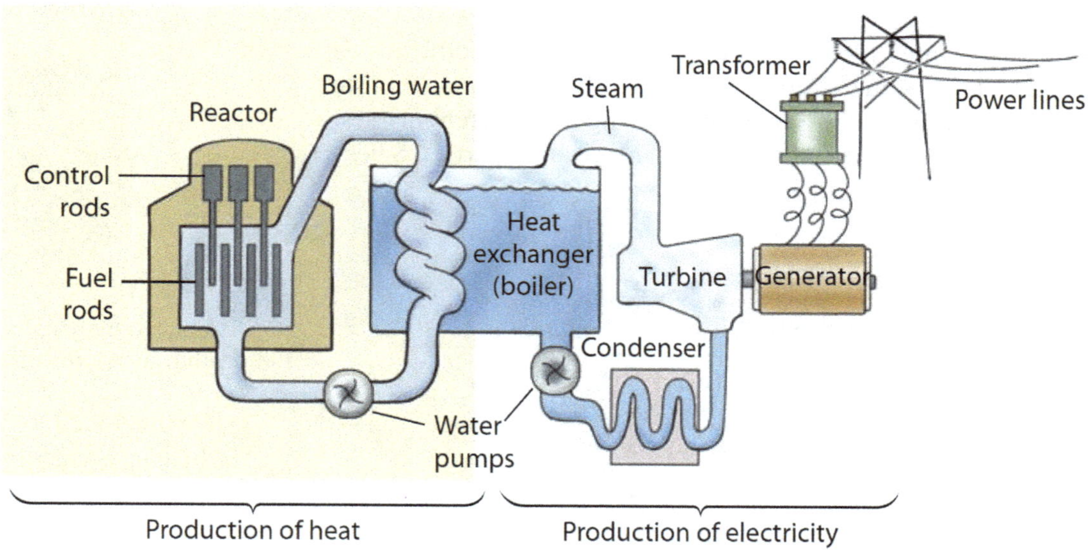

Figure 15.25
Diagram of a nuclear fission power plant. Note that the water in contact with the fuel rods is completely contained and that radioactive materials are not involved directly in the generation of electricity.

< Figure 15.26
The nuclear reactor is housed within a dome-shaped containment building designed to prevent the release of radioactive isotopes in the event of an accident. The Soviet-built Chornobyl nuclear power plant that reached meltdown in 1986 had no such containment building, leading to massive amounts of radiation being released into the surrounding environment.

generator in a conventional fashion. In this design, two separate water systems are used so that no radioactivity reaches the turbine or the outside environment. The entire setup resides inside a building like the one shown in **Figure 15.26**.

Fission power has benefits and drawbacks. The benefits include plentiful electricity, conservation of fossil fuels that presently are turned to heat and smoke (and in the long run may be far more precious as sources of organic molecules than as sources of heat), eliminating the megatons of sulfur oxides and other poisons as well as the greenhouse gas carbon dioxide released into the air each year by the burning of these fuels.

The drawbacks include the low-level release of radioactive materials into the air and groundwater, and, most important, the risk of an accidental release of large amounts of radioactivity. Then there's the problem of storing radioactive wastes. Some countries, such as France, dispose of wastes in deep ground repositories. Waste disposal nevertheless remains a problem.

Reasoned judgment requires not only that we examine fission power's benefits and drawbacks but that we compare its benefits and drawbacks with those of other power sources. We speak of "trade-offs," a comparison of benefits and drawbacks. The issue is not whether fission power is safe, but whether its dangers are greater or less than the dangers of supplying the same electric power in other ways. So, going nuclear isn't a long-term solution, but rather is a reasoned compromise.

FOR YOUR INFORMATION

Recent advances in fission power are smaller modular reactors (SMRs), slimmed-down versions of conventional reactors that take up 1% the space of a conventional reactor. Instead of water they use coolants such as molten salts.

CONCEPT**CHECK**

A 52-kilogram ball of uranium-235 has critical mass, but the same ball broken up into small chunks does not. Explain.

CHECK YOUR ANSWER

The small chunks have more combined surface area than the ball from which they came. Neutrons escape via the surface of each small chunk before a sustained chain reaction can build up.

15.8 Mass and Energy

Clearly, a lot of energy comes from every gram of nuclear fuel when it undergoes fission. What is the source of this energy? As we will see, it comes from nucleons losing mass through nuclear reactions.

In the early 1900s, Albert Einstein discovered that mass is actually "congealed" energy. Mass and energy are two sides of the same coin, as stated in his celebrated equation, $E = mc^2$. In this equation, E stands for the energy contained by any mass when at rest, m stands for mass, and c is the speed of light. This relationship between energy and mass is the key to understanding why and how energy is released in nuclear reactions.

READING CHECK

What do the symbols represent in the equation $E = mc^2$?

How easy might it be to pull a nucleon out of a nucleus? (Remember, "nucleon" is the generic name for either a proton or a neutron.) To do this, you would have to overcome the strong nuclear force, which holds the nucleon to the nucleus. So, a significant amount of energy would be required to pull this nucleon out of the nucleus. What we learn from Einstein's equation is that the energy you contribute to pull the nucleon out is not lost. Instead, this energy is absorbed by the nucleon, which thus becomes more massive as you pull it out. For example, if the mass of the nucleon were 1.00000 while in the nucleus, its mass might be a slightly greater 1.00728 after it has been pulled out of the nucleus. The energy you put into pulling the nucleon out was converted into mass. Energy and mass are two sides of the same coin. In other words, energy can become mass and mass can become energy as dictated by the equation $E = mc^2$.

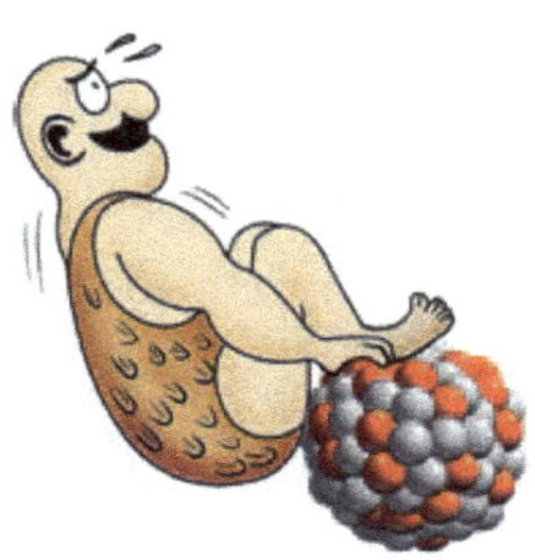

^ Figure 15.27
Work is required to pull a nucleon from an atomic nucleus. This work increases the energy and hence the mass of the nucleon outside the nucleus.

So, the mass of a nucleon depends upon where it is. In general, a nucleon's mass is greatest when it is free by itself outside of the nucleus. The nucleon's mass is smallest when it is tightly bound within the nucleus.

Not all nuclei are the same. In one nucleus, for example, a nucleon might find itself more tightly bound than in another. The mass of the nucleon, therefore, also depends upon which nucleus it is in. As illustrated in the graph of **Figure 15.28**, a nucleon has its greatest mass when in the hydrogen nucleus and its smallest mass when in the iron nucleus. The mass of a nucleon then gradually increases as it enters heavier nuclei, such as that of uranium.

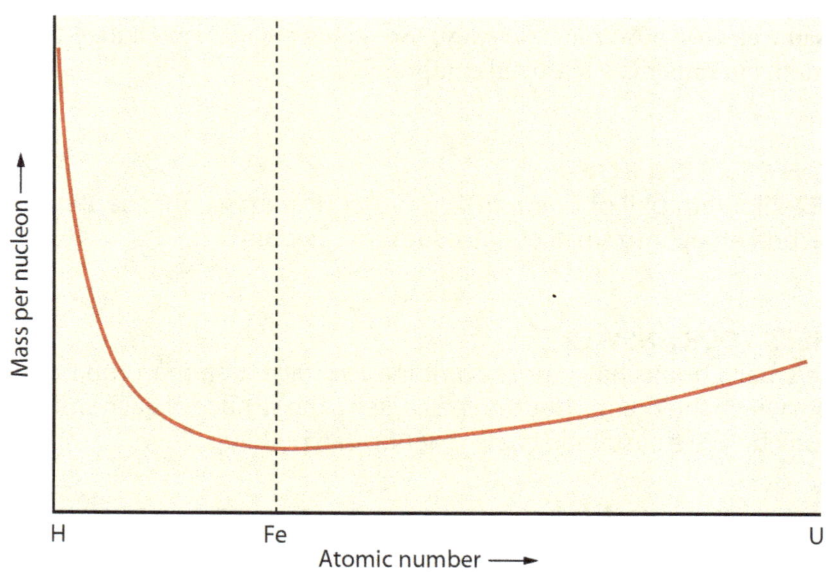

Figure 15.28 >
This graph shows that the average mass of a nucleon depends on which nucleus it is in. A nucleon has the most mass when in the lightest (hydrogen) nucleus. The nucleon has its least mass when in an iron nucleus and an intermediate mass when in the heaviest (uranium) nucleus.

The graph of Figure 15.28 is the key to understanding the energy released in nuclear processes. Uranium, being toward the right-hand side of the graph, is shown to have a relatively large amount of mass per nucleon. When the uranium nucleus splits in half, however, smaller nuclei of lower atomic numbers are formed. As shown in **Figure 15.29**, these nuclei are lower on the graph than uranium, which means that they have a smaller amount of mass per nucleon. Thus, nucleons lose mass in their transition from being in a uranium nucleus to being in one of its fragments. When this decrease in mass is multiplied by the speed of light squared (c^2 in Einstein's equation), the product is equal to the energy yielded by each uranium nucleus as it undergoes fission. Interestingly, Einstein's mass–energy relationship applies to chemical reactions as well as to nuclear reactions. For nuclear reactions, the energies involved are so great that the change in mass is measurable, corresponding to about 1 part in 1000. In chemical reactions, the energy involved is so small that the change in mass, about 1 part in 1,000,000,000, is not detectable. This is why chemistry's law of mass conservation is best stated as follows: There is no *detectable* change in the total mass of materials as they chemically react to form new materials. In truth, there are changes in the mass of atoms during a chemical reaction. These changes, however, are too small to be of any concern to the working chemist.

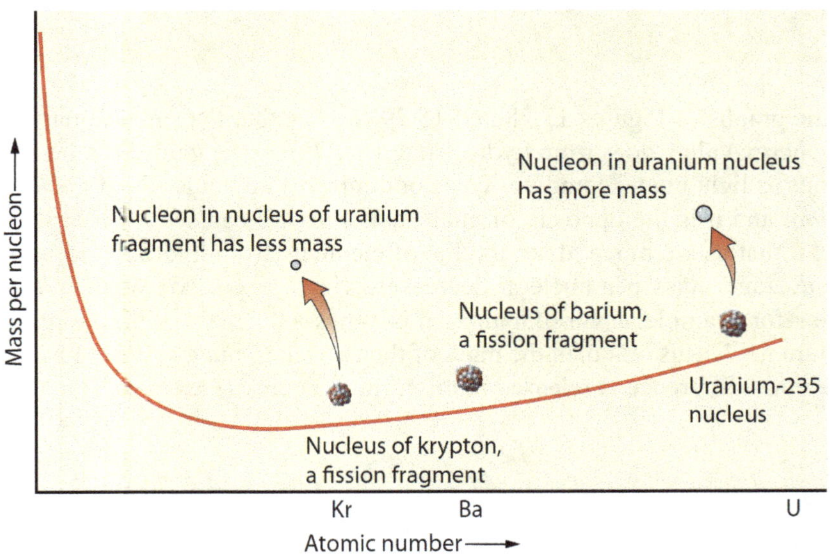

< Figure 15.29
The mass of each nucleon in a uranium nucleus is greater than the mass of each nucleon in any one of its nuclear fission fragments. This lost mass is mass that has been transformed into energy, which is why nuclear fission is an energy-releasing process.

CONCEPTCHECK
A nucleon has the least mass when in an iron nucleus. Does this mean that it is relatively easy to pull a nucleon out of an iron nucleus?

CHECK YOUR ANSWER
No! Just the opposite. The nucleon has the least mass when in an iron nucleus because that is where it is most tightly bound. To pull the nucleon from the iron nucleus would be much harder than pulling it from any other nucleus.

We can think of the mass-per-nucleon graph as an energy valley that starts at hydrogen (the highest point) and slopes steeply to iron (the lowest point), then slopes gradually up to uranium. Iron is at the bottom of the energy valley and has the most stable nucleus. It also has the most tightly bound nucleus; more energy per nucleon is required to separate nucleons from iron's nucleus than from any other element's nucleus. In this sense, iron, Fe, is a very special element.

All nuclear power today is produced by nuclear fission. A more promising long-range source of energy is to be found on the left side of the energy valley, in a process known as nuclear fusion.

CONCEPT**CHECK**

Correct the following incorrect statement: When a heavy element such as uranium undergoes fission, there are fewer nucleons after the reaction than before.

CHECK YOUR ANSWER

When a heavy element such as uranium undergoes fission, there are not fewer nucleons after the reaction. Instead, there is *less* mass in the same number of nucleons.

15.9 Nuclear Fusion

READING CHECK

How does the mass of a pair of hydrogen isotopes about to fuse compare with the mass of the resulting helium nucleus?

In the graphs of Figures 15.28 and 15.29, we see that the steepest part of the energy valley goes from hydrogen to iron. Energy is gained by the nucleons as light nuclei combine. This combining of nuclei is called *nuclear fusion*, and it is the opposite of nuclear fission. We can see from **Figure 15.30** that as we move along the list of elements from hydrogen to iron, the average mass per nucleon decreases. Thus, when two small nuclei fuse—for example, a pair of hydrogen isotopes—the mass of the resulting helium nucleus is less than the mass of the two small nuclei before fusion. The mass difference is released in the form of energy (**Figure 5.31**).

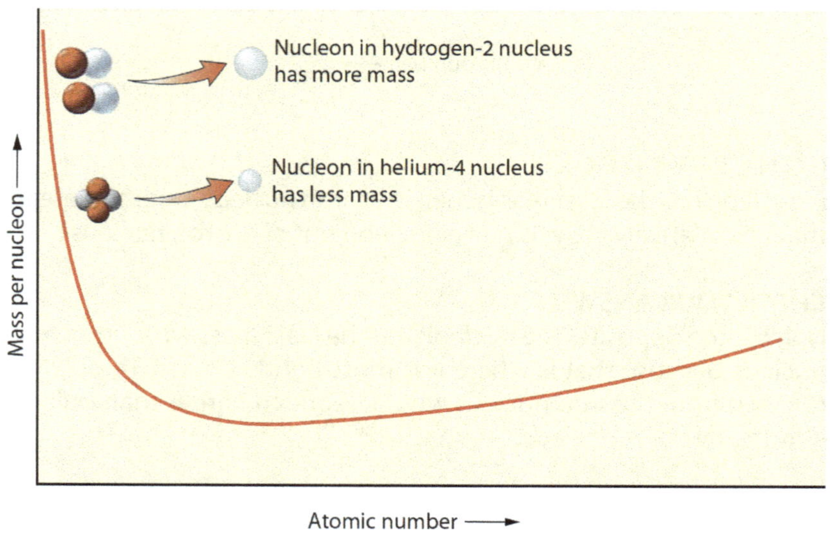

Figure 15.30 >
The further to the left on the graph, the greater is the mass of a nucleon. When fused (and therefore moved to the right), the resulting decrease in mass becomes mass that is converted to energy. Nuclear fusion is an energy-releasing process.

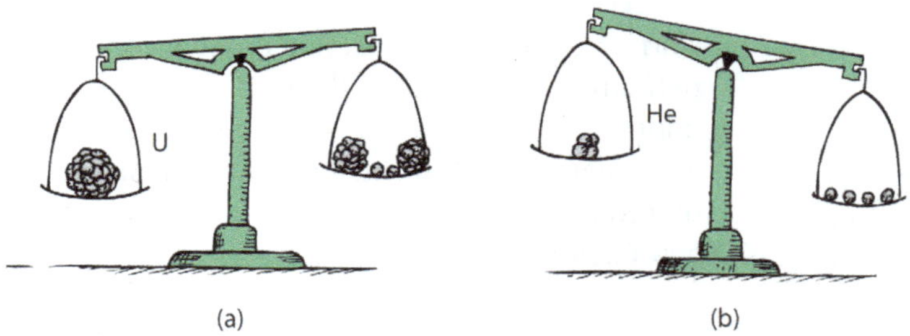

Figure 15.31
The mass of a nucleus is not equal to the sum of the mass of its parts. (a) The fission fragments of a uranium nucleus are less massive than the uranium nucleus. (b) Two protons and two neutrons are more massive in their free states than when they are combined to form a helium nucleus. Can you relate this to the graphs of Figure 15.29 and 15.30?

For a fusion reaction to occur, the nuclei must collide at a very high speed in order to overcome their mutual electric repulsion. The required speeds correspond to the extremely high temperatures found in the Sun and other stars. Fusion brought about by high temperatures is called **thermonuclear fusion**. In the high temperatures of the Sun, approximately 657 million tons of hydrogen is converted into 653 million tons of helium each second. The missing 4 million tons of mass is converted to energy—a tiny bit of which reaches our planet as sunshine. So thermonuclear fusion is the energy source of our Sun, which is, in turn, the ultimate energy source of life on Earth (**Figure 15.32**).

Such reactions are, quite literally, nuclear burning. Thermonuclear fusion is analogous to ordinary chemical combustion. In both chemical and nuclear burning, a high temperature starts the reaction; the release of energy by the reaction maintains a high enough temperature to spread the fire. The net result of the chemical reaction is the combination of atoms into more tightly bound molecules. In nuclear fusion reactions, the net result is more tightly bound nuclei. In both cases, mass decreases as the corresponding amount of energy is released.

Prior to the development of the atomic bomb, the temperatures required to initiate nuclear fusion on Earth were unattainable. When researchers found that the temperature inside an exploding atomic bomb is four to five times the temperature at the center of the Sun, the thermonuclear bomb was but a step away. This first thermonuclear bomb, a hydrogen bomb, was detonated in 1952. Whereas the critical mass of fissionable material limits the size of a fission bomb (atomic bomb), no such limit is imposed on a fusion bomb (thermonuclear, or hydrogen, bomb). Just as there is no limit to the size of an oil-storage depot, there is no theoretical limit to the size of a fusion bomb. Like the oil in the storage depot, any amount of fusion fuel can be stored with safety until ignited. Although a mere match can ignite an oil depot, nothing less energetic than an atomic bomb can ignite a thermonuclear bomb. We can see that there is no such thing as a "baby" hydrogen bomb. A typical thermonuclear bomb stockpiled by the United States today, for example, is about 1000 times more destructive than the atomic bomb detonated over Hiroshima at the end of World War II.

The hydrogen bomb is another example of a discovery used for destructive rather than constructive purposes. The potential constructive possibility is the controlled release of vast amounts of clean energy.

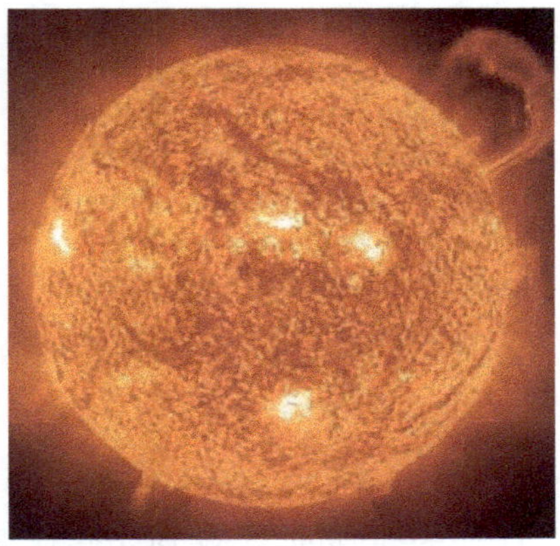

Figure 15.32
Thermonuclear fusion takes place in stars, such as the Sun. Some day, humans may produce vast quantities of energy through thermonuclear fusion, as the stars have always done.

Figure 15.33
Kenneth W Ford was part of the team that created the world's first hydrogen bomb. He wrote "Building the H Bomb: A Personal History," and mentored the author in textbook writing.

FOR YOUR INFORMATION

Elements are created in stars as smaller nuclei fuse to form larger nuclei. This process is energy releasing only up to iron. The manufacture of elements heavier than iron cannot be sustained within a shining star. So, where do heavier elements, such as gold, originate? The final stage of certain very large stars involves a mighty collapse, called a supernova. The energy of a supernova can outshine an entire galaxy, and it is this energy that makes the heavier-than-iron elements. The gold atoms in your jewelry were created using the abundant energy of a supernova that exploded very long ago.

Controlling Fusion

Carrying out fusion reactions under controlled conditions requires temperatures of millions of degrees. There are a variety of techniques for attaining high temperatures. No matter how the temperature is produced, a problem is that all materials melt and vaporize at the temperatures required for fusion. One solution to this problem is to confine the reaction in a nonmaterial container.

A nonmaterial container is a magnetic field, which can exist at any temperature and can exert powerful forces on charged particles in motion. "Magnetic walls" of sufficient strength provide a kind of magnetic straitjacket for hot ionized gases called plasmas. Magnetic compression further heats the plasma to fusion temperature.

Although there are no nuclear fusion power plants currently in operation, an international project now exists whose goal is to prove the feasibility of nuclear fusion power in the near future. This fusion power project is the International Thermonuclear Experimental Reactor (ITER) (**Figure 15.34**). The reactor will house electrically charged hydrogen gas (plasma) heated to about 150 million degrees Celsius, which is about 10 times hotter than the center of the Sun. In addition to producing about 500 MW of power, the reactor could be the energy source for the creation of hydrogen, H_2, which could be used to power fuel cells.

If people are one day to dart about the universe in the same way we jet about the Earth today, their supply of fuel is assured. The fuel for fusion—hydrogen—is found in every part of the universe, not only in the stars but also in the space between them. About 90 percent of the atoms in the universe are estimated to be hydrogen. For people of the future, the supply of raw materials is also assured, because all the elements known to exist result from the fusing of more and more hydrogen nuclei. Future humans might synthesize their own elements and produce energy in the process, just as the stars have always done.

▼ Figure 15.34

A cross-sectional view of the ITER (rhymes with "fitter") now underway in Cadarache, France. (Note for size the researcher in blue at the bottom of the reactor!)

CONCEPT**CHECK**

1. Fission and fusion are opposite processes, yet each releases energy. Isn't this contradictory?

2. To get a release of nuclear energy from the element iron, should iron undergo fission or fusion?

CHECK YOUR ANSWERS

1. No, no, no! This is contradictory only if the same element is said to release energy by both fission and fusion. Only the fusion of light elements and the fission of heavy elements result in a decrease in nucleon mass and a release of energy.

2. Neither, because iron is at the very bottom of the "energy valley." Fusing a pair of iron nuclei produces an element to the right of iron on the curve, in which the mass per nucleon is higher. If you split an iron nucleus, the products will lie to the left of iron on the curve and again have a higher mass per nucleon. So, no energy is released. For energy release, "decrease mass" is the name of the game—any game, chemical or nuclear.

Concept Review

Chapter 15

Summary of Terms

Alpha Particle A subatomic particle consisting of the combination of two protons and two neutrons ejected by a radioactive nucleus. The composition of an alpha particle is the same as that of the nucleus of a helium atom.

Beta Particle An electron emitted during the radioactive decay of a radioactive nucleus.

Carbon-14 Dating The process of estimating the age of once-living material by measuring the amount of radioactive carbon-14 present in the material.

Chain Reaction A self-sustaining reaction in which the products of one reaction event initiate further reaction events.

Critical mass The minimum mass of fissionable material needed for a sustainable chain reaction.

Gamma Rays High-frequency electromagnetic radiation emitted by radioactive nuclei.

Half-Life The time required for half the atoms in a sample of a radioactive isotope to decay.

Nuclear Fission The splitting of the atomic nucleus into two smaller halves.

Nuclear Fusion The combining of nuclei of light atoms to form heavier nuclei.

Radioactive A descriptive term for a material that contains nuclei that emit energetic particles and radiation.

Radioactivity The phenomenon exhibited by the spontaneous emission of energetic particles and radiation by an atomic nucleus.

Rem A unit for measuring the ability of radiation to harm living tissue.

Strong Nuclear Force The attractive force between all nucleons, effective only at very short distances.

Thermonuclear Fusion Nuclear fusion brought about by high temperatures.

Transmutation The conversion of an atomic nucleus of one element into an atomic nucleus of another element through a loss or gain in the number of protons.

Review Questions

1. Is radioactivity on Earth something relatively new? Defend your answer.
2. Why is cosmic radiation more intense in high-altitude locations?
3. Which type of radiation—alpha, beta, or gamma—results in the greatest change in mass number? The greatest change in atomic number?
4. Which of the three rays—alpha, beta, or gamma—has the greatest penetrating power?
5. What do flying in a commercial airliner and a visit to the doctors' office have in common?
6. What is a common source of radiation found within the basements of homes?
7. Which produces more radioactivity in the atmosphere, coal-fired power plants or nuclear power plants?
8. Why is the killing of cells by radioactivity preferable to damage of cells?
9. In what form is most of the energy released by ejected particles in nuclear processes?
10. When an element like thorium decays, does it become a completely different element, or a version of thorium?
11. What change in atomic number occurs when a nucleus emits an alpha particle? A beta particle?
12. What is the half-life of radium-226?
13. What is the half-live of uranium-238?
14. What changes occur to a nitrogen atom in the atmosphere that captures a neutron?
15. Why is carbon-14 dating not accurate for estimating the age of materials more than 50,000 years old?
16. Is a chain reaction more likely to occur in two separate pieces of uranium-235 or in the same pieces stuck together?
17. How is a nuclear reactor similar to a conventional fossil-fuel power plant? How is it different?
18. Who discovered that energy and mass are two different forms of the same thing?
19. How does the mass of a pair of atoms that have fused compare to the sum of their masses before fusion?
20. What kind of nuclear power is responsible for sunshine?

Answers to Odd-Numbered Questions

1. No. It has been around longer than the human race and has been a part of the environment as long as the Sun, rain and soil have been.
3. Alpha-radiation decreases the atomic number of the emitting element by 2 and the atomic mass number by 4. Beta-radiation increases the atomic number of an element by 1 and does not affect the atomic mass number. Gamma-radiation does not affect the atomic number or the atomic mass number. So alpha-radiation results in the greatest change in atomic number, and hence charge, and mass number as well.
5. In both cases you are taking on the risks of radiation exposure to enjoy certain benefits. For the commercial airliner, you are exposing yourself to gamma radiation from space, but you get the benefit of high-speed travel. For a doctor visit, you may be exposed to X rays (a form of radiation), but you get the benefit of the X ray's diagnostic powers.
7. Global combustion of coal produces about 13,000 tons of radioactive substances; nuclear plants produce about 10,000 tons.
9. Most is KE of the ejected particle and some is the KE of the recoiling nucleus.
11. Emission of an alpha particle decreases the atomic number of the nucleus by 2. Emission of a beta particle increases the atomic number of the nucleus by 1.
13. 4.5 billion years.
15. Because of the relatively short half-life of carbon-14, only trace amounts would be left after 50,000 years—too little to be statistically accurate.
17. A nuclear reactor and a fossil fuel power plant both boil water to produce steam for turbines. The main difference between them is the kind and amount of fuel involved. Nuclear reactors are much more efficient—one kilogram of uranium fuel yields more energy, for example, than 30 freight car loads of coal.
19. The pair has less mass after fusion, which is converted to energy.

Index

Photo Credits

Page 6	Cristo Vlahos
Ch 1 Opening	Hansueli Krapf
Figure 1.7	Brian Snelson
Figure 1.12	Andrew Dunn
Ch 2 Opening	Blackswan1923
Figure 2.18	Angie Spuc
Figure 3.16	Kelly Martin
Figure 4.11	Kmtextor
Ch 4 Opening	Nicolas Halftermeyer
Figure 4.6	Mark Harkin
Figure 4.10	Almonroth
Ch 5 Opening	MusikAnimal
Figure 5.10	中竹たけし
Ch 6 Opening	NASA
Figure 6.20	NASA
Ch 7 Opening	MGA73bot2
Figure 7.4	GreaterPonce665
Figure 7.32	Matthew T Rader
Ch 8 Opening	Alasdair
Figure 8.12	Soph556
Figure 8.13	Jarek Tuszyński
Figure 8.18	Pep.per de Ré
Figure 8.21	Alexey Kljatov
Figure 8.38	Ph0705
Ch 9 Opening	Jon Westra
Figure 9.8	Dwight Burdette
Figure 9.12	Togabi
Figure 9.19	KMJ
Figure 10.14	Paul Broeker
Figure 10.12	Geek3
Figure 10.12	OpenStax
Figure 11.12	FASTILY
Figure 12.17	Nicola Angeli
Figure 12.24	Damian Gadal
Figure 12.29	Geolina163
Figure 12.44b	Alvesgaspar
Figure 12.45	Tomascastelazo
Figure 12.46	Flystock
Figure 13.9b	Unukorno
Figure 13.9c	Agnez Bewer
Figure 13.9d	Arne-kaiser
Figure 13.23	MikeRun
Figure 13.25	Harvard College
Page 253 Activity	Yaroslav Ulyushin
Figure 14.3	NanoscaleRanger
Figure 14.7	Auckland Museum
Figure 14.7	Deutsche Fotothek
Figure 14.7	Ludvig14
Figure 14.7	Peellden
Figure 14.7	Daler Boi
Figure 14.7	Wilco Oelen
Figure 14.8	Gaius Cornelius
Ch 15 Opening	Martin Mergili
Figure 15.4	Kyle McDonald
Figure 15.7	BLK Hospital
Figure 15.16	Haruhiko Okumura
Figure 15.26	Photorush
Figure 15.34	Oak Ridge National Laboratory

www.ingramcontent.com/pod-product-compliance
Lightning Source LLC
LaVergne TN
LVHW082117200926
848484LV00008B/1965
* 9 7 9 8 9 8 6 2 7 0 6 2 3 *